2nd Edition

Algebraic Analysis I

Sun Tae Soh

Department of Mathematics
Myong Ji University

KYOWOO

Kyowoo Co., Ltd.
8, Yangnyeongsi-ro, Dongdaemun-gu,
Seoul, Korea 130-861,

Phone: 82-2-925-2861
F a x: 82-2-925-2860
e-mail: kyowoo@kyowoo.co.kr
http://www.kyowoo.co.kr

Printed in the Republic of Korea.
ISBN 979-11-251-0329-5 (93410)

₩ 28,000

To Sun Young for her love and support,

and to our four children.

Contents

Preface to the Second Edition

This book, *Algebraic Analysis I*, was first published in 2006 and then was improved in the following years for some time. Since then more time has passed fixing some typos and errors of mine as much as I could, I restated more clearly or added, if necessary, several statements of the previous book, hoping to improve its readability. Thus, this edition would be called the final version of my book, *Algebraic Analysis I.*

In particular, as of Dec. 18, 2008, the Computer Algebra System, REDUCE is available, free of charge, for various platforms such as PCs, Notebooks, and Linux Machines, etc., so I decide to provide the readers of the 2nd Edition of this book with the htm files consisting of those algorithms/programs written by the author in REDUCE commands, already presented as Algorithms, Programs, and Examples, in the 1st Edition of this book. This approach will assist the readers of this book a lot because it may enable them to actually perform their own computational experiments very easily and conveniently under the Internet Environment now very widely available to us. Here is a list of my algorithms/programs whose companion is the well-prepared Index of this book:

http://ch.catholic.or.kr/pundang/4/soh/2006_aa_i/a_list_of_APs.htm

A few of those algorithms and programs presented in the above list are *Working Examples* of those that were discovered by the author. The author originally planned to write another book with the title *Algebraic Analysis II*, which describes among other topics how he had discovered them, but the plan was delayed, and postponed until this time because of another project under the Internet Environment, called Computation Related Human Activities(CRHA), that the author launched in the year

2000 and is still going on. One of the recent results of this project is, for example, an algorithm called "*A Standard Method of Interpretation of Chinese Poems, Based on AI (Artificial Intelligence).*" The author strongly believes that this algorithm can significantly and transcendentally enhances the promotion of our intellectual virtues. Those methods developed by the author for this algorithm "Standard Method" under the Internet Environment have already produced many new and important discoveries/results in broad areas of Humanity.[1]

As a remark of this preface, I would like to mention briefly what I had foreseen in 1999, 22 years ago, what will be the main aspect of Mathematics in the 21st century, that is included as an Appendix A: An Introduction to Algebraic Analysis,[2] of my book, because most of the people around the world nowadays recognize the importance of Computation Related Human Activities(CRHA), after watching AlphaGo's victory in a Go-game competition in Seoul, in the Spring of 2016. Since the main aspect of CRHA consists of (1) Symbolic Computation, i.e., Algebraic Analysis, (2) deductive reasoning or inductive reasoning for intellect's understanding, and (3) decision making process in order to reach a meaningful conclusion after finitely many steps in reasoning process.

I think it is very important for us, human-beings, to understand what each of these three actually means, and it is fairly well described in the Appendix A: An Introduction to Algebraic Analysis, of my book, although I did not mention them explicitly in terms of technical terminologies, because, as it is well-known that the Greek Philosophy of those philosophers such as Socrates (470 - 399 BC), Plato (428/427 - 348/347 BC), and Aristotle (384 - 322 BC) is based on the method of Mathematical proof, still nowadays only well-taught in the courses of Pure Mathematics, which had already well been developed by the Pythagorean School of the 6th century BC, CRHA are also definitely based on the method of Mathematical proof, still nowadays only well-taught in the courses of Pure Mathematics.

[1] http://ch.catholic.or.kr/pundang/4/soh/interpretations_of_chinese_poems.htm
[2] http://ch.catholic.or.kr/pundang/4/soh/mathnet_973071.htm

I believe that people well know that there is no contradictions among any facts, as long as they are all mathematically proved to be true, and hence this is the main reason why the method of the mathematical proof is very important to us, human-beings, as long as we reason why. Another thing is that people usually do not make a mistake in deriving a true conclusion through a deductive reasoning from sound hypotheses, but people sometimes derive a false conclusion through an inductive reasoning even from sound hypotheses that is NOT a mathematical inductive reasoning. One should also recognize that Artificial Intelligence (AI) is born to have such a capacity but it inevitably has, like us human-beings, a defect in its inductive reasoning process.

I think both the Appendix A: An Introduction to Algebraic Analysis and the first chapter of my book are in fact well-suited for the beginners who want to learn how to better think and understand CRHA in the Computer Age, the 21st century, because their main underlying theme is not about any particular subjects in Mathematics, but rather about how we should think and reason to derive meaningful conclusions in an axiomatic manner.

Toward this direction, some amount of my research works and papers on finding or establishing a new relationship among objects, i.e., concepts/ideas, in Computation Related Human Activities(CRHA), that are meaningful to us human-beings, existing generically or specially among the given ideas/concepts, under the environment of internet-based Artificial Intelligence, which have been produced by the author for a period of more than recent 16 years, are available to the readers through the internet access.[3]

Sun T. Soh
February 2021[4]

[3]http://ch.catholic.or.kr/pundang/4/soh/a_list_of_research_outputs.htm

[4]Anything, typo corrections, updates or added materials etc., will be available from: http://ch.catholic.or.kr/pundang/4/soh/2006_aa_i/update.htm

The rest of the preface is the original preface to the first edition of this book, which is in the heart of my soul:

Preface [to the First Edition]

The following short historical remark is prepared for those who do not know about what's really happened in the development of the theory of Groebner basis in its beginning stage, which I myself happed to know only in the middle of 1980.

Suppose that

(1) you are given a polynomial $f(x_1, ..., x_n)$,
(2) a polynomial ideal I in the polynomial ring $k[x_1, ..., x_n]$ in n variables with $n \geq 1$, and
(3) you want to know whether $f = f(x_1, ..., x_n)$ is a member of I or not.

The problem is that, in most cases, the ideal I will have infinitely many polynomials in it, so that there is no easy way of deciding its membership to I, because even if you are allowed to apply the division algorithm, Algorithm 5 directly, sometimes it will tell you that the remainder is zero so that f belongs to I. However, since the resulting remainder is really dependent on your own path of your actual division process, you will not be able to get the zero polynomial most of the time as a remainder of your division process, even though f is in fact a member of the ideal I.

If you do not mind whether you are able to compute something out completely or not, like the posed question, it may not be a problem to you at all. Otherwise, this problem is a really serious one. If mathematicians refuse to take care of computational business who else could figure out the underlying reason in computational phenomena around us? In the mean while the history of mathematics tells us that ever since Euclid's Algorithm for computing the greatest common divisor of two numbers, there has been no comprehensible generalization of Euclid's algorithm for more than, say, 2,300 years. In the year 1965, however, the situation was changed suddenly and completely.

First of all, it should be quite the case that such an incomplete division algorithm, Algorithm 5, was not well-known to us or even not well-formulated as it is, because itself alone is not computationally useful at all, although someone says that we have some traces of such a division process in the works of Gauss.

Secondly, in his famous Ph.D. thesis in 1964 prepared under the guidance of Oscar Zariski, a Japanese mathematician Heisuke Hironaka (1931 -) claimed the existence of the so-called "standard bases" in the domain of formal power series and proved its existence by a method of contradiction, but he never actually tried to construct such a basis explicitly.

On the other hand, after Bruno Buchberger (1942 -) computationally constructed his very famous and very important algorithm, Algorithm 5 for Groebner bases in his Ph.D. thesis in the year 1965 without knowing that Hironaka proved its existence in his thesis of the year 1964, such an incomplete algorithm, Algorithm 5 has somewhat remained as the current formulation that *no terms of the remainder r should be divisible by any of the leading terms of the divisors*, because it suddenly becomes to have a significant meaning of computing a remainder in a *unique* manner, *provided that the divisors are a Groebner basis for the given ideal I.*

Now going back to the posted problem, we know by Hilbert Basis Theorem (Theorem 1.5 or Theorem 2.2) a finite number of generating elements of I are going to be given to us.

If $n = 1$, then it is very easy because $k[x]$ is a *PID*, so the single generator of I is given to us, and hence we simply apply the division algorithm, Algorithm 3, to see if f is a pure multiple of the single generator of I.

Otherwise, let $f_1, ..., f_s$ be generators of the ideal I. Then under a monomial order on $k[x_1, ..., x_n]$ of our choice, we compute the (reduced) Groebner basis $G = \{g_1, ..., g_t\}$ for I from these generators, using Buchberger's Algorithm, Algorithm 6 or an improved version of this algorithm for computing the (reduced) Groebner basis. After that, we apply the division algorithm, Algorithm 5 to the given f in order to see whether the remainder r is the zero polynomial or not. If it turns out that r is the zero polynomial, this equivalently means that f is indeed an element of the ideal I. Otherwise, since the remainder is uniquely determined by a Groebner basis, f is not a member of the ideal I.

Nowadays, most Computer Algebra Systems, since 1970s, say, REDUCE, MAPLE, etc., do provide a command for computing the (reduced) Groebner basis which satisfies at least the second condition of the definition for the reduced Groebner basis.

Some Computer Algebra System, for instance, (not the "preduce" command but) the "greduce" command in REDUCE, does provide a command which computes the unique remainder via the (reduced) Groebner basis.

Another point we would like to share with the reader of this book is a smooth combination of Axiomatic Algebra with the Theory of Groebner basis. In fact, the Theory of Groebner basis is independent of any other parts of Algebra, so that whenever we need it we can learn it. This also means that it would be very beneficiary if the reader learns the Theory of Groebner basis as early as they could. Because of this reason, we tried hard to move into the main computational tool, Buchberger's Algorithm for computing a Groebner basis for a given polynomial ideal I, as early as we could.

Here is the outline of each chapter of this book. After developing the basic Theory of Groebner basis, we really tried hard to go through all the other materials of this book in order to make them sense with historical accounts of who did what.

In Chapter 1, we review the most basic materials for Computational Algebra. Since this book is written for Symbolic Computation, we first review Euclid's algorithm for computing the greatest common divisor of two positive integers. And then we review basic facts for polynomials in one variable over a field k. Then the concept of computable rings, computable fields, and computable vector spaces are discussed. We also discuss why we cannot easily figure out the gcd of two polynomials in more than one variables, and then we discuss when a polynomial is also regarded as a polynomial function. This interpretation of a polynomial as a polynomial functions is very important because it enables us to connect Algebra with Geometry. Then we review the very basics of affine varieties. As a last part of Chapter 1, we examine Hilbert's basis theorem, the ascending chain condition for polynomial ideals, and Hilbert's Nullstellensatz, and their consequences. In particular, we discuss the evaluation homomorphism in detail, because it is this evaluation homomorphism which actually connects Algebra with Geometry under a categorically contravariant manner.

In Chapter 2, we develop the Theory of Groebner basis very slowly. For this, we first define a monomial ordering on the set of monomials in several variables, and then we closely examine the division process in detail. After that, we define the definition of a Groebner basis for a polynomial ideal purely axiomatically in terms of its associated monomial ideal. Then, we discuss Buchberger's Criterion for a Groebner basis and Buchberger's Algorithm for computing a Groebner basis. We finally define a reduced Groebner basis. If one examines the whole materials of Chapter 2, one will easily find that they are independent of any other parts of Mathematics, so that this chapter could be taught quite early.

In Chapter 3, we have some immediate consequences of Groebner basis: Solving a system of polynomial equations, the role of the dictionary ordering, the ideal membership problem, a program for computing the quotients appearing in division algorithm, a criterion for polynomial ideals to be the same, computations in a quotient ring of a polynomial ring, and finally the implicitization problem. Thus, the first three chapters of this book cover the most basic materials of the Theory of Groebner basis. Based on them, we have the next three chapters.

In Chapter 4, we begin out discussion on the Invariant Theory of the 19th Century from the Theory of Groebner basis. After reviewing the basic materials of the Invariant Theory, we discuss our own algorithms or new programs, and one of them enables us to compute the reduced Groebner basis for a certain ideal very efficiently. After that, we continue our discussion on the computation of the numerical invariants, discriminants, which play a very important role in the Galois Theory. To go further, we then explain the correspondence between Algebra and Geometry in detail, and finally we discuss how we can compute the generators of the invariant subring of a polynomial ring under an action of a finite group.

In Chapter 5, we begin our discussion toward how to compute a generating set of the intersection of two polynomial ideals, which is all possible due to the Elimination Theorem. Based on this, we have several algorithms for computing generators for certain polynomial ideals in Computational Algebra. Algorithms for the kernel and the image of a (affine) k-algebra homomorphism are fully discussed, which now enable us to establish the *contravariant* one-to-one correspondence between Algebra and Geometry in a great detail. Then as a serious application of the theory developed in this chapter, we begin our discussion on how we can compute the minimal polynomial of an algebraic element in a finite field extension of a given field k, including a primitive element of such an extension especially when the characteristic of k is zero. The reader will surely find at least one important reason why we need to learn the Theory of Groebner basis as early as we could, once are studied those materials in this part of the book for computation of the minimal polynomials over the bottom field of elements in a finite field extension.

Finally, Chapter 6 of this book is a geometric counterpart of Chapter 5, but it is in fact more than that. In particular, the Theory of Resultants in Invariant Theory plays its role in this direction, because it enables us to prove the Extension Theorem. Once again, resultants are related to the gcd of two polynomials, and so we start our discussion on resultants by considering the gcd again. Once the Extension Theorem is established, we combine Groebner basis with resultants in order to have a better understanding of implicitization process for a parametrized geometric object, including computation of the defining equation for the Zariski Closure of the image of a polynomial or rational mapping.

In this book, almost all theorems eventually produce algorithms or programs in a constructive manner, and almost all algorithms or programs have at least one example with a working set of codes in the Computer Algebra System, REDUCE.[5] For this, see the index of this book under Algorithms, Programs, or Examples.

As a computational counterpart or twin of the book, *Algebraic Structures*, of my teacher, late Professor George R. Kempf (1944 - 2002) at JHU, this book focuses on the subject, **Algebraic Analysis**, that is, symbolic computation, and hence its title: *Algebraic Analysis I.* Unless we approximate numbers which is very common in Numerical Analysis, the nature of all the objects under our mathematical consideration are exact. A brief introduction to Algebraic Analysis is provided as an Appendix of this book. There, our service, called InetCompu Service, is also discussed very briefly.

For more about Computer Algebra Systems, one may visit our homepage:

http://trinitas.mju.ac.kr/intro2softwares.html.[6]

There, you will further find very useful informations for users on CAS, MAPLE, REDUCE, MATLAB, etc. Also,

http://trinitas.mju.ac.kr/intro2InetCompu.html[7]

describes an important aspect of the availability of Internet environment to everybody in this wonderful world of the 21st century.

[5]In order to run those programs given in Examples of this book, one is strongly recommened to download the Computer Algebra System, REDUCE, and install it, free of charge, from: http://ch.catholic.or.kr/pundang/archive/4/reduce-algebra.htm.

[6]an archive site: http://ch.catholic.or.kr/pundang/archive/4/intro2softwares.html

[7]an archive site: http://ch.catholic.or.kr/pundang/archive/4/intro2InetCompu.html

CHAPTER 1

Some Preliminaries

1. Which are usually called "Bases"?

Let I be an ideal in the polynomial ring $k[x_1, ..., x_n]$ in n variables over a field k, $n \geq 1$. We then know that it is a vector space over the field k with a "basis" $B = \{x_1^{e_1} x_2^{e_2} \cdots x_n^{e_n} : e_i \in N \cup \{0\}\}$ consisting of monomials $x_1^{e_1} x_2^{e_2} \cdots x_n^{e_n}$ of every degree. Thus, although there are infinitely many elements in the basis, we can write each element of $k[x_1, ..., x_n]$ as k-linear combination of finitely many basis elements of this basis in a unique manner.

Now let $n = 1$. Then we already know that the polynomial ring $k[x]$ is a *PID* (Principal Ideal Domain), which exactly means that every ideal of $k[x]$ is generated by a single element, that is, there is a polynomial $d(x)$ in every ideal I such that each element of an ideal $I = (d(x))$ is a polynomial multiple of $d(x)$. Traditionally the singleton $\{d(x)\}$ is called a "basis" of the given polynomial ideal I.

These two cases thus suggest that a subset of an algebraic object is called a **"basis"** if any other elements of the object can be expressed through a finite number of algebraic manipulations of finitely many elements of the subset, called a basis. We know that most objects in mathematics consist of infinitely many elements, and hence figuring out its basis is an immediate job most of the time. I mean, figuring out a basis of an algebraic object consisting of infinitely many elements is an urgent job.

We recall that, when $n > 1$, although the polynomial ring $k[x_1, ..., x_n]$ is not a *PID*, it is a *UFD* (Unique Factorization Domain). Since it is not a *PID*, we ask ourselves whether its ideals I are all finitely generated, that is, there is a finite subset S of polynomials in $k[x_1, ..., x_n]$ such that it is a subset of I and every other polynomials in I can be expressed as a polynomial combination of the elements of this subset S. If an ideal has such a generating set S, then we also call S a basis of I. In a later section of this chapter, we treat the so-called Hilbert's "basis" theorem which says that every polynomial ideal I in a polynomial ring $k[x_1, ..., x_n]$ is finitely generated over k. (See Theorem 1.4).

This means that whenever we are given a polynomial ideal in $k[x_1, ..., x_n]$, its finite number of generators are going to be given to us. But then the next immediate question we can think of is, for instance: what is a generating set of the intersection $I \cap J$ of two polynomial ideals I and J? Of course $I \cap J$ is an ideal, but it is generally very difficult to explicitly figure out a generating set, or a "basis" of the intersection $I \cap J$ from those given generators of J and generators of J.

Fortunately, thanks to Buchberger's discovery of an algorithm for Groebner bases at around 1965, we are now able to answer this very difficult question affirmatively. (See, Theorem 2.5 and Algorithm 6 of Chapter 2, and Theorem 11 of Chapter 5). Since explicitly figuring out a basis is really important, we strongly believe that his discovery will survive forever like Euclid's algorithm for computing the greatest common divisor of two numbers. (See Algorithm 2 in the next section).

In summary, in what follows from now on we will mainly focus how to compute a basis of a given algebraic object.

Exercises for Section 1.1

EXERCISE 1.1. *Show that the maximal ideal (x, y) in the polynomial ring $k[x, y]$ in two variables over a field k is not generated by a single element of $k[x, y]$.*

EXERCISE 1.2. *Let k be a field.*

(1) *Prove that the polynomial ring $k[x]$ in one variables over k is a vector space over k.*
(2) *Prove that the polynomial ring $k[x, y]$ in two variables over k is a vector space over k.*

(3) *Are they finite-dimensional vector spaces? Figure out their k-bases, that is, generating sets or spanning sets whose elements are linearly independent over* k. *You need to justify your answer.*

EXERCISE 1.3. *By the above Exercise, we know that* $k[x_1, ..., x_n]$ *is a* k*-vector space. Let* I *be a polynomial ideal of* $k[x_1, ..., x_n]$. *Prove that the quotient ring* $k[x_1, ..., x_n]/I$ *is also a* k*-vector space. What is a* k*-basis of the quotient ring? [Hint: See Proposition 3.2].*

EXERCISE 1.4. *Prove that any ring homomorphism from a field to a ring is injective.*

2. A Review on Integers

The main purpose of this chapter is a study of division in rings. Because it is modeled on properties of the ring $\mathbb{Z}$ of integers, we need to review these properties. The results in this particular section are a continuation of those in the second-half of the last chapter.

REMARK 1.1. *We note that as long as factorization of polynomials are concerned, treatment of those polynomials in one variable is completely different from treatment of those polynomials in more than one variable. This is mainly because a polynomial ring* $k[x]$ *in one variable over a field* k *is a UFD that is also a PID, whereas a polynomial ring* $k[x, y]$ *in two variables over a field* k *is a UFD but is not a PID.*

In this section, we will study polynomials in one variable only, because a proper treatment of polynomials in more than one variables does require the Theory of Groebner bases, discovered and developed by Bruno Buchberger in late 20th *century.*

We start our discussion with an elementary school arithmetic stuff:

THEOREM 1.1. *(Division with Remainder) If* a, b *are two distinct natural numbers and* $a \neq 0$, *then there exist unique integers* q, r *such that*

$$b = aq + r, \tag{1.1}$$

with $0 \leq r < a$. □

ALGORITHM 1. *Thus, whoever computes the quotient* q *and the remainder* r*, the results are going to be the same. Moreover, we have an* ***algorithm*** *for actually computing* q *and* r *for a given set of inputs:* b *and* $a \neq 0$*, which we learned it even when we are in an elementary school.*

One immediate consequence of this theorem is:

COROLLARY 1.1. $\mathbb{Z}$ *is a PID.* $\square$

We note that given non-zero natural numbers a, b, the set defined by

$$a\mathbb{Z} + b\mathbb{Z} := \{ar + bs : r, s \in \mathbb{Z}\}$$

is again an ideal in $\mathbb{Z}$, the smallest ideal containing both ideals $a\mathbb{Z}$ and $b\mathbb{Z}$, and by this corollary it is generated by a single element, say d:

$$d\mathbb{Z} = a\mathbb{Z} + b\mathbb{Z}.$$

The immediate question one can think of is: how to compute d explicitly.

LEMMA 1.1. $d = \gcd(a, b)$.

PROOF. Certainly, $d = ar + bs$ for some $r, s \in \mathbb{Z}$. Also, d clearly divides both a and b since $a = a \cdot 1 + b \cdot 0$ for instance. Now, if h divides both a and b, then a and b are elements of the ideal $h\mathbb{Z}$; hence $d\mathbb{Z} = a\mathbb{Z} + b\mathbb{Z} \subset h\mathbb{Z}$; hence h does divide d.This exactly means that d is the greatest common divisor of a and b. $\square$

COROLLARY 1.2. $\gcd(a, b) = ar + bs$ *for some* $r, s \in \mathbb{Z}$. $\square$

REMARK 1.2. *We note that the above treatments do not tell us how we compute* $\gcd(a, b)$*,* r *and* s*. To compute* $\gcd(a, b), r,$ *and* s*, we apply the very well-known* ***division algorithm of Euclid*** *of linear time efficiency, which was known to us since about BC* 500*. Here is the algorithm of Euclid for* $\gcd(a, b)$.

ALGORITHM 2. *(**Euclid's algorithm for*** $\gcd(a, b)$*) We may assume that* $q_0 = b > a_1 = a \neq 0$*. If* b *is a multiple of* $a \neq 0$*, then* $a = \gcd(a, b)$*.*

Otherwise, we continuously apply the above theorem to obtain:

$$\begin{aligned} b &= q_1 a + r_1, \text{ with } r_1 > 0, \\ a &= q_2 a_1 + r_2, \text{ with } a_1 = r_1 \text{ and } r_2 > 0, \\ a_1 &= q_3 a_2 + r_3, \text{ with } a_2 = r_2 \text{ and } r_3 > 0, \\ &\dots\dots\dots \\ a_{k-2} &= q_k a_{k-1} + r_k, \text{ with } a_{k-1} = r_{k-1} \text{ and } r_k > 0, \\ a_{k-1} &= q_{k+1} a_k + 0, \text{ with } a_k = r_k > 0 \text{ but } r_{k+1} = 0. \end{aligned}$$

Then $r_1 > r_2 > \ldots > r_k > 0$ *and hence* $\gcd(a,b) = \gcd(r_1, r_2) = \gcd(r_2, r_3) = \ldots = \gcd(r_k, 0) = r_k$. □

This algorithm does work since

$$\gcd(b,a) = \gcd(b - aq, a) = \gcd(r, a),$$

and by rewriting these equations backwards starting from $\gcd(a,b)$, we can find those integers r and s in $\gcd(a,b) = ar + bs$. We note that these r and s are not uniquely determined.

EXAMPLE 1.1. *We know by inspection that* $\gcd(6, 29) = 1$. *On the other hand,*

$$\begin{aligned} 29 &= 4 \cdot 6 + 5, \\ 6 &= 1 \cdot 5 + 1, \\ 5 &= 5 \cdot 1 + 0. \end{aligned}$$

Thus, $1 - 6 - 1 \cdot 5 = 6 - 1 \cdot (29 - 4 \cdot 6) = 5 \cdot 6 + (-1) \cdot 29$; *hence* $r = 5$ *and* $s = -1$. *Obviously, these* r *and* s *are not determined uniquely:* $1 = 5 \cdot 6 + \{(-29) \cdot 6 + 29 \cdot 6\} + (-1) \cdot 29 = (-24) \cdot 6 + 28 \cdot 29$, *where* $6 \cdot 29 = \mathrm{lcm}(6, 29)$, *etc.*

LEMMA 1.2. *Let* p *be a prime number and let* a, b *be numbers. If* p *divides their product* ab, *then* p *divides at least one of them.*

PROOF. Suppose that the prime p does not divide a. Then $\gcd(p, a) = 1$; hence by the above corollary we have

$$1 = pr + as$$

for some $r, s \in \mathbb{Z}$. Multiplying both sides by b we see that p does divide b. □

THEOREM 1.2. *(**Fundamental Theorem of Arithmetic**) Every non-zero integer a can be written as a product*

$$a = cp_1 \cdots p_k, \tag{1.2}$$

where $c = \pm 1$, the p_i are prime numbers, and $k \geq 0$. This expression is unique except for the ordering of the prime factors p_i.

PROOF. First, we prove that a prime factorization exists. To prove this, it is enough to consider the case that a is greater than 1. By induction on a, we may assume that the existence is proved for all positive integers $b < a$. Either a is prime, in which case the product has one factor, or there is a proper divisor $b \neq a$. Then $a = bb'$, where both b and b' are smaller than a, and hence by induction hypothesis they can be factored into primes. Setting their factorization side by side gives a factorization of a.

Second, we prove that the factorization is unique. For this we suppose that

$$\pm p_1 \cdots p_n = \pm q_1 \cdots q_m.$$

The signs should certainly agree. Then we apply the above lemma with $p = p_1$: Since p_1 divides the product $q_1 \cdots q_m$ it divides some q_i, say q_1. Since q_1 is also prime, we must have that $p_1 = q_1$. We cancel p_1 from both sides and we proceed by induction, noting that 1 is never equal to a product of primes. □

REMARK 1.3. *Although the above factorization is unique up to units $c = \pm 1$, it is usually very difficult to compute those prime factors p_i when* *a is large. Moreover, the above proof does not tell us how one can compute those prime factors p_i efficiently.*

As we mentioned earlier, the structure of the ring $\mathbb{Z}$ of integers is closely analogous to that of a polynomial ring $k[x]$ in one variable x over a field k. Whenever a property of one of these two rings is derived, we should try to find an analogous property of the other.

EXAMPLE 1.2. *We have already seen that both $\mathbb{Z}$ and $k[x]$ are a PID in the previous chapter.*

Exercises for Section 1.2

EXERCISE 1.5. *Compute the* gcd *of* 48 *and* 35 *using Algorithm 2.*

EXERCISE 1.6. *Prove that there are infinitely many prime numbers.*

EXERCISE 1.7. *Prove that $\sqrt{2}$ is not a rational number.*

EXERCISE 1.8. *What is the maximal ideal of $\mathbb{Q}$?*

3. A Review on Polynomials in One Variable

3.1. Field Extensions. Here is our first lemma.

LEMMA 1.3. *An algebraically closed field is an infinite field.*

PROOF. Let k be a finite field consisting of s elements $a_1, ..., a_s$. Then the following polynomial

$$f(x) = 1 + \prod_{i=1}^{s}(x - a_i)$$

has no roots in k at all. Therefore, a finite field cannot be algebraically closed. □

In order to have a strong motivation, we first review the role of irreducible polynomials in one variable over the field $\mathbb{R}$ of real numbers. We start with an example, a famous construction of Leopold Kronecker (1823 - 1891):

EXAMPLE 1.3. *(Kronecker) We obviously know that the equation* $x^2 + 1 = 0$ *for an unknown* x *has no real solutions; hence* x^2+1 *is an irreducible polynomial over* $\mathbb{R}$*. The following arguments show that this is not a dead end. Rather, it opens a whole new beginning of the so-called Field Theory, provided that we are well equipped with some basic abstract Algebra:*

Since $\mathbb{R}[x]$ *is a PID, it is a UFD; hence irreducible elements are prime elements, and vice versa. Thus, the principal ideal* $(x^2 + 1)$ *in* $\mathbb{R}[x]$ *generated by* $x^2 + 1$ *is a maximal ideal; hence equivalently the quotient ring* $\mathbb{R}[x]/(x^2 + 1)$ *is a field.*

We then recall that there is a natural ring homomorphism

$$\varphi\colon \mathbb{R}[x] \longrightarrow \mathbb{R}[x]/(x^2 + 1),$$

which sends $f(x) \longmapsto \bar{f}(x) := f(x) \operatorname{mod}(x^2 + 1)$*. In particular, the unknown* x *is sent to* $x \operatorname{mod}(x^2 + 1) = x$ *under this map and the polynomial* x^2+1 *is sent to* 0 *under this map. Then since* φ *is a ring homomorphism, the image* $\bar{x}$ *of* x *under* φ*, which is* x*, in* $\mathbb{R}[x]/(x^2 + 1)$ *is a solution of* $(x)^2 + 1 = 0$*. That is,* x *in the field* $\mathbb{R}[x]/(x^2 + 1)$ *is no longer a relation free object, unlike its preimage,* x *in* $\mathbb{R}[x]$*.*

On the other hand, the composition of it after the natural injection: $\mathbb{R} \hookrightarrow \mathbb{R}[x]$ *gives a natural ring homomorphism*

$$\mathbb{R} \hookrightarrow \mathbb{R}[x] \longrightarrow \mathbb{R}[x]/(x^2 + 1).$$

Then since any ring homomorphism from a field is injective, which is the case because there are only two ideals $(0), (1)$ *in a field, and moreover since, by construction, any real number* r *is sent to itself under this composition, we have constructed a field* $\mathbb{R}[x]/(x^2 + 1)$ *containing* $\mathbb{R}$ *as its subfield. In other words, we have successfully extended the field* $\mathbb{R}$ *of real numbers to a newly constructed field* $\mathbb{R}[x]/(x^2 + 1)$*, where the equation* $x^2 + 1 = 0$ *indeed has a solution! We customarily write* i *for such an* x *in the field* $\mathbb{R}[x]/(x^2 + 1)$*: That is,* i *is by definition a number if ever exists whose square is equal to* -1*.*

Our construction explicitly shows that $\bar{x} := x \operatorname{mod}(x^2+1) = x$ *in the field* $\mathbb{R}[x]/(x^2+1)$ *is such an* i*. Then certainly* $-i \in \mathbb{R}[x]/(x^2+1)$ *is another solution of the same equation* $x^2+1=0$ *for an unknown* x*.*

Moreover, $\mathbb{R}[x]/(x^2+1)$ *is obviously a real vector space of dimension* 2*, since* $\{1, i\}$ *is its basis. This is because* $\mathbb{R}[x]$ *is a real vector space of infinite dimension with a basis* $\{1, x, x^2, ...\}$ *and every power* x^j *for* $j > 1$ *is reduced to* ± 1 *or* $\pm x$ *under modulo* (x^2+1)*.*

This allows us to write any element of the extended field $\mathbb{R}[x]/(x^2+1)$ *as a* $\mathbb{R}$*-linear combination of* 1 *and* i*,*

$$\mathbb{R}[x]/(x^2+1) = \{a+bi : a, b \in \mathbb{R}\}.$$

In this way, we have constructed the field

$$\mathbb{C} = \{a+bi : a, b \in \mathbb{R}\} = \mathbb{R}[x]/(x^2+1)$$

of complex numbers as a 2*-dimensional real vector space by "algebraically" adjoining an element* i *to* $\mathbb{R}$*.*

Since our construction of the field $\mathbb{C}$ *of complex numbers is canonical, whoever construct it, it must be isomorphic to ours.*

The main point of this example is that we extend the field $\mathbb{R}$ of real numbers in such a way that the irreducible polynomial x^2+1 over $\mathbb{R}$ has a root in it, and the desired field is the quotient ring:

$$\mathbb{R}[x]/(x^2+1),$$

where the image $\bar{x} = x$ under $\operatorname{mod}(x^2+1)$ is a root of $x^2+1=0$ and it contains $\mathbb{R}$ as a subfield. Thus, replacing x^2+1 with other irreducible polynomials over $\mathbb{R}$, we immediately see that irreducible polynomials over $\mathbb{R}$ provides extension fields of $\mathbb{R}$.

Then Gauss (1777 - 1855) proved

THEOREM 1.3. *(**Fundamental Theorem of Algebra**) The field* $\mathbb{C}$ *of complex numbers is algebraically closed.* □

Thus, whichever irreducible polynomial over $\mathbb{R}$ we choose to figure out an extension field K of $\mathbb{R}$ where it has its root, Gauss says that it is in fact a subfield of $\mathbb{C}$. Moreover, we have seen that

$$[\mathbb{C} : \mathbb{R}] = 2$$

which means that any irreducible polynomial over $\mathbb{R}$ has degree 1 or 2. Moreover, since $\mathbb{R}[x]/(x^2+1)$ is an $\mathbb{R}$-vector space of dimension 2 with a basis $\{1, i\}$, we can always express its elements in terms of these basis elements in a unique manner.

Similarly, if we start with an irreducible polynomial over the field $\mathbb{Q}$ of rational numbers, then its extension field, i.e., the one where it has its root is in fact a subfield of $\mathbb{C}$. But, there is something else here.

For example, $\mathbb{Q}[x]/(x^2+1)$ is a field extension of $\mathbb{Q}$ of degree 2 where i is. On the other hand, we have an interesting example:

EXAMPLE 1.4. *Let* $f(x) = x^4 + 2x^3 + 6x^2 + x + 9$. *Then since* $f(x) \bmod 2 = x^4 + x + 1$ *is irreducible over* $\mathbb{Z}_2$, *it is irreducible over* $\mathbb{Q}$. *Let* $\alpha = \sqrt[3]{2}$ *and* γ *be a root of* $f(x)$.

REMARK 1.4. *If one actually solves* $f(x)$ *for* x, *the* 4 *solutions are all non-real complex numbers and they are not real multiples of* $i = \sqrt{-1}$.

Then we easily see that $\alpha \notin \mathbb{Q}(\gamma)$ *since* $3 = [\mathbb{Q}(\alpha) : \mathbb{Q}]$ *never divides* $4 = [\mathbb{Q}(\gamma) : \mathbb{Q}]$. *Thus,* $\mathbb{Q}(\alpha)$ *is never a subfield of* $\mathbb{Q}(\gamma)$.

On the other hand, it is not easy to check whether i *is in* $\mathbb{Q}(\gamma)$, *although* i *has degree* 2 *over* $\mathbb{Q}$.

CLAIM 1.1. $i \notin \mathbb{Q}(\gamma)$.

One way of verifying this claim is the following:

Suppose $i \in \mathbb{Q}(\gamma)$. *Then since* $\mathbb{Q}(\gamma)$ *is a vector space with a basis* $\{1, \gamma, \gamma^2, \gamma^3\}$, $i = a + b\gamma + c\gamma^2 + d\gamma^3$ *for uniquely determined rational numbers* $a, b, c, d \in \mathbb{Q}$. *Then*

$$-1 = i^2 = (a + b\gamma + c\gamma^2 + d\gamma^3)^2 = A + B\gamma + C\gamma^2 + D\gamma^3,$$

where A, B, C *and* D *are obviously quadratic polynomials in* a, b, c, d, *and all* γ^j *for* $j > 3$ *are reduced to lower degree terms using* $\gamma^4 + 2\gamma^3 + 6\gamma^2 + \gamma + 9 = 0$.

Then since $\{1, \gamma, \gamma^2, \gamma^3\}$ *is a* $\mathbb{Q}$*-basis of* $\mathbb{Q}(\gamma)$, $(A+1) + B\gamma + C\gamma^2 + D\gamma^3 = 0$ *implies* $A, B, C,$ *and* D *must satisfy a system of polynomial equations for* a, b, c, d*:*

$$(*)\begin{cases} A+1 &= 0 \\ B &= 0 \\ C &= 0 \\ D &= 0 \end{cases}.$$

That is, this system must have rational solutions if $i \in \mathbb{Q}(\gamma)$.

Therefore, if we can show that this system does not have any rational solutions, then it just means that $i \notin \mathbb{Q}(\gamma)$.

From actual symbolic computations (using the Computer Algebra System REDUCE), the above procedure gives us the following system of polynomial equations for a, b, c, d*:*

$$(*)\begin{cases} a^2 - 18bd - 9c^2 + 36cd - 36d^2 + 1 &= 0 \\ 2ab - 2bd - c^2 - 14cd + 14d^2 &= 0 \\ 2ac + b^2 - 12bd - 6c^2 + 22cd - 31d^2 &= 0 \\ 2ad + 2bc - 4bd - 2c^2 - 4cd + 3d^2 &= 0 \end{cases}.$$

If one solves this system of quadratic equations in a, b, c, d, *using* ***the theory of Groebner bases****, one can easily check that it has no rational solutions. Therefore,* $i \notin \mathbb{Q}(\gamma)$*; hence* $\mathbb{Q}(i)$ *is not a subfield of* $\mathbb{Q}(\gamma)$.

In fact, if we examine the following solution set carefully, we recognize that d *is a solution of an irreducible polynomial over* $\mathbb{Q}$*; hence* d *is not a rational number; hence so are others.*

————————

```
answer := {
{d=root_of(222751263710994018575*d_**14
+ 327248046817526299 20*d_**12
+ 1024333138422735392*d_**10 - 6646593532140800*d_**8
+ 100778855020288*d_**6 + 742952112128*d_**4
- 1618280448*d_**2 - 16777216,d_,tag_4),
c=(d*( - 8374876905282633486301972 5\
```

```
115741370601846002617736357021762\
346872149077225*d**12-11290968900\
828018089493846300962145523331020\
460782200955244042491473112560*d**10
- 243571978407542681652414605202265\
431864678998131101052507725339652\
2745056*d**8 + 60827206431398169979\
393925549908850762335924500870169\
64936927101866496*d**6
- 103487770448543458846949576351951\
29995735833220025739745630479809152O*d**4
+ 305377420121353933353783686617421\
653272989311661892922663630770176*d**2
+3251572265947714084433951122616941117\
699119138907956589158878412 8))/10687223\
425590051434195585689901160826183925 23\
378061991375011840,
b=(d*(3396745222644812332803072264984 33\
736958200024095604577412257987889496382175*d**12
+ 502244508468234119645341028809204O895\
177839532673066779120617304562294048O*d**10
+ 1609285853130262313539396952071705271095O\
008873240506740743320221085324 48*d**8
-866634042895308806579010925914339591 35\
691814184877107105535643570260 48*d**6
+ 142938377726950567850459422368677 38\
395921770629740184932911167081344O*d**4
+ 130092449352402613432539374341841864\
089231867685964485644454091571 2*d**2
+4295453052452291492274867116715020356 3\
120654409865761800775925 76))/10687223425590051\
434195585689901160826183925233780619913750118 40,
```

a=(d*(23407577850156652015755239720127050 78\
4998336919351934069102976168784876675*d**12
+ 349970769207384193768859123983839 4352274\
27237927951188061022738319278268 0*d**10
+ 1151733389485529089562801717531 05011\
3217957243960397979630883274794 55168*d**8
-6185568810024089885246470906167464 9382421\
7194979720362629372379387008*d**6
+ 30458426905531619358808852805900 5435962\
787561210667259734018274944 0*d**4
+ 15586036997419414144587973317287 135426\
3289850109985233499157901312*d**2
-55201348856723044104108201208662058997041617\
35250198989752893 44))/9293237761382653 42103963\
973034883550102950020328749557717 40160},
{d=0,c=0,b=0,a=i},
{d=0,c=0,b=0,a= - i}
}$

End of Example. □

REMARK 1.5. *This example shows us usefulness of exact computation, i.e., symbolic computation in studying Abstract Algebra. Starting in next chapter, we are going to study the theory of Groebner bases.*

Thus, we have explicitly seen that there is an extension field of $\mathbb{Q}$ which does not contain i:

$$\mathbb{Q}[x]/(x^4 + 2x^3 + 6x^2 + x + 9).$$

Its degree over $\mathbb{Q}$ is thus 4. Thus, the degree of an extension field of $\mathbb{Q}$ could be very large if you choose an irreducible polynomial over $\mathbb{Q}$ of large degree. This means that the degree of the $\mathbb{Q}$-vector space $\mathbb{C}$

$$[\mathbb{C} : \mathbb{Q}]$$

is not finite. Then how about the degree of the $\mathbb{Q}$-vector space $\mathbb{R}$, $[\mathbb{R} : \mathbb{Q}]$? Is it a finite-dimensional vector space over $\mathbb{Q}$? For example, Eisenstein criterion says that it is not:

PROPOSITION 1.1. *(Eisenstein Criterion) Let* $f(x) = a_n x^n + \cdots + a_0$ $\in \mathbb{Z}[x]$ *be an integral polynomial, and let* p *be a prime integer. Suppose that the coefficients of* $f(x)$ *satisfies the following conditions:*

(1) p *does not divide the leading coefficient* a_n*;*
(2) p *divides all other coefficients* $a_{n-1}, ..., a_0$*;*
(3) p^2 *does not divide the constant term* a_0*.*

Then f *is irreducible in* $\mathbb{Q}[x]$*. If moreover* f *is primitive, then it is irreducible in* $\mathbb{Z}[x]$*.*

EXAMPLE 1.5. $f(x) = x^n - 2$ *is irreducible over* $\mathbb{Q}$ *for any positive integer* n*, and we have*

$$\mathbb{Q} \subset \mathbb{Q}(\sqrt[n]{2}) \subset \mathbb{R}.$$

The main reason why we have this is that the construction of the field $\mathbb{R}$ of real numbers from the field $\mathbb{Q}$ of rational numbers is not done as a way of finding a root of a polynomial equation in one variable over $\mathbb{Q}$. In fact, it took a very long time for mathematicians to figure out a mathematically legitimate method of constructing $\mathbb{R}$ from $\mathbb{Q}$, say, in terms of the so-called "completion of Cauchy sequences" via the limiting process in Real Analysis. (You should recall that every convergent sequence is Cauchy, but not conversely in general). In fact, we may say that it has taken more or less two millenniums from the Greek period until Cauchy and others finalized the construction of $\mathbb{R}$ from $\mathbb{Q}$.

3.2. Computable Rings, Fields, and Vector Spaces. We note that limiting process is not the main subject of Algebra, and moreover as long as computations are involved, our computations are supposed to be exact, i.e., there should be no approximation at all. This means that a computer, a machine to compute of finite nature, can handle the data exactly only with finitely many executions of finitely many binary operations, like exact integer arithmetic: addition, subtraction, multiplication, and division (when possible). That is, computers basically cannot have an ability to handle limiting process directly, as if it were a binary operation on numbers, like addition of two numbers. Thus, it may be a right place to introduce some terminology on computable stuffs.

DEFINITION 1.1.

a: *A (commutative) ring R is called a* ***computable ring*** *if its addition, multiplication, subtraction, and division (when possible) can be effectively performed on a computer.*

b: *A field k is call a* ***computable field*** *if it is a computable ring and the inverse of any non-zero element in F can be effectively computed.*

c: *A vector space V over a field k is a* ***computable vector space*** *if k is a computable field and each element of V can be represented on a computer, addition and subtraction in V and scalar multiplication can be effectively computed.*

EXAMPLE 1.6. *A prototypical example of a computable ring is the ring $\mathbb{Z}$ of integers. This is because the usual* 4 *exact integer arithmetic can be implemented on a computer.*

EXAMPLE 1.7. *A prototypical example of a computable field is the field $\mathbb{Q}$ of rational numbers. This is because it is a computable ring and we additionally can compute the* gcd *of any two numbers and hence we can reduce any "ratio" a/b into a non-reducible one by cancelling out* $\gcd(a,b)$ *from both a and b. Here, we note that* $\gcd(a,b)$ *can be computed only using the usual* 4 *integer arithmetic, thanks to* Euclid.

EXAMPLE 1.8. $\mathbb{Z}/m\mathbb{Z}$ *is a computable ring, because its operations are all inherited from those in $\mathbb{Z}$ and because each element $a+m\mathbb{Z}$ of $\mathbb{Z}/m\mathbb{Z}$ has a* "unique representation", $r+m\mathbb{Z}$, *where r is the unique remainder of a upon division by m, by the usual division algorithm in $\mathbb{Z}$.*

EXAMPLE 1.9. *For a prime number p, the finite field $\mathbb{Z}/p\mathbb{Z}$ is a computable field, because it is a computable ring as a ring and because one* *can find the inverse of any non-zero element of* $\mathbb{Z}/p\mathbb{Z}$ *by trial and error since it is a finite field.*

EXAMPLE 1.10. *A prototypical example of a computable vector space is* k^n *where* k *is a computable field.*

Now, going back to our main discussion, we know that $\{1, x, ..., x^{n-1}\}$ is a $\mathbb{Q}$-basis of the field,

$$\mathbb{Q}[x]/(f(x)),$$

where $f(x)$ is an irreducible monic polynomial over $\mathbb{Q}$ of degree n; hence everything in this field can be uniquely written as a $\mathbb{Q}$-linear combination of the basis elements, $1, x, ..., x^{n-1}$. Here we particularly note that the basis elements $1, x, ..., x^{n-1}$ are monomials not contributing to the leading term, i.e., the highest degree term, $LT(f)$ of $f(x)$. Here, we particularly note that the solution set of $f(x) = 0$ consists of only finite number of points in k.

From now on, we will not explicitly mention whether our rings, fields, or vector spaces are computable or not unless it is absolutely necessary to mention it clearly.

When we work with polynomials in one variables over a field k, the application of the following division algorithm is unavoidable:

ALGORITHM 3. ***(Division Algorithm in*** $k[x]$***)*** *Let* k *be a (computable) field and let* g *be a non-zero polynomial in* $k[x]$. *Then there is an* ***algorithm*** *for uniquely finding* $q, r \in k[x]$ *such that*

$$f = qg + r,$$

with $0 \leq \deg(r) < \deg(g)$.

PROOF. Since this is basically a high school algebra, we only write down the algorithm in pseudocode:

Input: g, f
Output: q, r
$q := 0; r := f$

WHILE $r \neq 0$ AND $LT(g)$ divides $LT(r)$ DO
$q := q + LT(r)/LT(g)$
$r := r - (LT(r)/LT(g)) \cdot g$

We first note that when we write down a polynomial in one variable x, we automatically assume that

$$x^n > x^{n-1} > \cdots > x > 1.$$

In this algorithm, $LT(g)$, for instance, means the highest-degree term of $g = a_n x^n + a_{n-1}x^{n-1} + ...$, so $LT(g) = a_n x^n$, etc.

We first note that, since we have the identity:

$$qg + r = (q + (LT(r)/LT(g)) \cdot g + (r - (LT(r)/LT(g)) \cdot g)$$

we all the time have $f = qg + r$ for variables q and r during each step of the algorithm.

Then the WHILE loop will terminate after finitely many steps, since the degree of $r - (LT(r)/LT(g)) \cdot g$ is either 0 or smaller than that of r. Thus, as the step in WHILE loop is increased, the degree drops to zero or $LT(g)$ does not divide $LT(r)$, i.e., $\deg(r) < \deg(g)$.

Finally, we need to show that q and r are uniquely determined. For this, let $f = q'g + r'$ be another expression. If $r \neq r'$, then

$$r' - r = (q - q')g,$$

and hence $q - q' \neq 0$; so $\deg(r' - r) = \det(q - q') + \deg(g) \geq \deg(g)$, which is impossible. Thus, $r = r'$; and hence $q = q'$. □

Since any field is an integral domain, we have:

PROPOSITION 1.2. *A polynomial $f(x)$ of degree n with coefficients in a field k has at most n roots in k.*□

Moreover,

PROPOSITION 1.3. *If k is a field, then $k[x]$ is a PID.*

PROOF. Let I be an ideal in $k[x]$. If $I = (0)$, then there is nothing to prove. If I contains a non-zero constant, then $I = (1)$, so we are done.

Otherwise, there exists a non-zero and non-constant polynomial in I, and hence let g be a non-zero polynomial of minimal degree in I. Then all the polynomial multiples of $g(x)$ are elements of I. That is, the principal ideal (g) generated by g is automatically contained in I. Now for the converse, we apply the above algorithm: let f be a polynomial in I. Then by the above algorithm, we have

$$f = qg + r,$$

with $0 \leq \deg(r) < \deg(g)$. If $\deg(r) \neq 0$, then $r = f - qg \in I$ has less degree than g, which contradicts to our choice of g. Therefore, $r \equiv 0$; hence $f \in (g)$. □

COROLLARY 1.3. *If two non-zero principal ideals (f) and (g) in $k[x]$ are the same, then g is a non-zero constant multiple of f.*

PROOF. By hypothesis, $f = hg$ for some $h \in k[x]$; hence

$$\deg(f) = \deg(h) + \deg(g).$$

Thus, $\deg(f) \geq \deg(g)$. Similarly, $\deg(g) \geq \deg(f)$. Thus, $\deg(h) = 0$, which is exactly the same as saying h is a non-zero constant in k. □

3.3. Polynomial Equations in One Variable. Now, we consider how to find the solution set of a given system of polynomial equations in one variable x over an algebraically closed field k:

$$\begin{cases} f_1(x) &= 0, \\ f_2(x) &= 0, \\ \quad\cdots & \\ f_q(x) &= 0. \end{cases}$$

Let $a \in k$ be a solution of the system. Then for any $f \in (f_1, ..., f_q)$, the ideal generated by $f_1, ..., f_q$, there exist $h_i \in k[x]$ such that

$$f = h_1 f_1 + \cdots + h_q f_q,$$

so that $f(a) = h_1(a)f_1(a) + \cdots + h_q(a)f_q(a) = 0 + \cdots + 0 = 0$; hence a is a common zero, i.e., a solution, of all the polynomials in the ideal

$(f_1, ..., f_q)$. Conversely, if $a \in k$ is a common zero of all the polynomials in the ideal $(f_1, ..., f_q)$, then certainly it is a solution of the polynomials f_i in the above system.

Therefore, as long as figuring out the solution set of a system of polynomial equations is concerned, both the system and the ideal generated by the polynomials in the system have the same solution set. Namely, the problem of figuring out the solution set of the system is changed to that of the ideal generated by the polynomials in the system.

Then, since $k[x]$ is a *PID*, we know that there must exist a single polynomial $g \in k[x]$ such that

$$(g) = (f_1, ..., f_q),$$

and consequently, the problem of solving the system is then changed to the problem of solving a single equation for x:

$$g = 0.$$

Therefore, the only thing remained is: how to figure out the polynomial g from the given polynomials $f_1, ..., f_q$.

Now, only for the simplicity of our explanation, let's assume that we are asked to solve a system of two polynomial equations $f_1(x) = 0$ and $f_2(x) = 0$ in one variable x over an algebraically closed field k. Then according to our discussion, it is reduced to find a polynomial $g \in k[x]$ such that

$$(g) = (f_1, f_2).$$

If this equality happens, then g certainly divides both f_1 and f_2, and $g = h_1 f_1 + h_2 f_2$ for some $h_1, h_2 \in k[x]$; hence if h divides both f_1 and f_2 then it also divides g. Thus, what we are looking for g is the $\gcd(f_1, f_2)$. Since $k[x]$ is a *PID*, it is a *UFD*; hence the $\gcd(f_1, f_2)$ certainly exists in $k[x]$. We thus have proved

PROPOSITION 1.4. *Let $f_1, f_2 \in k[x]$. Then*

$$(f_1, f_2) = (\gcd(f_1, f_2)). \quad \square$$

Moreover, if we have more than two polynomials, $f_1, f_2, ..., f_s$, then since we are in a *UFD* $k[x]$, an easy inductive argument shows that

$$\gcd(f_1, ..., f_s) = \gcd(f_1, \gcd(f_2, ..., f_s)).$$

Then since we are in a *PID* $k[x]$,

$$(f_1, \gcd(f_2, ..., f_s)) = (f_1, ..., f_s)$$

holds. Thus, we have

$$(f_1, ..., f_s) = (\gcd(f_1, ..., f_s)).$$

Therefore, by inductively computing

$$\begin{aligned} \gcd(f_1, ..., f_s) &= \gcd(f_1, \gcd(f_2, ..., f_s)) = \gcd(f_1, \gcd(f_2, \gcd(f_3, ..., f_s)) \\ &= \cdots \end{aligned}$$

we can compute a single generator for

$$(f_1, ..., f_s) = (\gcd(f_1, ..., f_s)),$$

provided that we have an algorithm for computing $\gcd(f_1, f_2)$:

ALGORITHM 4. ***(Euclid's Algorithm)*** *For any $f, g \neq 0 \in k[x]$, the following procedure computes $\gcd(f, g)$:*

Input: f, g

Output: h

Initialization: $h := f, s := g$

WHILE $s \neq 0$ ***DO***

$rem :=$ *remainder*(h, s)

$h := s$

$s := rem$

Then this algorithm computes $\gcd(f, g)$: For this we first write $f = qg+r$ with uniquely determined q, r with $0 \leq \deg(r) < \deg(g)$.Then since

$k[x]$ is a *PID*, $(f, g) = (f - qg, g)$ holds, and hence

$$\gcd(f, g) = \gcd(g, r).$$

Now, if $r \neq 0$, we apply the same argument to g this time replacing the role of g with r: $g = q'r + r'$ with $0 \leq \deg(r') < \deg(r)$, inductively to obtain

$$\gcd(f, g) = \gcd(g, r) = \gcd(r, r') = \gcd(r'.r") = ...$$

where $\deg(g) > \deg(r) > \deg(r') > \deg(r") > \cdots$. Thus, after finitely many steps we will stop, and this is exactly what the above WHILE loop says.

REMARK 1.6. *If we compare this algorithm with Algorithm 2, then we see that they are exactly the same, as algorithm.*

EXAMPLE 1.11. *Thus,*

$$\begin{aligned} \gcd\left(\begin{array}{c} x^3 - 3x + 2, x^4 - 1, \\ x^6 - 1 \end{array}\right) &= \gcd(x^3 - 3x + 2, \gcd(x^4 - 1, x^6 - 1)) \\ &= \gcd(x^3 - 3x + 2, x^2 - 1) \\ &= x - 1. \end{aligned}$$

In summary, when we are in $k[x]$ where k is an algebraically closed field, we need not to solve a system of polynomial equations over k, instead, we compute their gcd, and try to find its solutions in k. Because of this reason, Gauss basically settled down everything for the case when we are in $k[x]$, where k is an algebraically closed field, by proving his very famous theorem, Theorem 1.3: (**Fundamental Theorem of Algebra**) The field $\mathbb{C}$ of complex numbers is algebraically closed.

3.4. Polynomial Equations in More Than One Variables. But,

what happens when we consider the quotient ring

$$k[x, y]/(f(x, y)),$$

where $f(x, y)$ is an irreducible polynomial over k in two variables x, y?

First of all, we recall that $k[x,y]$ is not a *PID*. Then since irreducible elements and prime elements are the same in a *UFD*, the quotient ring is an integral domain. For example, if $f(x,y) = x+y$ then $f(x,y) \in (x,y)$ a maximal ideal in $k[x,y]$, because $k[x,y]/(x,y)$ is the bottom field k. Thus, $(x+y)$ is not a maximal ideal of $k[x,y]$; hence the quotient ring associated to an irreducible polynomial $f(x)$ is an integral domain that is not a field, in general. In this direction, we note that $(f(x,y))$ is the kernel of the natural ring homomorphism $k[x,y] \longrightarrow k[x,y]/(f(x,y))$ which is always surjective.

On the other hand, we know that it is a k-vector space, so there must be a k-basis for it. What is a k-basis of it? Is it going to be those monomials not contributing to the leading term $LT(f(x,y))$ of $f(x,y)$? It was the case when we are dealing with polynomials in one variable. We will later see that it is so if the solution set consists of only finite number of points. Of course, $f(x,y) = 0$ has infinitely many solutions, so that we are far away from this in the case of $k[x,y]/(f(x,y))$.

The second question is, an ideal membership question: Suppose you are given an ideal, say, generated by two polynomials $f(x,y), g(x,y) \in k[x,y]$, and let $h(x,y)$ be a polynomial in $k[x,y]$. The question is whether $h(x,y)$ is a member of the ideal $(f(x,y), g(x,y))$ or not.

When we deal with polynomials in one variable, it is very easy to answer this question: Since $k[x]$ is a *PID*,

$$(f(x), g(x)) = (d(x)),$$

where $d(x) = \gcd(f(x), g(x))$ and hence we simply divide a given $h(x) \in k[x]$ by $d(x)$ to see if it belongs to $(f(x), g(x))$. If the remainder is zero, it is a member of the ideal $(f(x), g(x))$; otherwise, it is not.

But starting with $k[x,y]$, since a polynomial ring with more than one unknown is never a *PID*, the ideal membership question suddenly becomes a very difficult problem.

In the next chapter, we are going to study the Theory of Groebner basis and it is this theory which enables us to computationally answer these difficult questions affirmatively.

Exercises for Section 1.3

EXERCISE 1.9. *Show that the polynomial* $x^3 - 3$ *is irreducible over* $\mathbb{Q}$.

EXERCISE 1.10. *Show that the polynomial* $x^4 + 2x^3 + 6x^2 + x + 9$ *is irreducible over* $\mathbb{Z}$. *[Hint: Show first that it is irreducible over the finite field* $\mathbb{Z}_2 = \mathbb{Z}/2\mathbb{Z}$, *and then try to use the canonical ring homomorphism* $\mathbb{Z} \longrightarrow \mathbb{Z}/2\mathbb{Z}$.

EXERCISE 1.11. *Compute a generator for the ideal* $(x^4 - 1, x^6 - 1)$ *in* $k[x]$. *[Hint: Use Algorithm 4].*

EXERCISE 1.12. *Compute a generator for the ideal* $(x^3 - 3x + 2, x^4 - 1, x^6 - 1)$ *in* $k[x]$.

EXERCISE 1.13. *Let* I *be an ideal in a ring* R. *Prove that* I *is a prime ideal if and only if* R/I *is an integral domain.*

EXERCISE 1.14. *Let* I *be an ideal in a ring* R. *Prove that* I *is a maximal ideal if and only if* R/I *is a field.*

4. Why Geometry and Polynomial Functions?

The 3-dimensional $\mathbb{R}$-vector space $\mathbb{R}^3$, which is usually called the *Euclidean 3-space*, is particularly very special to us human beings, because we reside there. We can draw the graph of a real-valued function $z = f(x, y)$ in two variables within this space.

In particular, a solution set of a system of polynomial equations over a field k

$$\begin{cases} f_1(x_1, ..., x_n) &= 0, \\ f_2(x_1, ..., x_n) &= 0, \\ \quad .... & \\ f_s(x_1, ..., x_n) &= 0 \end{cases}$$

is a subset of k^n, $n \geq 1$. Thus, they are a geometric object in k^n. That is,

DEFINITION 1.2. *Any subset of k^n is called a* ***geometric object*** *in k^n.*

In particular,

DEFINITION 1.3. *The n-dimensional k-vector space k^n is called the* ***affine n-space****.*

When $n = 1$, we reviewed what we need in the previous section. For a general $n \geq 2$, we have already seen that $k[x, y]$ is not a *PID*, and this usually causes us a real difficulty in solving a system of polynomial equations.

On the other hand, we cannot draw the graph of a function if the number of its (independent) variables is bigger than 3. But, it is quite often the case that geometric observations tell us which direction we should go ahead, even when we are dealing with algebraic objects.

EXAMPLE 1.12. *Suppose we are given a system of linear equations in n unknowns. Then, the so-called elementary row operations in Linear Algebra are nothing but those algebraic manipulations we apply to the given set of equations,* without changing the solution set of the given system, *and hence simplifying the given equations, we finally judge whether the system has a non-empty solution set or not. If so, then we can have a means to find all the solutions of the system. Here, the crucial point is that the solution set must remain* ***invariant*** *under algebraic manipulations. In what follows, this important point that a solution set be remained invariant will be always kept in out mind even when we try to solve a system of non-linear, i.e., polynomial equations in n unknowns.*

4.1. When Polynomials Are Called Polynomial Functions? To make our discussions more clear, we would like to introduce a few geometric terminologies.

We know that the terminologies

"*functions*" and "*maps*"

are set-theoretically the same. But, historically they have different usage so they mean to have different meanings: **When we want to "compare" one set with another we are supposed to use the terminology "map", but when we want to "evaluate" points of k^n, we are supposed to use the terminology "function".** More precisely, when we compare two sets, we usually use the horizontal arrows, but

DEFINITION 1.4. *Let k^n be the affine n-space. A* vertical *map*

$$\begin{array}{c} k^n \\ \downarrow \\ k \end{array}$$

is called a ***function*** *(or an* ***evaluation map****).*

It is called a function because its values in k are assumed to be well known to us.

Among functions from k^n to k, **polynomial functions** play a very important role in this book because it basically connects geometry with algebra. First of all, everybody knows that polynomial functions are one of the simplest functions, and they are continuous, repeatedly differentiable everywhere, etc. Other usefulness of polynomial functions needs further explanation.

Elements of $k[x_1, ..., x_n]$ are called polynomials in n unknowns with coefficients in k and they are algebraic objects, but each element of $k[x_1, ..., x_n]$ is also a function from k^n to k. We recall that a map consists of three things; its *domain*, its *target*, and an *assignment*. Then since we assume that we know everything very well about elements of the target, k, and since properties of polynomials, as algebraic objects, are well studied in Algebra, by considering enough *fibres*, i.e., inverse images of various points in k, we really hope that we can have a better understanding of the properties of geometric objects in the domain, k^n. As we recalled already, the solution set of a single polynomial equation $f(x_1, ..., x_n) = 0$ is in fact the fibre of the polynomial function at $0 \in k$:

$$\begin{array}{c} k^n \\ \downarrow f \\ k \end{array} .$$

Similarly, the solution set of a system of polynomial equations

$$f_i(x_1, ..., x_n) = 0, \;\; i = 1, 2, ..., s$$

is in fact the intersection of the fibres $f_i^{-1}(0)$ of the polynomial functions

$$f_i(x_1, ..., x_n)$$

at $0 \in k$.

EXAMPLE 1.13. *Let $k = \mathbb{F}_2 = \{0, 1\}$. Then as a polynomial in $\mathbb{F}_2[x, y]$,*

$$f(x, y) = x(x^{2-1} - 1) + y(y^{2-1} - 1)$$

is not the zero polynomial 0 at all, but as a function it is the zero function, that is, for all possible choices $(0,0), (1,0), (0,1), (1,1)$ of points of $\mathbb{F}_2^2$, their values are all zero.

This was not a problem, say until the end of the 19th century, because people did not even think of finite fields. But, to make it precise we definitely need the following proposition.

PROPOSITION 1.5. *Let $f \in k[x_1, ..., x_n]$. If k is an infinite field, then $f = 0$ as a polynomial in $k[x_1, ..., x_n]$ if and only if*

$$\begin{array}{c} k^n \\ \downarrow f \\ k \end{array}$$

is the zero function

$$\begin{array}{c} k^n \\ \downarrow 0 \\ k \end{array} .$$

PROOF. Obviously, the zero polynomial is the zero function. Now, suppose that evaluation of every point of k^n under a polynomial function f is zero: $f(a_1, ..., a_n) = 0$ at each point $(a_1, ..., a_n)$ of k^n. We are going to inductively prove that f is the zero polynomial.

If $n = 1$, then we would have infinitely many solutions of a polynomial equation $f(x) = 0$ unless $f(x) \equiv 0$, because we already know that any polynomial with coefficients in an integral domain of degree d can have at most d solutions by Proposition 1.2.

Now assume that the converse holds for $n-1$, and let f be a polynomial function in n variables which vanishes at each point of k^n. We can write is as a polynomial in one variable in x_n:

$$f = \sum_{i=0}^{m} g_i(x_1, ..., x_{n-1})x_n^i$$

For a fixed $(a_1, ..., a_{n-1}) \in k^{n-1}$, f is a polynomial in one variable x_n, and by hypothesis it has infinitely many solutions $a_n \in k$; hence f must be the zero polynomial in x_n; consequently, each $g_i(a_1, ..., a_{n-1}) = 0$. Then since $(a_1, ..., a_{n-1})$ was an arbitrary point in k^{n-1}, the induction hypothesis says that each polynomial $g_i(x_1, ..., x_{n-1})$ is the zero polynomial 0 in $k[x_1, ..., x_{n-1}]$. Therefore, f is the zero polynomial in $k[x_1, ..., x_n]$. $\square$

COROLLARY 1.4. *Let $f, g \in k[x_1, ..., x_n]$. If k is an infinite field, then $f = g$ as polynomials in $k[x_1, ..., x_n]$ if and only if*

$$\begin{array}{ccc} k^n & & k^n \\ \downarrow f & = & \downarrow g \\ k & & k \end{array}$$

as polynomial functions on k^n. $\square$

In Remark 1.10 of this Chapter, the next material to the current subject is presented.

4.2. Algebraic Varieties. We review a few terminologies for the solution set of a system of polynomial equations.

DEFINITION 1.5. *Let $f_1, ..., f_s \in k[x_1, ..., x_n]$.*

$$V(f_1, ..., f_s) = \{(a_1, ..., a_n) \in k^n : f_i(a_1, ..., a_n) = 0, \ i = 1, 2, ..., s\}$$

is called the ***(affine) variety*** *defined by the polynomials $f_1, ..., f_s$ in n variables $x_1, ..., x_n$, and polynomials $f_1, ..., f_s$ are called the* ***defining equations*** *of the variety.*

More generally,

DEFINITION 1.6. *Let I be an ideal in $k[x_1, ..., x_n]$.*

$$V(I) = \{(a_1, ..., a_n) \in k^n : f(a_1, ..., a_n) = 0, \ \textit{for each } f \in I\}$$

is called the (***affine***) ***variety*** *defined by the ideal I.*

DEFINITION 1.7. *Let S be a subset of k^n. Then*

$$I(S) = \{f \in k[x_1, ..., x_n] : f(a_1, ..., a_n) = 0 \textit{ for each } a \in S\}$$

is called the ***ideal vanishing*** *at S or the* ***ideal of*** *S.*

COROLLARY 1.5. *$I(k^n) = \{0\}$ if and only if the zero polynomial $f = 0$ in $k[x_1, ..., x_n]$ is the only polynomial function vanishing everywhere. In particular, if the field k is infinite then $I(k^n) = \{0\}$.* □

COROLLARY 1.6. *$I(\varnothing) = k[x_1, ..., x_n]$.*

PROOF. This is because $k[x_1, ..., x_n] = (1)$ as an ideal, and the constant polynomial 1 never vanishes at any points, that is, 1 vanishes at no points of k^n, i.e., 1 vanishes at the empty subset; hence so are all of its polynomial multiples $= k[x_1, ..., x_n]$. □

LEMMA 1.4. *If $V, W \subseteq k^n$ are affine varieties, then so are their finite union and finite intersection.*

PROOF. Write $V = V(f_1, ..., f_s)$ and $W = V(g_1, ..., g_t)$. Then obviously, $V \cap W = V(f_1, ..., f_s, g_1, ..., g_t)$, so it is an affine variety. We note that $V \cup W = V(f_i g_j : 1 \leq i \leq s, 1 \leq j \leq t)$. The $\subseteq$ inclusion is again obvious. For another inclusion, suppose $(a_1, ..., a_n) \in V(f_i g_j)$. If it belongs to $V(f_i)$, then we are done. Otherwise, there exists one j such that f_j does not vanish at this point $(a_1, ..., a_n)$, but the t products $f_i g_j$, $1 \leq j \leq t$, vanishes at the point; hence it must be an element of $V(g_j)$. □

LEMMA 1.5. *$(f_1, ...f_s) \subseteq I(V(f_1, ..., f_s))$, where $f_1, ..., f_s \in k[f_1, ..., f_s]$.*

PROOF. Let $f \in (f_1, ..., f_s)$, so there exists $h_1, ..., h_s \in k[x_1, ..., x_n]$ such that $f = h_1 f_1 + \cdots + h_s f_s$. Since each f_i vanishes at each point $(a_1, ..., a_n)$ of $V(f_1, ..., f_s)$, f obviously vanishes at each point of the variety $V(f_1, ..., f_s)$. □

REMARK 1.7. *The converse of this lemma does not hold in general, because, for instance, since $V(x^2, y^2) = V(x, y) = \{(0,0)\}$, we have*

$$(x^2, y^2) \subsetneq (x, y) = I(V(x, y)) = I(V(x^2, y^2)).$$

This is because, the set of all polynomials vanishing the origin $(0,0)$ consists of those polynomials with no non-zero constant terms; hence it is generated by x and y.

PROPOSITION 1.6. *Let V and W be affine varieties in k^n. Then:*

(1) *$V \subseteq W$ if and only if $I(V) \supseteq I(W)$.*
(2) *$V = W$ if and only if $I(V) = I(W)$.*

PROOF. If $V \subseteq W$, then any polynomial function vanishing on W vanishes on V. Conversely, since $W = V(f_1, ..., f_s)$ for some $f_1, ..., f_s \in k[x_1, ..., x_n]$, and since these polynomials are in $I(V(f_1, ..., f_s))$ by the above lemma, we see that they also vanish on V by hypothesis. Therefore, V is contained in the solution set W of the polynomial equations

$$f_1 = 0, ..., f_s = 0.$$

The second statement is an immediate consequence of the first statement. □

PROPOSITION 1.7. *Let $V = V(I), W = V(J)$ be two affine varieties in k^n. Then*

(1) *$V(I) \cup V(J) = V(IJ) = V(I \cap J)$.*

(2) $V(I) \cap V(J) = V(I+J)$.

PROOF. In the next section, it is given as exercises that the two subsets IJ and $I+J$ of $k[x_1, ..., x_n]$ defined by

$$\begin{aligned} IJ &= \left\{ \sum_{i=1}^{\text{finite}} f_i g_i : f_i \in I, g_i \in J \right\}, \\ I+J &= \{f+g : f \in I,\ g \in J\}, \end{aligned}$$

are ideals of $k[x_1, ..., x_n]$. Then, we obviously have

$$IJ \subseteq I \cap J,$$

and $I+J$ is the smallest ideal containing both I and J, i.e., containing the *subset* $I \cup J$.

Part (1). Suppose $x \in V(IJ)$. In particular, we have $g(x)h(x) = 0$ in the field k, which is an integral domain, for each $g \in I$ and $h \in J$. Thus, either $g(x) = 0$ or $h(x) = 0$; hence $x \in V(I) \cup V(J)$. Conversely, suppose $x \in V(I) \cup V(J)$. Then either $g(x) = 0$ in k for each $g \in I$ or $h(x) = 0$ in k for each $h \in J$; hence $g(x)h(x) = 0$ for any $g \in I$ and any $h \in J$; hence $\sum_{i=1}^{\text{finite}} f_i(x)g_i(x) = 0$ for any finite sum $\sum_{i=1}^{\text{finite}} f_i g_i$; hence $x \in V(IJ)$. Thus, we have proved that

$$V(IJ) = V(I) \cup V(J).$$

Now, suppose $x \in V(I) \cup V(J)$. Then either $x \in V(I)$ or $x \in V(J)$; hence either $g(x) = 0$ in k for each $g \in I$ or $h(x) = 0$ for each $h \in J$; hence $f(x) = 0$ in k for any $f \in I \cap J$. Thus, $x \in V(I \cap J)$. Thus, $V(I) \cup V(J) \subseteq V(I \cap J)$ holds. Then since $IJ \subseteq I \cap J$, we have $V(I \cap J) \subseteq V(IJ)$, so we have

$$V(I) \cup V(J) \subseteq V(I \cap J) \subseteq V(IJ) = V(I) \cup V(J).$$

Part (2). Suppose $x \in V(I+J)$. Then $x \in V(I)$ because $I \subseteq I+J$; similarly, $x \in V(J)$. Thus, $x \in V(I) \cap V(J)$. For the reverse inclusion, suppose $x \in V(I) \cap V(J)$, and let h be *any* polynomial in $I+J$. Thus, there exists $f \in I$ and $g \in J$ such that $h = f + g$. Then, $h(x) = f(x) + g(x) = 0 + 0$ in k; hence $x \in V(I + J)$. □

Exercises for Section 1.4

EXERCISE 1.15. *Show that the projection map $\pi : \mathbb{R}^2 \longrightarrow \mathbb{R}$ is not a closed map, that is, the image of an affine variety is not an affine variety. [Hint: Consider the affine variety defined by $xy = 1$].*

EXERCISE 1.16. *What is the vanishing ideal $I(V(x^2, y^2))$ of the variety $V(x^2, y^2)$?*

EXERCISE 1.17. *Let k be a field. Prove the following equalities:*

(1) $(x + y, x - y) = (x, y)$ *in* $k[x, y]$

(2) $V(x + xy, y + xy, x^2, y^2) = V(x, y)$ *in* k^2. *[Hint:* $x = (x + xy) - \frac{1}{2}(x^2 + y^2 - (x + xy - (y + xy))^2$*].*

EXERCISE 1.18. *Let $V \subset \mathbb{R}^3$ be the space curve parametrically given by (t^2, t^3, t^4).*

(1) *Prove that V is an affine variety in $\mathbb{R}^3$.*

(2) *What is a defining polynomial equation of V?*

(3) *Determine $I(V)$.*

EXERCISE 1.19. *Let S be a subset of k^n. Prove that $I(S)$ is an ideal of $k[x_1, ..., x_n]$.*

EXERCISE 1.20. *Compute a generator of the vanishing ideal $I(V)$ in $k[x]$ of the affine variety V defined by two polynomial equations in one variable, $x^5 - 2x^4 + 2x^2 - x = 0$ and $x^5 - x^4 - 2x^3 + 2x^2 + x - 1 = 0$.*

EXERCISE 1.21. *Prove that a finite set on k^n is an affine variety.*

EXERCISE 1.22. *Prove that the x-axis in $\mathbb{R}^2$ with removed origin $(0, 0)$ is not an affine variety. [Hint. Proposition 1.5].*

5. Hilbert's Basis Theorem and Ascending Chain Condition

We recall the definition of an ideal:

DEFINITION 1.8. *An **ideal** I of a commutative ring R is an additive subgroup of R such that whenever $a \in I$, we have the following absorption property: for each $r \in R$,*

$$ra \in I.$$

DEFINITION 1.9. *Let I be an ideal in a commutative ring R.*

a: *I is **generated** by a subset S of I if, for each element a of I, there exist $s_1, ..., s_q \in S$ such that de*

$$a = h_1 s_1 + \cdots + h_q s_q,$$

where s_j's are elements of S and h_j are elements of R.

b: *I is called **finitely generated** by $f_1, ..., f_s \in R$ if $I = (f_1, ..., f_s)$, where*

$$(f_1, ..., f_s) = \{h_1 f_1 + \cdots + h_s f_s : h_i \in R\}.$$

It is very easy to prove the next lemma, so it is left as an exercise.

LEMMA 1.6. *$(f_1, ..., f_s)$ is an ideal of R.* □

LEMMA 1.7. *Let I be a polynomial ideal of the polynomial ring $k[x_1, ..., x_n]$ in n variables over a field k. If I is generated by $f_1, ..., f_s$, then $V(I) = V(f_1, ..., f_s)$.*

PROOF. Here, $V(f_1, ..., f_s)$ is by definition the solution set of the system of polynomial equations,

$$\begin{cases} f_1 &= 0, \\ f_2 &= 0, \\ \cdots & \\ f_s &= 0. \end{cases}$$

Since $\{f_1, ..., f_s\}$ is a subset of I,we clearly have $V(I) \subseteq V(f_1, ..., f_s)$. Now for another inclusion, let $(a_1, ..., a_n)$ be a point in $V(f_1, ..., f_s)$. Then, for each f_i, we have $f_i(a_1, ..., a_n) = 0$. Then since $I = (f_1, ..., f_s)$, any $f \in I$ is written as $f = h_1 f_1 + \cdots + h_s f_s$; hence

$$\begin{aligned} f(a_1, ..., a_n) &= h_1(a_1, ..., a_n) f_1(a_1, ..., a_n) + \cdots \\ &\quad + h_s(a_1, ..., a_n) f_s(a_1, ..., a_n) \\ &= 0 + \cdots + 0 \\ &= 0. \end{aligned}$$

This exactly means that $(a_1, ..., a_n) \in V(I)$. □

DEFINITION 1.10. *Let I, J be two polynomial ideals of $k[x_1, ..., x_n]$.*

(1) The **sum** of I and J, denoted by $I + J$, is the subset

$$I + J = \{f + g : f \in I, g \in J\}.$$

(2) The **product** of I and J, denoted by IJ, is the subset

$$IJ = \left\{ \sum_{i=1}^{\text{finite}} f_i g_i : f_i \in I, g_i \in J \right\}.$$

LEMMA 1.8. (1) *$I + J$ is the smallest ideal of $k[x_1, ..., x_n]$ containing the subset $I \cup J$.*
(2) *IJ is an ideal of $k[x_1, ..., x_n]$.*

PROOF. We leave their proofs as an exercise. □

PROPOSITION 1.8. *If $I = (f_1, ..., f_s)$ and $J = (g_1, ..., g_t)$ are finitely generated ideals, then*

(1) $I + J = (f_1, ..., f_s, g_1, ..., g_t)$.
(2) $IJ = (f_i \cdot g_j : 1 \leq i \leq s, 1 \leq j \leq t)$.

PROOF. Part (1). Since $(f_1, ..., f_s, g_1, ..., g_t)$ is an ideal containing both I and J, we have $I + J \subseteq (f_1, ..., f_s, g_1, ..., g_t)$, and the reverse inclusion is obvious.

Part (2). We clearly have $(f_i \cdot g_j : 1 \leq i \leq s, 1 \leq j \leq t\} \subseteq IJ$. The reverse inclusion is very routinely verified. □

PROBLEM 1. *Then, what is the generating set of the ideal $I \cap J$ when both polynomial ideals $I = (f_1, ..., f_s)$ and $J = (g_1, ..., g_t)$ of $k[x_1, ..., x_n]$ with $n \geq 2, s \geq 2, t \geq 2$, are finitely generated ideals?*

When $n = 1$, since $k[x]$ is a PID, every ideal is principal, so that $I = (f(x))$ and $J = (g(x))$ and hence we have an obvious answer,

$$(f(x)) \cap (g(x)) = (\mathrm{lcm}(f(x), g(x)),$$

because $f(x)$ divides $h(x)$ and $g(x)$ divides $h(x)$ if and only if $\mathrm{lcm}(f(x), g(x))$ divides $h(x)$.

Even when $n \geq 2$, if $I = (f)$ and $J = (g)$ are principal ideals, then by Lemma 5.2, for instance, we still have the answer:

$$(f) \cap (g) = (\mathrm{lcm}(f, g)).$$

But, when $n \geq 2$ and the polynomial ideals I and J are non-principal, it has been generally considered to be very difficult to answer this "simple" question, at least until 1965. *In order to answer this basic question, we definitely need the Theory of Groebner bases of which elementary concepts and results are very clearly expounded in Chapter 2 of this book.*

5.1. Hilbert's Basis Theorem And Ascending Chain Condition. For an easiness of our exposition, we are now going to prove an important theorem. As long as constructibility of a mathematical objects is concerned, this part of mathematics is the core material we should fully understand.

PROPOSITION 1.9. *The following two statements are equivalent in a commutative ring R.*

- **a:** *For any ideal I in R, there exists elements $f_1, ..., f_s \in R$ such that $I = (f_1, ..., f_s)$.*
- **b:** *For any ascending chain of ideals of R,*

$$I_1 \subseteq I_2 \subseteq \cdots \subseteq I_n \subseteq \cdots,$$

there exists a number N such that $I_N = I_j$ for any $j \geq N$. That is, every increasing chain of ideals in R becomes stationary.

PROOF. The $\Rightarrow$ direction. We simply note that the union

$$\bigcup_{j=1}^{\infty} I_j$$

of all these ideals I_j, $j = 1, 2, ...$, is an ideal; hence it is finitely generated by part a, say by $f_1, ..., f_s$ in R. Thus, each f_i belongs to some I_{ji} appearing in the chain; hence all of the f_i's belong to the biggest one I_N among these ideals I_{j_i}; hence $I_j \subseteq I_N$ if $j \geq N$.

Now if part a is not the case, i.e., if there exists an ideal I in R that is not finitely generated, then for any $f_1 \in I$, (f_1) does not contain I; hence there exists $f_2 \in I - (f_1)$ such that $(f_1) \subsetneq (f_1, f_2)$. Then since (f_1, f_2) does not contain I, there exists $f_3 \in I - (f_1, f_2)$ such that $(f_1) \subsetneq (f_1, f_2) \subsetneq (f_1, f_2, f_3)$. We inductively continue the same process to obtain a strictly increasing chain of ideals in R, which is the negation of part b. □

DEFINITION 1.11. *The second statement of this proposition is called the* ***Ascending Chain Condition*** *(ACC), and any commutative ring satisfying this condition is called a* ***Noetherian ring***.

REMARK 1.8. *It is well known that Emmy Neother, a daughter of Max Noether, a famous Algebraic Geometer in the* 19th *century, was the first mathematician who recognized the importance of this condition; hence in order to celebrate her works, her name is cited whenever a (commutative) ring satisfies the ACC condition.*

Here is a generalized version of Hilbert Basis Theorem:

THEOREM 1.4. ***(Hilbert's Basis Theorem)*** *If R is a Noetherian ring, then so is $R[x]$.*

THE FIRST PROOF OF THEOREM 1.4. Let J be an ideal in $R[x]$. By Proposition 1.9, it is enough to show that J is finitely generated. We may assume that J is a proper non-zero ideal of $R[x]$. For $n \geq 0$, define

$$I_n = \{LC(f) : f \in J \text{ such that } \deg(f) = n\} \cup \{0\},$$

where $LC(f)$ is the leading coefficient of f. In particular, I_0 consists of all constant polynomials in J, i.e., $I_0 = J \cap R$, and $I_0 \subseteq I_n$ for each n because $I_0 x^n \in J$. Then one can easily show that for each $n \geq 0$, I_n is an ideal in R, satisfying $I_n \subseteq I_{n+1}$. (For, if $a \in I_n$, there exists $f = ax^n + ...$ in J and hence $f \cdot x = ax^{n+1} + ... \in J$; thus $a \in I_{n+1}$). Then since R is Noetherian, there exists N such that $I_N \supseteq I_n$ for each $n \geq N$. Again since R is Noetherian, each I_i is finitely generated, say by $r_{i1}, ..., r_{it_i}$:

$$I_i = (r_{i1}, ..., r_{it_i}).$$

Now, let f_{ij} be the polynomial in J of degree i with leading coefficient r_{ij} for each $j, 1 \leq j \leq t_i$ where $i = 0, 1, 2, ..., N$. Then obviously, we have an ideal in $R[x]$,

$$J^* = (f_{ij} \in J : 1 \leq j \leq t_i,\ i = 0, 1, ..., N),$$

generated by these f_{ij}. Then by construction, it is contained in J. Thus, if we can show that $J \subseteq J^*$ then we are done.

For this, let $f \in J$ be a polynomial in x of degree n. If $n = 0$, then f is a constant polynomial in I_0; hence it is in J^*. Now for $n > 0$, we assume that all the elements of J of degree at most $n - 1$ are in J^* (induction hypothesis). Let $r = LC(f)$. We then have two cases: either $n \leq N$ or not.

If $n \leq N$, then $r \in I_n$; hence

$$r = \sum_{j=1}^{t_n} s_j r_{nj}$$

for some $s_j \in R$. Then obviously, the polynomial

$$g = \sum_{j=1}^{t_n} s_j f_{nj}$$

is a polynomial in J of degree n and whose leading coefficient $LC(g)$ is r because each $f_{nj} = r_{nj}x^n + ...$ Moreover, it is in J^*. Since f and g have the same leading coefficient r, their difference $f - g$ is a polynomial in J

of degree $\leq n-1$. Thus, by induction hypothesis, it is in J^*; consequently $f = (f-g) + g \in J^*$.

Otherwise, i.e., if $n > N$, then the leading coefficient $r = LC(f) \in I_n = I_N$; hence

$$r = \sum_{j=1}^{t_N} s_j r_{Nj}$$

for some $s_j \in R$. Then the polynomial

$$g = \sum_{j=1}^{t_N} s_j x^{n-N} f_{Nj}$$

in J has degree n and its leading coefficient is again r. Moreover, it is in J^*. Thus, $f - g$ is a polynomial in J of degree $\leq n-1$, and by induction hypothesis, $f - g \in J^*$; thus, $f = (f-g) + g \in J^*$. □

Now, here is the second proof of Hilbert Basis Theorem, Theorem 1.4. It is known that this "indirect" proof by a method of contradiction is the shortest proof for his theorem.

The Second Proof of Theorem 1.4. In what follows, we are going to prove:

If $R[x]$ is not noetherian, then R is not noetherian ring.

Let I be an ideal of $R[x]$ that is not finitely generated. Let $f_1 \in I$ be a polynomial of least degree. If f_k, $k \geq 1$, has already been chosen, let f_{k+1} be a polynomial of least degree in $I - (f_1, ..., f_k)$. For these polynomials f_k, let n_k be the degree of f_k and $a_k \in R$ be the leading coefficient of f_k, for $k = 1, 2, ...$ Then by our choice of f_k, we see that

$$n_1 \leq n_2 \leq \cdots.$$

Moreover, the increasing chain of ideals in R,

$$(a_1) \subset (a_1, a_2) \subset \cdots$$

must be a strictly increasing chain of ideals in R. For, if $(a_1, ..., a_k) = (a_1, ..., a_{k+1})$, then

$$a_{k+1} = \sum_{i=1}^{k} b_i a_i, \qquad b_i \in R$$

and hence the polynomial

$$p := f_{k+1} - \sum_{i=1}^{k} b_i x^{n_{k+1}-n_i} f_i \in I - (f_1, ..., f_k)$$

is of lower degree than f_{k+1}, contradicting our choice of f_{k+1}, and this finishes the second proof of Hilbert Basis Theorem. □

As an immediate corollary of Theorem 1.4, we obtain the original version of Hilbert Basis Theorem:

THEOREM 1.5. ***(Hilbert's Basis Theorem)*** *$k[x_1, x_2, ..., x_n]$ is Noetherian.*

PROOF. In a field $k, (0)$ and (1) are the only ideals, so k is finitely generated as as ideal. Thus, by this theorem $k[x_1]$ is Noetherian. Then again by this theorem, $k[x_1, x_2]$ is noetherian, and so on. □

REMARK 1.9. *Once the ACC condition is satisfied, we do not need the Axiom of Choice, or equivalently, the falsely called Zorn's Lemma any more, this is exactly the reason why Emmy Noether's notion on finiteness through the ACC is so important. Thus, Algebra-oriented readers must be far away from this Axiom, as long as we are concerned how to actually construct mathematical objects, although whose existence may be proved based on this Axiom.*

PROPOSITION 1.10. *Let R be a Noetherian ring, and let I be an ideal of R. Then the quotient ring R/I is a Noetherian ring.*

PROOF. Let J' be the ideal of R/I, and $J = \pi^{-1}(J')$ is the corresponding ideal of R, where $\pi\colon R \longrightarrow R/I$ is the canonical projection. Then J is finitely generated by, say, $J = (a_1, ..., a_m)$, by hypothesis; hence the finite subset $(a_1 + I, ..., a_m + I)$ will generate J'. □

Exercises for Section 1.5

EXERCISE 1.23. *Prove that any ring which is a quotient of a polynomial ring over $\mathbb{Z}$ or a field is Noetherian.*

EXERCISE 1.24. *Find a ring R and an ideal of R which is not finitely generated.*

6. Hilbert's Nullstellensatz

In this section, we are going to review certain properties of maximal ideals in a (commutative) ring R, and through the so-called "evaluation homomorphism", this will eventually lead us to meet the proof of Hilbert's famous theorem, Hilbert's weak Nullstellensatz, which then in turn allows us to prove Hilbert's (strong) Nullstellensatz. These two theorems are in fact equivalent and we will see that one of these two connects solution sets of systems of polynomial equations and ideals in the polynomial ring $k[x_1, ..., x_n]$ in a one-to-one correspondence manner, *provided* that the bottom field k is algebraically closed. The following is a short historical remark on the subject we are going to study in this book, in a heavily computation-oriented manner.

Ever since the proof of David Hilbert (1862 - 1943) of his very famous theorem, Nullstellensatz, Algebra and Geometry, which were more or less studied separately until his proof the Nullstellensatz, were successfully put together coherently. In other words, without his Nullstellensatz, we would not be able to have the so-called "Modern Algebraic Geometry", whose foundation was provided by Oskar Zariski (1899 - 1986) during his stay at The Johns Hopkins University (1928 - 1946), based on the works of Emmy Noether (1882 - 1935) and Wolfgang Krull (1899 - 1971). And without Zariski's work, we would not be able to have the so-called "Theory of Schemes", the foundation of Algebraic Geometry which was developed by Alexander Grothendieck (1928 - 2014) with the help of the notion of Jean-Pierre Serre (1926 -) on Algebraic Coherent Sheaf (1955), during his stay at Institut des Hautes Etudes Scientifiques, an advanced mathematical research instituted in France which was originally established in 1959 for mainly helping Grothendieck's fundamental research works on establishing the foundation of Algebraic Geometry during 1959 - 1970, and since then it is more well known to us simply as IHES. We note that it was Hilbert who announced 23 problems in mathematics at the Second International Congress of Mathematicians in Paris at the turn of the 20th century, and one of them was about the foundation of Algebraic Geometry. Because of his successful development for the foundation of Algebraic Geometry, Grothendieck received Fields Medal in 1966.

As we shall see it later in this book, Hilbert's Weak Nullstellensatz can be thought of as the "**Fundamental Theorem of Algebra for polynomials in several variables**", because it is a "natural" generalization of Gauss's "**Fundamental Theorem of Algebra (for polynomials in one variable)**" in a computational sense.

6.1. Maximal Ideals in Polynomial Rings over $\mathbb{C}$. We first recall

DEFINITION 1.12. *An ideal M of a ring R is called a **maximal ideal** if it is a proper ideal of R not contained in any other ideal, other than R and itself.*

LEMMA 1.9. *A non-trivial ring R is a field if and only if it has no non-trivial proper ideals.*

PROOF. If R is a field, for any non-zero element a of an ideal I of R, since it is invertible, we have

$$1 = a^{-1}a \in I,$$

and consequently, every non-zero ideal I of R is R itself. Conversely, if R is a ring with no proper ideals, then for any non-zero element of R, the principal ideal (a) generated by a must be R itself; hence it contains the unit 1, so that $1 = ba$ for some $b \in R$. That is, every non-zero element of R is invertible. □

PROPOSITION 1.11. *For any ideal I of a non-trivial ring R, the following statements are equivalent:*

(1) *If α, β are elements of R such that $\alpha\beta \in I$, then either $\alpha \in I$ or $\beta \in I$.*
(2) *If J, K are ideals of R such that $JK \subseteq I$, then either $J \subseteq I$ or $K \subseteq I$.*
(3) *The quotient ring R/I is an integral domain.*

PROOF. Part (2) clearly implies Part (1), if we take $J = (\alpha)$ and $K = (\beta)$.

The condition for $\bar{R} = R/I$ to be an integral domain are that $\bar{R} \neq 0$ and $\bar{\alpha}\bar{\beta} = 0$ implies either $\bar{\alpha} = 0$ or $\bar{\beta} = 0$. We can translate these conditions back to: $I \neq R$ and if $\alpha\beta \in I$ then either $\alpha \in I$ or $\beta \in I$, and conversely. Thus, Part (1) and Part (3) are equivalent to each other.

Now, assume that Part (1) holds and let J, K be ideals of R such that $JK \subseteq I$. Suppose that J is not contained in I. Then there exists an element $\alpha \in J$ not contained in I. Let β be any element of K. Then $\alpha\beta \in JK$; hence $\alpha\beta \in I$. Thus, by hypothesis, $\beta \in I$. That is, we have shown that $K \subseteq I$. □

DEFINITION 1.13. *An ideal of R which satisfies one of these conditions is called a* ***prime ideal****.*

PROPOSITION 1.12. *For any ideal I of a non-trivial ring R, we have:*
(1) *I is a maximal ideal if and only if R/I is a field.*
(2) *I is a prime ideal if and only if R/I is an integral domain.*

PROOF. Part (1). By Lemma 1.9, we know that R is a field if and only if it has no non-trivial proper ideals. Then since there is a natural one-to-one correspondence between the set of ideals containing I and the set of ideals of the quotient ring R/I, we see that R/I is a field if and only if there are no ideal of R properly contained between R and I, namely, I is a maximal ideal of R.

The proof of Part (2) is immediate from the definition of an ideal I being prime: $ab \in I \Longrightarrow$ either $a \in I$ or $b \in I$, because this definition exactly says the cancellation law in the quotient ring: $(a + I)(b + I) = I \Longrightarrow$ either $a + I = I$ or $b + I = I$. □

PROPOSITION 1.13. *The maximal ideals of the ring $\mathbb{Z}$ of integers are the principal ideals generated by prime integers.*

PROOF. This is true because every ideal of $\mathbb{Z}$ is principal. □

Similarly, we know by Proposition 1.3 that the polynomial ring $k[x]$ in one variable over a field is a *PID*.

In addition to this, if the field k is moreover algebraically closed, for instance, $k = \mathbb{C}$, the field of complex numbers, then we can say more. For example, the maximal ideals of the ring $\mathbb{C}[x]$ of complex polynomials in one variable are described very simply:

PROPOSITION 1.14. *Let $\mathbb{C}$ be the field of complex numbers. Then:*

(1) *The maximal ideals of the polynomial ring $\mathbb{C}[x]$ in one variable are principal ideals generated by the linear polynomials $x - \alpha$.*
(2) *The ideal M_α generated by $x - \alpha$ is the kernel of the* ***evaluation homomorphism*** *ε_α:*

$$\begin{array}{c} \mathbb{C}[x] \\ \downarrow \varepsilon_\alpha \\ \mathbb{C} \end{array}$$

defined by

$$\varepsilon_\alpha(f(x)) = f(\alpha).$$

(3) *There is a one-to-one correspondence between maximal ideals*

$$M_\alpha = (x - \alpha)$$

of $\mathbb{C}[x]$ and complex numbers $\alpha \in \mathbb{C}$.

PROOF. Part (1). Let M be a maximal ideal in the *PID* $\mathbb{C}[x]$. Thus, it is generated by a single monic polynomial $f(x)$ of least degree over $\mathbb{C}$. Then since every complex polynomial of positive degree has a root α, $f(x)$ is divisible by the linear factor $x - \alpha$; hence $f(x)$ is contained in the principal ideal $(x - \alpha)$. Since M is a maximal ideal, we must have $M = (x - \alpha)$.

Part (2). The kernel, $\ker(\varepsilon_\alpha)$ is an ideal of $\mathbb{C}[x]$ obviously containing the linear polynomial $x - \alpha$, and other elements of $\ker(\varepsilon_\alpha)$ are clearly divisible by $x - \alpha$, and hence $\ker(\varepsilon_\alpha) = (x - \alpha)$. We denote this maximal ideal by M_α. Part (3) then does not need a proof. □

REMARK 1.10. *Here is our comment on the role of the evaluation homomorphism*

$$\begin{array}{c}\mathbb{C}[x]\\ \downarrow \varepsilon_\alpha\\ \mathbb{C}\end{array}$$

and the evaluation function

$$\begin{array}{c}\mathbb{C}^1\\ \downarrow f\\ \mathbb{C}\end{array}$$

defined in Definition 1.4.

When we consider a function $g(x)$, we usually assume that x is a variable for a fixed g. But, on the other hand, we can also think of a situation where g is a variable for a fixed x, whenever it makes sense mathematically.

Thus, when $f(x)$ is a given polynomial over $\mathbb{C}$, we can consider the evaluation function:

$$\begin{array}{c}\mathbb{C}^1\\ \downarrow f\\ \mathbb{C}\end{array}$$

which does make sense. And moreover, when α is a given point in $\mathbb{C}^1$, we can consider the evaluation homomorphism

$$\begin{array}{c}\mathbb{C}[x]\\ \downarrow \varepsilon_\alpha\\ \mathbb{C}\end{array}$$

which also makes sense. Therefore, in other words, the evaluation homomorphism:

$$\begin{array}{c} \mathbb{C}[x] \\ \downarrow \varepsilon_\alpha \\ \mathbb{C} \end{array}$$

and the evaluation function:

$$\begin{array}{c} \mathbb{C}^1 \\ \downarrow f \\ \mathbb{C} \end{array}$$

are ***dual*** *to each other, related by the same value*

$$f(\alpha) = \varepsilon_\alpha(f) \in \mathbb{C}.$$

Exactly because of this reason, we sometimes define $\alpha(f)$ to be $\varepsilon_\alpha(f)$, i.e.,

$$\alpha(f) \equiv \varepsilon_\alpha(f) = f(\alpha), \tag{1.3}$$

so that we can easily move from one side of the duality to another side of the duality through this "local point evaluation process" in the bottom field $\mathbb{C}$. We thus have: "The point α in geometry corresponds to the evaluation k-algebra homomorphism ε_α to the bottom field k, and its converse also holds."

In the above argument, we can safely replace $\mathbb{C}$ with any other ring R, without destroying the concept of the duality of the evaluation function and the evaluation homomorphism.

REMARK 1.11. *In Remark 1.10, we in fact have discussed the most important but very subtle fundamental concept that direcly and contravariantly connects Algebra with Geometry in a natual and canonical manner. In particular, about those books that had helped the author of this book a lot better understand this fundamental notion are the following: He had read, in the Spring of the year1981, a related subject in Bertrand Russell's famous book,* Principia Mathematica *(2nd Ed. 1923), Introduction to the Second Edition, VI. Functions as variables, page xxviii,*[1] *and then, in the Spring of the year 1989, he had suddenly found a very short explanation on a subject of the "same" kind in his dear Professor, George R. Kempf's unpublished Lecture Note,* Algebraic Variety, *Chapter 1. initial concepts, Section 1.1. functions on sets, page 1.*[2] *Therefore, the author decided to use in this book his dear Professor's notation, ε_α, for the evaluation k-algebra homomorphism to the bottom field k that was used in his Professor's Lecture Note,* Algebraic Variety.

REMARK 1.12. *In Remark 4.15 of Chapter* 4, *the next material to the current subject is presented.*

The extension of Proposition 1.14 to more than one variables is called the Null-stellen-satz (= Zero- Places-Theorem). Before we state and prove this important theorem, we need some lemmas and their consequences.

LEMMA 1.10. *Let V be a vector space which is spanned by a countable family $\{v_1, ..., v_n, ...\}$ of vectors. Then every set L of linearly independent vectors in V is either finite or countably infinite.*

PROOF. Let L be a linearly independent subset of V, V_n be the span of the first n vectors $v_1, ..., v_n$, and let

$$L_n = L \cap V_n.$$

Then L_n is a linearly independent subset in a finite-dimensional subspace V_n and hence it is a finite set. In addition to this, we easily see that

$$L = \cup_{n=1}^{\infty} L_n.$$

On the other hand, we recall that the union of countably many finite sets is either finite or countably infinite. □

EXAMPLE 1.14. *$k[x]$ is a k-vector space with a countable basis*

$$\{x^j : j \in \mathbb{N} \cup \{0\}\}.$$

[1] http://ch.catholic.or.kr/pundang/archive/4/functions_as_variables.htm
[2] http://ch.catholic.or.kr/pundang/4/soh/2006_aa_i/functions_as_variables.htm

Similarly, $k[x_1, ..., x_n]$ *is a* k*-vector space with a countable basis*

$$\{x_1^{\alpha_1} \cdots x_n^{\alpha_n} : \alpha_j \in \mathbb{N} \cup \{0\}\}.$$

We recall that the set $\mathbb{Q}$ of rational numbers is a countable infinite set, but the set $\mathbb{R}$ of real numbers are uncountable infinite set; hence so is the set $\mathbb{C}$ of complex numbers.

LEMMA 1.11. *The* uncountably *many rational functions*

$$\frac{1}{(x-\alpha)}, \qquad \alpha \in \mathbb{C},$$

are linearly independent over $\mathbb{C}$.

PROOF. We suppose that a linear combination of such rational functions:

$$\sum_{i=1}^{n} \frac{c_i}{x - \alpha_i},$$

where $\alpha_1, ..., \alpha_n$ are distinct complex numbers with undetermined coefficients. Then by clearing out the denominators, and then by putting $x_i = c_i$ for each $i = 1, 2, ..., n$, we easily see that each c_i must be zero. Thus, the rational functions $1/(x-\alpha)$ are linearly independent over $\mathbb{C}$. □

That is, we have equivalently shown that:

COROLLARY 1.7. *The field of fractions* $\mathbb{C}(x)$ *of the polynomial ring* $\mathbb{C}[x]$ *has a linearly independent subset consisting of uncountably many elements of* $\mathbb{C}(x)$. □

COROLLARY 1.8. *Let* M *be a maximal ideal of* $\mathbb{C}[x_1, ..., x_n]$, K *the field* $\mathbb{C}[x_1, ..., x_n]/M$, *and let* π_1 *be the restriction of the canonical projection*

$$\pi : \mathbb{C}[x_1, ..., x_n] \longrightarrow K$$

to the subring $\mathbb{C}[x_1]$ *of polynomials in one variable:*

$$\pi_1\colon \mathbb{C}[x_1] \overset{i}{\hookrightarrow} \mathbb{C}[x_1, ..., x_n) \overset{\pi}{\longrightarrow} K, \qquad \pi_1 = \pi \circ i.$$

Then the kernel $\ker(\pi_1)$ *of* π_1 *is either* 0 *or else it is a maximal ideal of* $\mathbb{C}[x_1]$.

PROOF. Suppose that its kernel, $\ker(\pi_1)$, is not zero, and let $f(x_1)$ be a non-zero polynomial in $\ker(\pi_1)$. Since K is not the zero ring, $\ker(\pi_1)$ is not the whole ring. Then since we have the natural injective ring homomorphism $\mathbb{C} \longrightarrow K$, and any non-zero constant polynomial in $\mathbb{C}[x_1]$ remains the same under π_1; hence $f(x_1)$ is not a non-zero constant polynomial. Thus, it is divisible by a linear polynomial over $\mathbb{C}$, say $f(x_1) = (x_1 - \alpha_1)g(x_1)$ in $\mathbb{C}[x_1]$, since $\mathbb{C}$ is an algebraically closed field. Thus, $0 = \pi_1(x_1 - \alpha_1)\pi_1(g(x_1))$ in K. Then since K is an integral domain, one of $x_1 - \alpha_1$ or $g(x_1)$ is an element of $\ker(\pi_1)$. Thus, by induction on the degree of $f(x_1)$, we conclude that $\ker(\pi_1)$ contains a linear polynomial. Therefore, by Proposition 1.14 (1), we conclude that ker (π_1) is a maximal ideal of $\mathbb{C}[x_1]$. □

The following version of Hilbert's Nullstellensatz is not the original version Hilbert proved. In the proof of the next theorem, we note that the proof of the $\Longrightarrow$ direction of (1) *only* requires that the bottom field k be algebraically closed.

THEOREM 1.6. ***(Hilbert's Nullstellensatz)***

(1) *The maximal ideals of the polynomial ring*

$$\mathbb{C}[x_1, ..., x_n]$$

are in a one-to-one correspondence with points of the complex n-dimensional space $\mathbb{C}^n$.

(2) *A point $\alpha = (\alpha_1, ..., \alpha_n)$ in $\mathbb{C}^n$ corresponds to the kernel of the evaluation homomorphism*

$$\begin{array}{c} \mathbb{C}[x_1, ..., x_n] \\ \downarrow \varepsilon_\alpha \\ \mathbb{C} \end{array}$$

defined by

$$\varepsilon_\alpha(f(x_1, ..., x_n)) = f(\alpha_1, ..., \alpha_n).$$

(3) *The kernel M_α of this map ε_α is the ideal generated by the linear polynomials*

$$x_1 - \alpha_1, x_2 - \alpha_2, ..., x_n - \alpha_n.$$

PROOF. Let $\alpha = (\alpha_1, ..., \alpha_n)$ be a point in $\mathbb{C}^n$, and let M_α be the kernel of the evaluation homomorphism

$$\begin{array}{c} \mathbb{C}[x_1, ..., x_n] \\ \downarrow \varepsilon_\alpha \\ \mathbb{C} \end{array} .$$

Since this homomorphism is surjective and $\mathbb{C}$ is a field, M_α is a maximal ideal of $\mathbb{C}[x_1, ..., x_n]$.

Now, in order to show that M_α is generated by the linear polynomials, $x_1 - \alpha_1, ..., x_n - \alpha_n$, we first observe that we can always expand a polynomial $f(x) \in \mathbb{C}[x_1, ..., x_n]$ in powers of $x_1 - \alpha_1, ..., x_n - \alpha_n$:

$$f(x) = f(\alpha) + \sum_i c_i(x_i - \alpha_i) + \sum_{i,j} c_{ij}(x_i - \alpha_i)(x_j - \alpha_j) + \cdots.$$

The existence of such an expression of a finite sum can be derived by first substituting $x = \alpha + u$ into $f(x)$, and expanding in powers of the variables $u = (u_1, ..., u_n)$, and then substituting $u = x - \alpha$ back into the result. Thus, every term on the right side except $f(\alpha)$ is divisible by at least one of the linear polynomials $(x_i - \alpha_i)$. Thus, if $f(x) \in \ker(\varepsilon_\alpha)$, then $f(\alpha) = 0$ so that $f(x)$ is an element of the ideal generated by these linear polynomials $x_i - \alpha_i$. Since we have already verified that $\ker(\varepsilon_\alpha)$ is a maximal ideal, it must be the same as the ideal $(x_1 - \alpha_1, ..., x_n - \alpha_n)$:

$$\ker(\varepsilon_\alpha) = M_\alpha = (x_1 - \alpha_1, ..., x_n - \alpha_n).$$

This proves part (2) and part (3). It also proves half of part (1), namely,

$$\alpha \longmapsto M_\alpha = (x_1 - \alpha_1, ..., x_n - \alpha_n)$$

is one-to-one. Until now, we did not apply the hypothesis that the bottom field k is algebraically closed. Thus to finish the whole proof, we are left to to prove that every maximal ideal M of $\mathbb{C}[x_1, ..., x_n]$ is equal to M_α for some point $\alpha \in \mathbb{C}^n$, and this requires the hypothesis that he bottom field k is the algebraically closed field $\mathbb{C}$.

Let M be a maximal ideal of $\mathbb{C}[x_1, ..., x_n]$. Then by Corollary 1.8, we already know that the kernel $\ker(\pi_1)$ of π_1 is either 0 or else it is a maximal ideal of $\mathbb{C}[x_1]$,

$$\pi_1 \colon \mathbb{C}[x_1] \overset{i}{\hookrightarrow} \mathbb{C}[x_1, ..., x_n] \xrightarrow{\pi} K = \mathbb{C}[x_1, ..., x_n]/M, \qquad \pi_1 = \pi \circ i.$$

CLAIM 1.2. $\ker(\pi_1)$ *cannot be the zero ideal.*

Suppose that we have proved that $\ker(\pi_1)$ can not be the zero ideal, then the maximal ideal $\ker(\pi_1) = (x_1 - \alpha_1)$ for some $\alpha_1 \in \mathbb{C}$, and hence working with another variable x_i we see that the maximal ideal M contains linear polynomials $x_i - \alpha_i$ for $i = 1, ..., n$. Thus, we must have $M = (x_1 - \alpha_1, ..., x_n - \alpha_n)$ for some $\alpha = (\alpha_1, ..., \alpha_n) \in \mathbb{C}^n$. Then since we have already verified that $(x_1 - \alpha_1, ..., x_n - \alpha_n) = \ker\ (\varepsilon_\alpha)$, M is the kernel of an evaluation homomorphism ε_α for some $\alpha = (\alpha_1, ..., \alpha_n) \in \mathbb{C}^n$. Thus, we need to prove the claim.

PROOF OF THE CLAIM. Suppose on the contrary that $\ker(\pi_1) = (0)$. Then π_1 maps $\mathbb{C}[x_1]$ isomorphically to its image, which is a subring of K; and hence we can extend this map to the field $\mathbb{C}(x_1)$ of fractions of $\mathbb{C}[x_1]$ by the universal property of the field of fractions, so that the field K has a subfield which is isomorphic to $\mathbb{C}(x_1)$. Then by Corollary 1.7, we know that $\mathbb{C}(x_1)$ has a linearly independent subset consisting of uncountably many elements; consequently, we have a basis of K consisting of uncountably many elements. But, this is impossible because, since $\mathbb{C}[x_1, ..., x_n]$ is a complex vector space with a countably infinite basis consisting of monomials, the quotient ring K is spanned as a complex vector space by a countable family, consisting of the residues of those monomials. Thus, K has a basis consisting of countably many linearly independent elements. Then since the dimension of the vector space K can not be both countably infinite and uncountably infinite, we must have $\ker(\pi_1) \neq (0)$. □

This finishes the proof of Hilbert's Nullstellensatz. □

REMARK 1.13. *For the consequence of Hilbert's Nullstellensatz in bijectively establishing Geometry and Algebra correspondence, one may refer to Remark 4.15.*

Hilbert's Nullstellensatz provides us with an important connection between Geometry and Algebra. It tells us that the maximal ideals in the polynomial ring $\mathbb{C}[x_1, ..., x_n]$ corresponds to the points in $\mathbb{C}^n$. This correspondence can also be used to relate affine varieties to quotient rings of the polynomial ring:

THEOREM 1.7. *Let V be the affine variety in $\mathbb{C}^n$ defined by a system of polynomial equations,*

$$\begin{cases} f_1 &= 0, \\ f_2 &= 0, \\ \dots & \\ f_s &= 0. \end{cases},$$

and let $I = (f_1, ..., f_s) \subset \mathbb{C}[x_1, ..., x_n]$ be the ideal generated by these polynomials. Then the maximal ideals of the quotient ring

$$R = \mathbb{C}[x_1, ..., x_n]/I$$

are in one-to-one correspondence with the points of V.

PROOF. We first recall that the maximal ideals of the quotient ring R correspond to those maximal ideals M of $\mathbb{C}[x_1, ..., x_n]$ which contain I. On the other hand, by Hilbert's Nullstellensatz, Theorem 1.6, we know that the maximal ideal M of $\mathbb{C}[x_1, ..., x_n]$ is equal to $M_\alpha = (x_1 - \alpha_1, ..., x_n - \alpha_n)$ for some point $\alpha \in \mathbb{C}^n$. Thus, combining these two results together, we have: $I \subset M_\alpha$ if and only if each $f_i \in M_\alpha = \ker(\varepsilon_\alpha)$ if and only if for each i

$$0 = \varepsilon_\alpha(f_i) = \alpha(f_i) = f_i(\alpha)$$

if and only if $\alpha \in V$. □

DEFINITION 1.14. *The quotient ring R in Theorem 1.7 is called the* ***coordinate ring*** *of the affine variety defined by the polynomial equations $f_1 = f_2 = \cdots = f_s = 0$.*

Theorem 1.7 explains that the algebraic properties of the ring R are closely related with the geometry of V. In principle, all properties of the system of polynomial equations

$$\begin{cases} f_1 &= 0, \\ f_2 &= 0, \\ \dots & \\ f_s &= 0. \end{cases}$$

are reflected in the structure of the ring $R = \mathbb{C}[x_1, ..., x_n]/(f_1, ..., f_s)$, and the theory of this relationship is called **Algebraic Geometry**.

The following theorem says that the zero ring is the only ring with no maximal ideals.

THEOREM 1.8. *(Krull) Let R be a ring. Every ideal I of R which is not the unit ideal is contained in a maximal ideal.*

This theorem is usually proved using the Axiom of Choice, or equivalently, Zorn's Lemma, so we leave it as an exercise. □

COROLLARY 1.9. *The only ring R with no maximal ideals is the zero ring.* □

If we combine Theorem 1.7 and Theorem 1.8 together, we obtain another important corollary:

COROLLARY 1.10. *(**Hilbert's Weak Nullstellensatz**) If the system of polynomial equations,*

$$\begin{cases} f_1 &= 0, \\ f_2 &= 0, \\ \dots & \\ f_s &= 0. \end{cases}$$

over $\mathbb{C}$ has no solution in $\mathbb{C}^n$, then we can express 1 as a linear combination of these f_i's,

$$1 = \sum g_i f_i,$$

with polynomial coefficients in $\mathbb{C}[x_1, ..., x_n]$, and conversely.

That is, $V(f_1, ..., f_s) = \varnothing$*, then* $(f_1, ..., f_s) = \mathbb{C}[x_1, ..., x_n]$*, and conversely.*

PROOF. The converse is trivial. Now, if the system has no solution, then Hilbert's Nullstellensatz, Theorem 1.7 says that there is no maximal ideal containing the ideal $I = (f_1, ..., f_s)$; hence by Theorem 1.8, I must be a unit ideal (1). $\square$

REMARK 1.14. *Thus, Hilbert's Weak Nullstellensatz is a generalization of Gauss's Fundamental Theorem for Algebra because it states that* "Every system of polynomial equations that generates an ideal which is smaller than $\mathbb{C}[x_1, ..., x_n]$ has a common solution in $\mathbb{C}^n$". *Thus, Hilbert's Weak Nullstellensatz is also called* ***"Fundamental Theorem of Algebra for polynomials in several variables"***. *Then Corollary 3.1 tells us how we can computationally verify whether* $1 \in (f_1, ..., f_d)$ *or not, by the Theory of Groebner bases.*

The Nullstellensatz has been reformulated in many apparently different ways, and the one we proved above is not its original form. Here is the original form of Nullstellensatz:

THEOREM 1.9. ***(Hilbert's Strong Nullstellensatz)*** *Let*

$$V = V(f_1, ..., f_s)$$

be the affine variety defined by polynomials $f_1, ..., f_s \in \mathbb{C}[x_1, ..., x_n]$*. If a polynomial* $g \in \mathbb{C}[x_1, ..., x_n]$ *vanishes identically zero on* V*, then there exists a positive integer* N *such that*

$$g^N \in I = (f_1, ..., f_s),$$

and conversely.

PROOF. The converse is trivial since $g^N(a_1, ..., a_n) = 0$ if and only if $g(a_1, ..., a_n) = 0$ holds for any point $(a_1, ..., a_n) \in \mathbb{C}^n$.

We consider the $s + 1$ polynomials $f_1, ..., f_s, gy - 1$ in the variables $x_1, ..., x_n$, and y, where $gy = g(x_1, ..., x_n)y$ is a polynomial with one more variable y. Then the system of these polynomials does not have a solution, because if $(\alpha_1, ..., \alpha_n, b)$ is a solution of them, then each $f_i(\alpha_1, ..., \alpha_s) = 0$ and $g(\alpha_1, ..., \alpha_n) = 0$ by hypothesis, and hence $gy - 1 = -1$, which is impossible. Thus, by Corollary 1.10 (Hilbert's Weak Nullstellensatz), we know that they generate the unit ideal (1) of $\mathbb{C}[x_1, ..., x_n, y]$. Thus, we have

$$\begin{aligned} 1 &= \sum_i p_i(x_1, ..., x_n, y) f_i(x_1, ..., x_n) \\ &\quad + q(x_1, ..., x_n, y)(g(x_1, ..., x_n)y - 1), \end{aligned}$$

for some $p_i(x_1, ..., x_n, y)$ and $q(x_1, ..., x_n, y) \in \mathbb{C}[x_1, ..., x_n, y]$. Thus, by substituting $y = 1/g(x_1, ..., x_n)$ into this expression, and clearing the denominators of the resulting expression,

$$1 = \sum_i p_i(x_1, ..., x_n, 1/g(x_1, ..., x_n)) f_i(x_1, ..., x_n)$$

we find that there exists a positive integer N such that

$$g(x_1, ..., x_n)^N = \sum_i h_i(x_1, ..., x_n) f_i(x_1, ..., x_n) \in I,$$

where $h_i(x_1, ..., x_n) = g(x_1, ..., x_n)^N p_i(x_1, ..., x_n, 1/g(x_1, ..., x_n))$. □

REMARK 1.15. *The method we applied in the above proof is usually called* ***Rabinowitch's trick.***

In fact, Hilbert's Strong Nullstellensatz is equivalent to Hilbert's Weak Nullstellensatz:

THEOREM 1.10. ***(Hilbert's Weak Nullstellensatz)*** *Let I be an ideal of $\mathbb{C}[x_1, ..., x_n]$. Then $V(I) = \varnothing$ if and only if $I = \mathbb{C}[x_1, ..., x_n]$.*

PROOF. The $\Longleftarrow$ direction it trivial because, the constant polynomial 1 has no solution at all.

Conversely, suppose that I is a proper ideal of $\mathbb{C}[x_1, ..., x_n]$, that is, $I \subsetneq \mathbb{C}[x_1, ..., x_n]$. Then it is contained in a maximal ideal M (Theorem 1.8). Then by Theorem 1.6 (Hilbert's Nullstellensatz),

$$M = (x_1 - \alpha_1, ..., x_n - \alpha_n)$$

for some point $(\alpha_1, ..., \alpha_n) \in \mathbb{C}^n$; hence the point

$$(\alpha_1, ..., \alpha_n) \subseteq V(M) \subseteq V(I).$$

That is, $V(I) \neq \varnothing$. □

Another way of describing Hilbert's Strong Nullstellensatz is:

DEFINITION 1.15. *Let I be an ideal of a Noetherian ring R.*

(1) *Then the set defined by*

$$\sqrt{I} = \{r \in R : r^N \in I \text{ for some } N \geq 1\},$$

is called the ***radical*** *of I.*

(2) *If $\sqrt{I} = I$ holds, then I is called a* ***radical ideal****.*

COROLLARY 1.11. *Let I be an ideal of a Noetherian ring R. Then radical $\sqrt{I}$ of I is an ideal of R containing I.*

PROOF. We leave its proof as an exercise. □

COROLLARY 1.12. *Let V be an affine variety. Prove that $I(V)$ is a radical ideal.*

PROOF. We leave its proof as an exercise □

COROLLARY 1.13. *Let I be an ideal of the polynomial ring $\mathbb{C}[x_1, ..., x_n]$ over the field $\mathbb{C}$ of complex numbers. Then*

$$I(V(I)) = \sqrt{I}.$$

REMARK 1.16. *Thus, we have proved both Weak and Strong Versions of Hilbert's Nullstellensatz when the bottom field is $\mathbb{C}$. When the bottom field k is an algebraically closed field in general, we still have the Hilbert's Nullstellensatz and one may find its proof in, for instance, Hungerford's Book, Algebra.*

Before we move into the next Chapter, we want to closely examine what's going on when we attempt to solve a system of linear equations in several variables. The following example actually suggests a few very important notions which were nevertheless neither noticed nor well-understood until 1965.

EXAMPLE 1.15. *Let* $f_1 = y-z$, $f_2 = x+2y+3z$, *and* $f_3 = 3x-4y+2z$ *be linear polynomials in* $\mathbb{Q}[x]$. *We consider the ideal* $I = (f_1, f_2, f_3)$ *and the variety* $V(f_1, f_2, f_3)$, *i.e., the solution set of the system of linear equations:*

$$\begin{cases} y - z &= 0, \\ x + 2y + 3z &= 0, \\ 3x - 4y + 2z &= 0. \end{cases}$$

Applying elementary row operations to the following matrix we solve the system:

$$\begin{bmatrix} 0 & 1 & -1 \\ 1 & 2 & 3 \\ 3 & -4 & 2 \end{bmatrix} \longrightarrow \begin{bmatrix} 0 & 1 & -1 \\ 1 & 2 & 3 \\ 0 & -10 & -7 \end{bmatrix} \longrightarrow \begin{bmatrix} 1 & 2 & 3 \\ 0 & 1 & -1 \\ 0 & 0 & -17 \end{bmatrix}.$$

We recall that any one of the three elementary row operations never change the solution set at all. That is, the solution set must be invariant under these elementary row operations. In this example, we thus obtain: $V(f_1, f_2, f_3) = V(f_1, f_2, -17z)$. *In fact, we have more than that:*

$$\begin{aligned} &(f_1, f_2, f_3), \\ &(f_1, f_2, f_3 - 3f_2), \\ &(f_2, f_1, f_3 - 3f_2 + 10f_1), \end{aligned}$$

where $f_3 - 3f_2 + 10f_1 = -17z$. *We note that they are the same ideal* I, *but its generators are transformed by elementary row operations to the next ones in such a way that we can easily figure out its solution set!*

REMARK 1.17. *There are a few important but usually not-mentioned notions in the business of elementary row operations for solving a system of linear equations.*

First of all, we psychologically think that x *is greater than* y *is greater than* z; *and then try to eliminate the term with* x *in* f_3; *then try to eliminate the term with* y *in* f_3. *More precisely, we try to eliminate the leading term of* f_3 *by the leading term of* f_2*; then we try to eliminate the leading term of the resulting polynomial,* $f_3 - 3f_2$ *by the leading term of* f_1.

Secondly, we stopped because the resulting term $-17z = f_3 - 3f_2 + 10f_1$ *is "small enough" so that none of the leading terms of either* f_1 *or* f_2 *can eliminate, or reduce further, the resulting term,* $-17z$.

Exercises for Section 1.6

EXERCISE 1.25. *Let* $I = (x^2 + y^2 - 1, y - 1)$. *Show that* I *is properly contained in the ideal* $I(V(I))$.

EXERCISE 1.26. *Prove that the radical of an ideal* I *is again an ideal.*

EXERCISE 1.27. *Prove that the radical of a radical ideal* I *is equal to* I *itself.*

EXERCISE 1.28. *Let* V *be an affine variety. Prove that* $I(V)$ *is a radical ideal.*

CHAPTER 2

Groebner Bases

1. Monomial Orderings on $k[x_1, ..., x_n]$

If one examines in detail the division algorithm in $k[x]$ and the elementary row-reduction (or Gaussian elimination) algorithm for systems of linear equations (or matrices) in Linear Algebra, one can recognize, as we have seen Example 1.15 of the previous section, that the notion of ordering terms in polynomials is an important step, although it is not often mentioned at all. Namely, if we examine the division algorithm in $k[x]$ more carefully, we easily see that we write polynomials in a non-increasing manner, and we try to get rid of the leading term of the target polynomial, i.e., the one that is going to be divided into pieces, with that of one of the dividing polynomials in order to obtain a part of the final quotient in a repeated manner in such a way that the left-over, i.e., the remainder must satisfy that it is never divisible by the leading term of any of the dividing polynomials. Of course, when we apply this algorithm, we always assume that $x^{n+1} > x^n > ...$ Moreover, when we had Example 1.15, we have also seen that we psychologically regard that $x > y > z$.

This means that when we have more than one variables it is inevitable to introduce an "order" among unknowns, a very special kind of monomials i.e., monomials of degree 1.

Then what are we supposed to do in order to properly handle monomials in several variables of degree > 1? How are we going to divide out one polynomial in several variables with a set of finite number of other polynomials, also in several variables?

In this section, we first introduce the notion of an order on the set M of monomials in $k[x_1, ..., x_n]$, and then we are going to define the most commonly used and most easily understandable monomial orders on $k[x_1, ..., x_n]$.

1.1. Orderings on the set of Monomials in $k[x_1, ..., x_n]$.

We start with a definition:

DEFINITION 2.1. *Let $k[x_1, ..., x_n]$ be a polynomial ring over a field k. A **monomial** in n variables $x_1, ..., x_n$ is a product*

$$x^\alpha = x_1^{\alpha_1} x_2^{\alpha_2} \cdots x_n^{\alpha_n},$$

where $\alpha = (\alpha_1, ..., \alpha_n)$, where each $\alpha_i \in \mathbb{Z}_{\geq 0} = \{n \in Z : n \geq 0\}$, i.e., each α_i is a non-negative integer.

We recall that the set of all monomials in $k[x_1, ..., x_n]$ form a k-basis of the k-vector space $k[x_1, ..., x_n]$, that is, every polynomial f in $k[x_1, ..., x_n]$ is written uniquely as a k-linear combination of finitely many of these monomials:

$$f = \sum_\alpha a_\alpha x^\alpha, \; a_\alpha \in k.$$

In this expression for f, we call a_α the **coefficient** of the monomial x^α; if moreover $a_\alpha \neq 0$, then $a_\alpha x^\alpha$ is called a **term** of f; and the maximum $|\alpha| = \alpha_1 + \cdots + \alpha_n$ among the exponents of the monomials of f is called the **total degree** of f, denoted by

$$\deg(f).$$

In the next section, we will introduce more terminologies in order to handle polynomials in several variables neatly.

Let M be the set of all monomials in $k[x_1, ..., x_n]$. Then we easily see that

(1) it is closed under multiplication of $k[x_1, ..., x_n]$,
(2) the multiplicative identity $1 \in k[x_1, ..., x_n]$ is also in M, and
(3) the associativity for multiplication holds in M.

On the other hand, if we denote the set of all exponents of monomials in M by

$$\mathbb{Z}^n_{\geq 0} = \{\alpha = (\alpha_1, ..., \alpha_n) : \text{ each } \alpha_i \text{ is a non-negative integer}\},$$

then we easily see that

(1) it is closed under the usual vector addition,
(2) the additive identity $(0, 0, ..., 0)$ is there, and
(3) the associativity law for addition holds in $\mathbb{Z}^n_{\geq 0}$.

Such things are called a *monoid.* We start our discussion with the following lemma.

LEMMA 2.1. *There exists a one-to-one correspondence between M and $\mathbb{Z}^n_{\geq 0}$.*

PROOF. Just send $x_1^{\alpha_1} \cdots x_n^{\alpha_n} \longmapsto (\alpha_1, ..., \alpha_n)$. □

Via this correspondence, an order relation between two elements in M naturally corresponds to that between the corresponding two elements in $\mathbb{Z}^n_{\geq 0}$.

DEFINITION 2.2. *A* ***monomial ordering*** *on $k[x_1, ..., x_n]$ is a relation $>$ on $\mathbb{Z}^n_{\geq 0}$, or equivalently on the set M of monomials $x^\alpha, \alpha \in \mathbb{Z}^n_{\geq 0}$, satisfying:*

(1) $>$ is a linear (i.e., total) ordering on $\mathbb{Z}^n_{\geq 0}$.
(2) If $\alpha > \beta$, then $\alpha + \gamma > \beta + \gamma$ for any $\gamma \in \mathbb{Z}^n_{\geq 0}$.
(3) $0 < \alpha$ for each $\alpha \neq 0$ in $\mathbb{Z}^n_{\geq 0}$.

REMARK 2.1. *Under the first two conditions, we will prove that the third condition in this definition is equivalent that $>$ is a well-ordering (see Corollary 2.4).*

We are going to see later: Dickson's Lemma (Theorem 2.1) is implied by Hilbert's Basis Theorem (Theorem 1.5), and conversely, Hilbert's Basis Theorem is also implied by Dickson's Lemma (Theorem 2.2); moreover, since our proofs of these two implications do not require a particular monomial ordering $>$ *on* $k[x_1, ..., x_n]$, *it is a lot better to use the above third condition in defining a monomial ordering* $>$ *on* $k[x_1, ..., x_n]$.

For example, under the first two conditions, it is (computationally) a lot easier to verify the above third condition than to verify the order is a well-ordering.

LEMMA 2.2. *For* $\alpha, \beta \in \mathbb{Z}^n_{\geq 0}$, *if there exists* $\gamma \in \mathbb{Z}^n_{\geq 0}$ *such that* $\beta = \alpha + \gamma$ *then* $\beta \geq \alpha$. *That is, if* x^α *divides* x^β *then* $x^\alpha \leq x^\beta$.

PROOF. This is very easy, since $\gamma \geq 0$, so by part 2 of the definition $\beta = \alpha + \gamma \geq \alpha + 0 = \alpha$. □

REMARK 2.2. *Definition 2.2 may need some remarks. The first requirement is needed in order to compare any of two monomials. That is, we want the so-called* trichotomy law *holds for a given pair of monomials,* x^α *and* x^β*:*

$$x^\alpha > x^\beta, \quad x^\alpha = x^\beta, \quad \text{or } x^\alpha < x^\beta.$$

The second requirement is also necessary because, when we attempt to rearrange the given polynomial into another, for instance, by applying a division algorithm, Algorithm 5, which will be soon introduced, to obtain the so-called quotient polynomial and remainder polynomial, we always attempt to get rid of the leading term, which is of course a monomial, of the given polynomial by the leading term of one of dividing polynomials, the second requirement is really needed. Otherwise, a systematic treatment of such a division is virtually impossible.

The third requirement is also needed in order to terminate our division algorithm, Algorithm 5, after finitely many number of steps.

We recall that an order $>$ on a set S is a *well-ordering* if every nonempty subset of S has a "least element" in it, i.e., there exists $s_0 \in S$ such that $s_0 \leq s$ for each element $s \in S$, with respect to the given order $>$.

LEMMA 2.3. *An ordering relation $>$ on $\mathbb{Z}^n_{\geq 0}$ is a well-ordering if and only if every decreasing sequence,*

$$\alpha_1 \geq \alpha_2 \geq \alpha_3 \geq \cdots,$$

in $\mathbb{Z}^n_{\geq 0}$ eventually terminates.

PROOF. If $>$ is not a well-ordering, then there exists a nonempty subset S of $\mathbb{Z}^n_{\geq 0}$ with no least element. Since S is nonempty, we have $\alpha_1 \in S$. Then since S has no least element, there exists α_2 that is strictly smaller than α_1: $\alpha_1 > \alpha_2$, and we just repeat this argument indefinitely to obtain a strictly decreasing infinite sequence:

$$\alpha_1 > \alpha_2 > \alpha_3 > \cdots.$$

Conversely, given such an infinite sequence, the nonempty subset of $\mathbb{Z}^n_{\geq 0}$ consisting of these α_i's certainly does not have a least element. □

PROPOSITION 2.1. *Every monomial order on M is a well-ordering.*

PROOF. Suppose there exists a monomial order $>$ on M that is not a well-ordering. Then under this order $>$ there exists a nonempty subset S of M, having no least element. Thus, we have an element x^{α_1} in S which is not a least element, so we have another x^{α_1} in S which is smaller than x^{α_1}: $x^{\alpha_1} > x^{\alpha_2}$, and we can repeat this argument indefinitely to obtain a strictly decreasing infinite sequence of monomials of S:

$$x^{\alpha_1} > x^{\alpha_2} > x^{\alpha_3} > \cdots. \tag{2.1}$$

This then gives rise to an increasing infinite chain of (monomial) ideals:

$$(x^{\alpha_1}) \subseteq (x^{\alpha_1}, x^{\alpha_2}) \subseteq (x^{\alpha_1}, x^{\alpha_2}, x^{\alpha_3}) \subseteq \cdots. \tag{2.2}$$

If we can show that each inclusion in the chain is proper, then this chain does not satisfy the ACC condition, so that $k[x_1, ..., x_n]$ is not Noetherian, which is a contradiction to Hilbert's basis theorem, Theorem 1.5.

Now, suppose that one inclusion in the chain is not proper. Then there exists i such that

$$(x^{\alpha_1}, ..., x^{\alpha_i}) = (x^{\alpha_1}, ..., x^{\alpha_i}, x^{\alpha_{i+1}}),$$

so that we have

$$x^{\alpha_{i+1}} = \sum_{j=1}^{i} h_j x^{\alpha_j},$$

for some polynomials $h_j \in k[x_1, ..., x_n]$. But, since the set of monomials forms a k-basis of the k-vector space $k[x_1, ..., x_n]$, each h_j can be uniquely written as a k-linear combination of monomials in M; hence the last equation simply means that you can express one basis element $x^{\alpha_{i+1}}$ as a k-linear combination of others, and such an expression is impossible unless $x^{\alpha_{i+1}}$ does appear among the $h_j x^{\alpha_j}$. That is, *there must be cancellation among the terms $h_j x^{\alpha_j}$, only leaving us the term $x^{\alpha_{i+1}}$*. In particular, this means that $x^{\alpha_{i+1}}$ is divisible by some $x^{\alpha_j}, 1 \le j \le i$; hence $x^{\alpha_{i+1}} \ge x^{\alpha_j}$. But, this obviously contradicts to our choice of the x^{α_i}'s in Eq. (2.1). Therefore, the chain in Eq. (2.2) is a strictly increasing chain of ideals. □

1.2. Examples of Monomial Orderings. Here are three basic examples of monomial orderings in $k[x_1, ..., x_n]$.

DEFINITION 2.3. *(**Lexicographic Order** or **Dictionary Order**) Let $\alpha = (\alpha_1, ..., \alpha_n)$ and $\beta = (\beta_1, ..., \beta_n) \in \mathbb{Z}^n_{\ge 0}$. We define $\alpha >_{lex} \beta$ if the left-most non-zero entry of $\alpha - \beta$ in the vector difference is positive.*

One can easily check that it is a monomial ordering on $k[x_1, ..., x_n]$. In short, this ordering is also called, "*lex order*".

EXAMPLE 2.1. *As its another name indicates, it is this ordering we actually apply when we want to find a word in English Language Dictionary: All the words in English Dictionary are listed alphabetically i.e., according to the Dictionary Order. For example,*

$$studio >_{lex} study\ .$$

EXAMPLE 2.2. $x_1^3 x_2 x_3^2 x_4^4 >_{lex} x_1^3 x_2 x_3 x_4^6$, *because*

$$(3, 1, 2, 4) - (3, 1, 1, 6) = (0, 0, 1, -2)$$

and the left-most non-zero $= 1 > 0$.

EXAMPLE 2.3. *We note that the variables $x_1, x_2, ...x_n$ are ordered "in the usual way":*

$$(1, 0, ..., 0) >_{lex} (0, 1, 0, ..., 0) >_{lex} \cdots >_{lex} (0, ..., 0, 1),$$

i.e., $x_1 >_{lex} x_2 >_{lex} \cdots >_{lex} x_n$, provided that we define in advance

$$x_1 = (1, 0, ..., 0), x_2 = (0, 1, 0, ..., 0), ..., x_n = (0, ..., 0, 1).$$

As this example indicates, given any ordering of the variables of our choice

$$x_1, x_2, ..., x_n,$$

we are going to have the lex order as defined in the above definition. That is, when we are given more than one unknown, say, x and y, we psychologically decide which variable is bigger than the other, i.e., either $x > y$ or $y > x$, etc.

LEMMA 2.4. *Given n variables, there are $n!$ lex orders on $k[x_1, ..., x_n]$ in total.*

PROOF. This is because we have $n!$ number of listing n objects x_i's. □

REMARK 2.3. *Under the lex order with $x > y > z$, we note that*

$$x >_{lex} y^6 z^5.$$

In some naive sense, the "degree of a polynomial" should reflect the "level of difficulty" in handling it, but the lex order sometimes does not agree with our "this" feeling or our intuition. We feel that since we do not know about x, another immediately obtainable unknown quantity, its square x^2, must be more difficult to know than x itself. Or, for instance, which algebraic process is going to take less time, multiplication of x by x itself or taking a square root of x^2 to get x? Here, the total degree of x is 1 but that of $y^6 z^5$ is $6 + 5 = 11$, so that we actually feel that the degree of difficulty in knowing $y^6 z^5$ is far greater, i.e., far complicated as is obvious from its writing, than that in knowing x, even though we assume that $x > y > z$ in advance.

The following two monomial orders indeed successfully reflect our this kind of feeling on the "degree of difficulties" in handling a polynomial = the (total) degree of the polynomial.

DEFINITION 2.4. *(**Graded Lex Order**) Let* $\alpha = (\alpha_1, ..., \alpha_n)$ *and* $\beta = (\beta_1, ..., \beta_n) \in \mathbb{Z}^n_{\geq 0}$. *We define* $\alpha >_{grlex} \beta$ *if either*

$$\text{“}|\alpha| = \sum_{i=1}^{n} \alpha_i > |\beta| = \sum_{i=1}^{n} \beta_i\text{”},$$

or

$$\text{“}|\alpha| = |\beta| \text{ and } \alpha >_{lex} \beta\text{”}.$$

One can easily check that it is a monomial ordering on $k[x_1, ..., x_n]$. Its full name must be "*Graded Lexicographic Order*". We see that grlex first tries to order by total degree first, then "break ties" using lex order as a second means.

EXAMPLE 2.4. $x^2y^2z^4 >_{grlex} x^2yz^5$ *because* $2+2+4 = 2+1+5$, *but* $(2,2,4)-(2,1,5) = (0,1,-1)$ *and* $1 > 0$.

EXAMPLE 2.5. *If we are under the graded lex order using the lex order with* $x_1 > x_2 > \cdots > x_n$, *we still have* $x_1 >_{grlex} x_2 >_{grlex} \cdots >_{grlex} x_n$.

LEMMA 2.5. *Given* n *variables, there are* $n!$ *grlex orders on* $k[x_1, ..., x_n]$ *in total.* □

DEFINITION 2.5. *(**Graded Reverse Lex Order**) Let* $\alpha = (\alpha_1, ..., \alpha_n)$ *and* $\beta = (\beta_1, ..., \beta_n) \in \mathbb{Z}^n_{\geq 0}$. *We define* $\alpha >_{grevlex} \beta$ *if either*

$$\text{“}|\alpha| = \sum_{i=1}^{n} \alpha_i > |\beta| = \sum_{i=1}^{n} \beta_i\text{”},$$

or "$|\alpha| = |\beta|$ *and the right-most non-zero entry of* $\alpha - \beta$ *in the vector difference is negative".*

EXAMPLE 2.6. $x^5yz >_{grevlex} x^4yz^2$, *because* $5+1+1 = 4+1+2$, *but* $(5,1,1)-(4,1,2) = (1,0,-1)$ *and* $-1 < 0$.

EXAMPLE 2.7. *With*

$$x_1 = (1, 0, ..., 0), x_2 = (0, 1, 0, ..., 0), ..., x_n = (0, ..., 0, 1),$$

we also have:

$$x_1 >_{grevlex} x_2 >_{grevlex} \cdots >_{grevlex} x_n.$$

LEMMA 2.6. *Given n variables, there are $n!$ grevlex orders on $k[x_1, ..., x_n]$ in total.* □

REMARK 2.4. *Its short name is "grevlex". We note, unlike others, that when grevlex with*

$$x_1 = (1, 0, ..., 0) >_{grevlex} \cdots >_{grevlex} x_n = (0, ..., 0, 1)$$

finds the same total degree, it looks at the right-most (i.e., smallest) variable and the one with smaller power is taken to be bigger one under this order. This may be one of the reasons why the grevlex ordering provides the most efficient computation for some operations, (especially related to Commutative Algebra stuffs, not Algebraic Geometry stuffs.) Algebraic Geometry stuffs usually requires us to use the grlex order or lex order.

There are many other monomial orders besides these three we have considered here, and most Computer Algebra Systems (CAS's) implement lex order, and some (such as Macaulay and REDUCE) also allow other orders, such as grlex and grevlex. Once such an order is chosen, these systems allow the user to specify any of the $n!$ orderings of the variables.

EXAMPLE 2.8. *Let $f = -5x^3 + x^2z^2 + 4xy^2z + z^2 \in k[x, y, z]$ Then under the above three monomial orders with $x = (1, 0, 0), y = (0, 1, 0)$, and $z = (0, 0, 1)$, this polynomial has three different expressions all in a non-increasing order, respectively:*

(1) *(Lex Order)* $f = -5x^3 + x^2z^2 + 4xy^2z + z^2$.
(2) *(Graded Lex Order)* $f = x^2z^2 + 4xy^2z - 5x^3 + z^2$.
(3) *(Graded Reverse Lex Order)* $f = 4xy^2z + x^2z^2 - 5x^3 + z^2$.

REMARK 2.5. *When we use a CAS (Computer Algebra System), we also need to know about the default order the system uses when it expresses a computed output, because this order is generally different from the order given to the computer by a user who wants his ordering to be applied to a specific computational job the user wants to perform. The former should be understood as an internal order only for output expressions. Of course, a user must be able to change this internal default order at least a little bit if the CAS he is using is well-developed. Among other things, a Computer Algebra System, REDUCE is very good in this sense.*

Exercises for Section 2.1

EXERCISE 2.1. *Show that the lex order is a monomial ordering on the polynomial ring* $k[x_1, ..., x_n]$.

EXERCISE 2.2. *Show that the graded lex order is a monomial ordering on* $k[x_1, ..., x_n]$.

EXERCISE 2.3. *Show that the graded reverse lex order is a monomial ordering on* $k[x_1, ..., x_n]$.

2. A Division Algorithm in $k[x_1, ..., x_n]$

Before we move into the next subject, we would like to introduce more terminologies in order to have a closer look at what's actually happening when we attempt to divide, over a field, a polynomial in several variables with a set of non-zero polynomials in several variables.

DEFINITION 2.6. *Let* $f = \sum a_\alpha x^\alpha$ *be a non-zero polynomial in the polynomial ring* $k[x_1, ..., x_n]$ *over a field* k *and let* $>$ *be a monomial order on* $k[x_1, ..., x_n]$.

(1) *The* ***multidegree*** *of* f *is*

$$multideg(f) = \max(\alpha \in \mathbb{Z}^n_{\geq 0} : a_\alpha \neq 0),$$

where the maximum is taken with respect to $>$.

(2) *The* ***leading coefficient*** *of* f *is*

$$LC(f) = a_{multi\deg(f)} \in k.$$

(3) *The **leading monomial** of f is*

$$LM(f) = x^{multi\deg(f)},$$

i.e., with coefficient 1.

(4) *The **leading term** of f is*

$$LT(f) = LC(f) \cdot LM(f).$$

EXAMPLE 2.9. *Let $f = -5x^3+x^2z^2+4xy^2z+z^2 \in k[x, y, z]$. Under the lex order, we have that $multi\deg(f) = (3, 0, 0), LC(f) = -5, LM(f) = x^3, LT(f) = -5x^3$.*

Here is a lemma, whose proof requires k to be an integral domain.

LEMMA 2.7. *Let $f, g \in k[x_1, ..., x_n]$ be non-zero polynomials. Then under any monomial order on $k[x_1, ..., x_n]$, we have*

(1) $multi\deg(fg) = multi\deg(f) + multi\deg(g)$.

(2) *If $f + g \neq 0$, then*

$$multi\deg(f+g) \leq \max(multi\deg(f), multi\deg(g)),$$

and if furthermore $multi\deg(f) \neq multi\deg(g)$, then the equality holds. □

Now, we are going to have two examples on how we may divide out a given polynomial in several variables with a set of polynomials in several variables, in order to see first of all what we can do, second what's actually happening during the division process involved. After that, we are going to state a division algorithm for polynomials in several variables. Unlike to the case of the division algorithm for polynomials in one variable, this algorithm is incomplete: both the quotient and remainder are never uniquely determined in general, even under the same monomial order on $k[x_1, ..., x_n]$.

EXAMPLE 2.10. *Let $f = x^2y+xy^2+y^2$ and $f_1 = y^2-1, f_2 = xy-1$. We want to divide f by first f_1 and then f_2 in this example under lex order with $x > y$, that is, we want to find a_1, a_2, and r such that $f = a_1f_1 + a_2f_2 + r$.*

Here, we can have at least two cases:

Case 1-1.

(1) *The leading term* $LT(f) = x^2y$ *of* f *is not divisible by* $LT(f_1) = y^2$, *so we move to the next largest term of* f.

(2) *The second largest term* xy^2 *of* f *is divisible by* $LT(f_1)$, *so its quotient* x *is saved as a part of the first quotient:*

$$\begin{aligned} & f \\ \longrightarrow\quad & f - xf_1 = x^2y + x + y^2. \end{aligned}$$

(3) *The third term of* $f - xf_1$ *is divisible by* $LT(f_1)$, *so its quotient* 1 *is saved as a part of the first quotient:*

$$\begin{aligned} & f - xf_1 \\ \longrightarrow\quad & f - xf_1 - f_1 = x^2y + x - 1, \end{aligned}$$

where none of the resulting terms is divisible by $LT(f_1)$, *so we are done with division by* $LT(f_1)$.

(4) *The leading term of* $f - xf_1 - f_1$ *is divisible by* $LT(f_2) = xy$, *so its quotient* x *is saved as a part of the second quotient:*

$$\begin{aligned} & f - (x+1)f_1 \\ \longrightarrow\quad & f - (x+1)f_1 - xf_2 = 2x + 1, \end{aligned}$$

where both $2x$ *and* 1 *in* $2x+1$ *are never divisible by either* $LT(f_2)$ *or* $LT(f_1)$, *so they are saved as a part of the remainder.*

(5) *Thus, we have*

$$f = (x+1) \cdot f_1 + x \cdot f_2 + (2x+1),$$

where $x + 1$ *and* x *are the first and second quotients, respectively and* $2x + 1$ *is the remainder.*

Case 1-2. *We again try to divide* f *by first* f_1 *and then* f_2. *But here, we stick to argue on the leading term of the intermediate polynomial* p *at each step.*

(1) *The leading term* $LT(f) = x^2y$ *of* f *is not divisible by* $LT(f_1) = y^2$.

(2) $LT(f)$ *is divisible by* $LT(f_2)$, *so its quotient* x *is saved as a part of the second quotient:*

$$\begin{aligned} & f \\ \longrightarrow \quad & f - xf_2 = xy^2 + x + y^2. \end{aligned}$$

(3) *The leading term* xy^2 *of* $xy^2 + x + y^2$ *is divisible by* $LT(f_1)$, *so its quotient* x *is saved as a part of the first quotient:*

$$\begin{aligned} & f - xf_2 \\ \longrightarrow \quad & f - xf_2 - xf_1 = 2x + y^2. \end{aligned}$$

(4) *The leading term* $2x$ *of* $2x+y^2$ *is never divisible by either* $LT(f_1)$ *or* $LT(f_2)$, *so it is saved as a part of the remainder:*

$$\begin{aligned} & f - xf_2 - xf_1 \\ \longrightarrow \quad & f - xf_2 - xf_1 - 2x = y^2 \end{aligned}$$

(5) *The leading term* y^2 *of* y^2 *is divisible by* $LT(f_1)$, *so its quotient* 1 *is saved as a part of the first quotient:*

$$\begin{aligned} & f - xf_2 - xf_1 - 2x \\ \longrightarrow \quad & f - xf_2 - xf_1 - 2x - f_1 = 1. \end{aligned}$$

(6) *The leading term* 1 *of* 1 *is never divisible by either* $LT(f_1)$ *or* $LT(f_2)$, *so it is saved as a part of the remainder:*

$$\begin{aligned} & f - xf_2 - xf_1 - 2x - f_1 \\ \longrightarrow \quad & f - xf_2 - xf_1 - 2x - f_1 - 1 = 0. \end{aligned}$$

(7) *Thus, we have*

$$f = (x+1) \cdot f_1 + x \cdot f_2 + (2x+1).$$

We note that in both cases, we have the same quotients and remainders.

EXAMPLE 2.11. *Let f, f_1, and f_2 be the same as before. Now this time, we want to divide f by first f_2 and then f_1 successively under lex $x > y$, that is, we want to find a_1, a_2, and r such that $f = b_1 f_1 + b_2 f_2 + r'$.*

Here, we can have two cases:

Case 2-1.

(1) *The first term of f is divisible by $LT(f_2) = xy$, so its quotient x is saved as a part of the second quotient:*

$$\begin{aligned} & f \\ \longrightarrow \quad & f - xf_2 = xy^2 + x + y^2. \end{aligned}$$

(2) *The first term of $f - xf_2$ is again divisible by $LT(f_2)$, so its quotient y is saved as a part of the second quotient:*

$$\begin{aligned} & f - xf_2 \\ \longrightarrow \quad & f - xf_2 - yf_2 = x + y^2 + y. \end{aligned}$$

(3) *Since $LT(f_2)$ never divides none of the terms in $x + y^2 + y$, we are done with f_2.*

(4) *The first term of $f - xf_2 - yf_2$ is not divisible by $LT(f_1) = y^2$.*

(5) *The second term of $f - xf_2 - yf_2$ is divisible by $LT(f_1)$, so its quotient 1 is saved as a part of the first quotient:*

$$\begin{aligned} & f - (x+y)f_2 \\ \longrightarrow \quad & f - (x+y)f_2 - f_1 = x + y + 1, \end{aligned}$$

where both x and y in $x + y$ is never divisible by either $LT(f_1)$ or $LT(f_2)$, so they are saved as a part of the remainder.

(6) *Thus, we have*

$$f = 1 \cdot f_1 + (x+y) \cdot f_2 + (x + y + 1).$$

Case 2-2. *We again try to divide f by first f_2 and then f_1. But here, we stick to argue on the leading term of the intermediate polynomial p at each step.*

(1) *The leading term of f is divisible by $LT(f_2) = xy$, so its quotient x is saved as a part of the second quotient:*

$$\begin{aligned} & f \\ \longrightarrow \quad & f - xf_2 = xy^2 + x + y^2. \end{aligned}$$

(2) *The leading term of $xy^2 + x + y^2$ is divisible by $LT(f_2) = xy$, so its quotient x is saved as a part of the first quotient:*

$$\begin{aligned} & f - xf_2 \\ \longrightarrow \quad & f - xf_2 - yf_2 = x + y^2 + y. \end{aligned}$$

(3) *The leading term x of $x+y^2+y$ is never divisible by either $LT(f_1)$ or $LT(f_2)$, so it is saved as a part of the remainder:*

$$\begin{aligned} & f - xf_2 - yf_2 \\ \longrightarrow \quad & f - xf_2 - yf_2 - x = y^2 + y. \end{aligned}$$

(4) *The leading term y^2 of $y^2 + y$ is not divisible by $LT(f_2)$, but is divisible by $LT(f_1)$ so its quotient 1 is saved as a part of the first quotient:*

$$\begin{aligned} & f - xf_2 - yf_2 - x \\ \longrightarrow \quad & f - (x + y)f_2 - x - f_1 = y + 1. \end{aligned}$$

(5) *The leading term y of $y + 1$ is not divisible by either $LT(f_2)$ or $LT(f_1)$, so it is saved as a part of the remainder:*

$$\begin{aligned} & f - xf_2 - yf_2 - x - f_1 \\ \longrightarrow \quad & f - xf_2 - yf_2 - x - f_1 - y = 1. \end{aligned}$$

(6) *Thus, we have*

$$f = 1 \cdot f_1 + (x + y) \cdot f_2 + (x + y + 1).$$

Note that both quotients and remainders are the same in these cases.

If we compare the results of these two examples, we immediately see that both quotients and remainders are not the same all the time, i.e., they are dependent on a particular sequence of division process. In other words, they are really dependent on who performs the division, so for instance this kind of division cannot be used in the process of proving something in mathematics. There are two problems here: the quotients are not uniquely determined and the remainder is not uniquely determined, under a given same monomial order, lex order, on $k[x_1, ..., x_n]$. We think this should be the main reason why virtually no progress had been made in the computation theory of divisions for polynomials in several variables for a long time at least since BC 300.

But, the following example should suggest how these two problems could be partially fixed:

EXAMPLE 2.12. *We note that the ideal*

$$I_1 = (f_1, f_2) = (y^2 - 1, xy - 1)$$

is the same as the ideal

$$I_2 = (x - y, y^2 - 1)\,,$$

since

$$x - y = -x \cdot f_1 + y \cdot f_2,$$

and

$$xy - 1 = y \cdot (x - y) + 1 \cdot (y^2 - 1).$$

That is, suppose we have somewhat figured out another generating set

$$\{x - y, y^2 - 1\}$$

of the same ideal $I = (f_1, f_2)$.

Write

$$\begin{aligned} g_1 &= x - y, \\ g_2 &= y^2 - 1, \end{aligned}$$

and again under lex order with $x > y$ *we first try to divide* f *by first* g_1 *and then* g_2*, and then secondly we try to divide* $f = x^2y + xy^2 + y^2$ *by* g_2 *and* g_1*. We still do this in order to see what differences could show up.*

Case 1-1. *We try to divide f by first g_1 and then g_2.*

(1) *$LT(g_1) = x$ divides the first term of f, so we save its quotient xy as a part of the first quotient:*

$$\begin{aligned} & f \\ \longrightarrow \quad & f - xyg_1 = 2xy^2 + y^2. \end{aligned}$$

(2) *$LT(g_1)$ divides the first term of $2xy^2 + y^2$, so we save its quotient $2y^2$ as a part of the first quotient:*

$$\begin{aligned} & f - xyg_1 \\ \longrightarrow \quad & f - xyg_1 - 2y^2g_1 = 2y^3 + y^2. \end{aligned}$$

(3) *Since none of the terms in $2y^3 + y^2$ are divisible by $LT(g_1)$, we are done with g_1.*

(4) *$LT(g_2) = y^2$ divides both terms of $2y^3 + y^2$, so we save its quotient as a part of the second quotient:*

$$\begin{aligned} & f - xyg_1 - 2y^2g_1 \\ \longrightarrow \quad & f - xyg_1 - 2y^2g_1 - (2y+1)g_2 = 2y + 1. \end{aligned}$$

(5) *Since $2y + 1$ is never divisible by either $LT(g_2)$ or $LT(g_1)$, we save it as a part of the remainder.*

(6) *Thus, we have*

$$f = (xy + 2y^2) \cdot g_1 + (2y + 1)g_2 + (2y + 1).$$

Case 1-2. *We again try to divide f by first g_1 and then g_2. But here, we stick to argue on the leading term of the intermediate polynomial p at each step.*

(1) *$LT(g_1) = x$ divides the leading term of f, so we save its quotient xy as a part of the first quotient:*

$$\begin{aligned} & f \\ \longrightarrow \quad & f - xyg_1 = 2xy^2 + y^2. \end{aligned}$$

(2) *$LT(g_2) = y^2$ divides the leading term of $2xy^2 + y^2$, so we save its quotient $2x$ as a part of the first quotient:*

$$\begin{aligned} & f - xyg_1 \\ \longrightarrow\ & f - xyg_1 - 2xg_2 = 2x + y^2. \end{aligned}$$

(3) *$LT(g_1)$ does divides the leading term of $2x + y^2$, so its quotient 2 is saved as a part of the second quotient:*

$$\begin{aligned} & f - xyg_1 - 2xg_2 \\ \longrightarrow\ & f - xyg_1 - 2xg_2 - 2g_1 = y^2 + 2y. \end{aligned}$$

(4) *$LT(g_2)$ divides the leading term y^2 of $y^2 + 2y$, so its quotient 1 is saved as a part of the second quotient:*

$$\begin{aligned} & f - (xy + 2)g_1 - 2xg_2 \\ \longrightarrow\ & f - (xy + 2)g_1 - 2xg_2 - g_2 = 2y + 1. \end{aligned}$$

(5) *None of $LT(g_1)$ and $LT(g_2)$ divides the leading term $2y$ of $2y+1$, so $2y$ is saved as a part of the remainder:*

$$\begin{aligned} & f - (xy + 2)g_1 - (2x + 1)g_2 \\ \longrightarrow\ & f - (xy + 2)g_1 - (2x + 1)g_2 - 2y = 1. \end{aligned}$$

(6) *None of $LT(g_1)$ and $LT(g_2)$ divides the leading term 1 of 1, so 1 is saved as a part of the remainder:*

$$\begin{aligned} & f - (xy + 2)g_1 - (2x + 1)g_2 - 2y \\ \longrightarrow\ & f - (xy + 2)g_1 - (2x + 1)g_2 - 2y - 1 = 0. \end{aligned}$$

(7) *Thus, we have*

$$f = (xy + 2)g_1 + (2x + 1)g_2 + (2y + 1).$$

Case 2-1. *Now we try to divide f by first $g_2 = f_1$ and then g_1.*

(1) *The leading term $LT(f) = x^2y$ of f is not divisible by $LT(g_2) = y^2$, so we move to the next largest term of f.*

(2) *The second largest term* xy^2 *of* f *is divisible by* $LT(g_2)$, *so its quotient* x *is saved as a part of the second quotient:*

$$\begin{aligned} & f \\ \longrightarrow\ & f - xg_2 = x^2y + x + y^2. \end{aligned}$$

(3) *The third term of* $f - xg_2$ *is divisible by* $LT(g_2)$, *so its quotient* 1 *is saved as a part of the second quotient:*

$$\begin{aligned} & f - xg_2 \\ \longrightarrow\ & f - xg_2 - g_2 = x^2y + x + 1, \end{aligned}$$

where none of the resulting terms is divisible by $LT(g_2)$, *so we are done with division by* $LT(g_2)$.

(4) *The first term of* $f - xg_2 - g_2$ *are divisible* $LT(g_1) = x$, *so its quotient* $xy + 1$ *is saved as a part of the first quotient:*

$$\begin{aligned} & f - (x+1)g_2 \\ \longrightarrow\ & f - (x+1)g_2 - (xy+1)g_1 = xy^2 + y + 1. \end{aligned}$$

(5) *The first and second terms of* $xy^2 + y + 1$ *is divisible by* $LT(g_1)$, *so its quotient* y^2 *is saved as a part of the first quotient:*

$$\begin{aligned} & f - (x+1)g_2 - (xy+1)g_1 \\ \longrightarrow\ & f - (x+1)g_2 - (xy+1)g_1 - y^2g_1 = y^3 + y + 1. \end{aligned}$$

(6) *Now no terms of* $y^3 + y + 1$ *are divisible by* $LT(g_1)$. *But we now note that the first term of* $y^3 + y + 1$ *is again divisible by* $LT(g_2)$, *so we save its quotient* y *as a part of the second quotient:*

$$\begin{aligned} & f - (x+1)g_2 - (xy+y^2+1)g_1 \\ \longrightarrow\ & f - (x+1)g_2 - (xy+y^2+1)g_1 - yg_2 = 2y + 1. \end{aligned}$$

(7) *Then* $2y + 1$ *is divisible by none of either* $LT(g_1)$ *or* $LT(g_2)$. *Thus, we have*

$$f = (xy + y^2 + 1) \cdot g_1 + (x + y + 1) \cdot g_2 + (2y + 1).$$

Case 2-2. *Now we try to divide f by first $g_2 = f_1$ and then g_1. But here, we stick to argue on the leading term of the intermediate polynomial p at each step.*

(1) *The leading term $LT(f) = x^2y$ of f is not divisible by $LT(g_2) = y^2$.*
(2) *$LT(f)$ is divisible by $LT(g_1) = x$, so its quotient xy is saved as a part of the first quotient:*

$$\begin{aligned} & f \\ \longrightarrow \quad & f - xyg_1 = 2xy^2 + y^2. \end{aligned}$$

(3) *$LT(g_2)$ divides $LT(2xy^2 + y^2) = 2xy^2$, so its quotient $2x$ is saved as a part of the second quotient:*

$$\begin{aligned} & f - xyg_1 \\ \longrightarrow \quad & f - xyg_1 - 2xg_2 = 2x + y^2. \end{aligned}$$

(4) *$LT(g_1)$ divides $LT(2x + y^2) = 2x$, so its quotient 2 is saved as a part of the first quotient:*

$$\begin{aligned} & f - xyg_1 - 2xg_2 \\ \longrightarrow \quad & f - xyg_1 - 2xg_2 - 2g_1 = y^2 + 2y. \end{aligned}$$

(5) *$LT(g_2)$ divides $LT(y^2 + 2y) = y^2$, so its quotient 1 is saved as a part of the second quotient:*

$$\begin{aligned} & f - xyg_1 - 2xg_2 - 2g_1 \\ \longrightarrow \quad & f - xyg_1 - 2xg_2 - 2g_1 - g_2 = 2y + 1. \end{aligned}$$

(6) *Both $LT(g_2)$ and $LT(g_1)$ do not divide $LT(2y + 1) = 2y$, so $2y$ is saved as a part of the remainder:*

$$\begin{aligned} & f - (xy + 2)g_1 - (2x + 1)g_2 \\ \longrightarrow \quad & f - (xy + 2)g_1 - (2x + 1)g_2 - 2y = 1. \end{aligned}$$

(7) *Both* $LT(g_2)$ *and* $LT(g_1)$ *do not divide* $LT(1) = 1$, *so* 1 *is saved as a part of the remainder:*

$$\begin{aligned} & f - (xy+2)g_1 - (2x+1)g_2 - 2y \\ \longrightarrow\ & f - (xy+2)g_1 - (2x+1)g_2 - 2y - 1 = 0. \end{aligned}$$

(8) *Thus, we have*

$$f = (xy+2)g_1 + (2x+1)g_2 + (2y+1).$$

We note that all Cases give the same remainder, and Case $1-2$ *and Case* $2-2$ *even give the same quotients.*

If we compare the results of the four computations in the last example, we see that the remainders of all of them are the same, $2y+1$, although we still get different quotients depending on the order of division.

We also note that in these examples, Example 2.10, Example 2.11, and Example 2.12, we have used the same division process in the sense that the remainder r satisfies the following condition, called the **reduction condition** mod F, that "*either* $r = 0$ *or no term of the remainder* r *is divisible by any of the leading terms of the divisors,* $F = \{f_1, ..., f_s\}$", but we also note that it is far from being complete in the sense that *the quotients and the remainder are not uniquely determined.*

The only point in the last example is that our divisors g_1 and g_2 are very special ones, and we obtain this by first considering a lot more complicated objects, ideals $I = (f_1, f_2) = (g_1, g_2)$, than elements, f_1, f_2. That is, *the new and important point is that the generating set* $\{f_1, f_2\}$ *of the same ideal* I *is changed to another generating set* $\{g_1, g_2\}$. But we do not know yet why this approach works generally (which requires axiomatic treatments) and how to obtain such a generating set that produces the same remainder, without depending on the division process applied (which requires an algorithm for finding such a generating set for the same ideal I).

One reason why the Taylor series expansion of a real-valued function, if exists, is so important is that, it does not matter which methods are used for what different manners. As long as one obtains the Taylor series expansion of a real-valued function, it is uniquely determined. That is, it is independent of the path for obtaining it, and this is the kind we like most of the time. We note that in the above example, the remainder is independent of the path for obtaining it.

We will see very soon that we do not accidentally have this nice result of having the same remainder. But before that, we officially describe the division algorithm applied in these examples.

ALGORITHM 5. ***(Division Algorithm in** $k[x_1, ..., x_n]$**)*** *Fix a monomial order* $>$ *on* $\mathbb{Z}^n_{\geq 0}$*, and let* $F = (f_1, f_2, ..., f_s)$ *be an* ordered *s-tuple of non-zero polynomials in* $k[x_1, ..., x_n]$*. Then every polynomial* $f \in k[x_1, ..., x_n]$ *can be written as*

$$f = a_1 f_1 + \cdots + a_s f_s + r,$$

where $a_i, r \in k[x_1, ..., x_n]$*, and* r *is reduced, i.e., "either* $r \equiv 0$ *or* r *is a* k*-linear combination of monomials, none of which is divisible by any of* $LT(f_1), ..., LT(f_s)$*." We call* r ***a remainder of** f **on division by** F*. *Furthermore, for those* $a_i f_i \neq 0$*, we have*

$$multi\deg(f) \geq multi\deg(a_i f_i).$$

PROOF. If we can prove the following algorithm terminates, then the existence of $a_1, ..., a_s$ and r is proved.

Input: $f_1, ..., f_s, f$
Output: $a_1, ..., a_s, r$

$a_1 := 0, ..., a_s := 0; r := 0$
$p := f$
WHILE $p \neq 0$ **DO**
$i := 1$
divisionoccurred := false
  WHILE $i \leq s$ **AND divisionoccurred** = **false DO**
    If $LT(f_i)$ **divides** $LT(p)$ **THEN**
      $a_i := a_i + \frac{LT(p)}{LT(f_i)}$

$p := p - \frac{LT(p)}{LT(f_i)} \cdot f_i$
divisionoccurred = true

ELSE

$i := i + 1$
divisionoccurred = false

IF divisionoccurred = false THEN

$r := r + LT(p)$
$p := p - LT(p)$

We first note that the above algorithm consists of two parts, Division Step and Remainder Step.

- (Division Step) If some $LT(f_i)$ divides $LT(p)$, the the algorithm proceeds as in the one-variable case.
- (Remainder Step) If none of $LT(f_i)$ divides $LT(p)$, ten the algorithm adds $LT(p)$ to the remainder.

Now, we can easily see that the form like

$$f - a_1 f_1 - \cdots - a_s f_s - r = p,$$

holds at both Division Step and Remainder Step within at every step.

We also note that the algorithm certainly terminates after finitely many steps. This is because at each step a new p is defined in such a way that *either* its multidegree drops *or* it becomes 0. Thus, if it does not stop, i.e., p does not become 0, after finitely many steps, then we would have an infinite decreasing sequence of multidegrees, which is impossible by the well-ordering property of our monomial order $>$ on $k[x_1, ..., x_n]$. Thus, after finitely many steps we must have $p = 0$. Thus, as we have seen Case $1-2$ and Case $2-2$ in the above examples, the algorithm stops when $p = 0$, and at this very moment, we have the final result:

$$f = a_1 f_1 + \cdots a_s f_s + r,$$

where r is either 0 or none of the terms of r is divisible by any $LT(f_i)$.

Finally, it is easy to see $multi\deg(a_i f_i) \leq multi\deg(f)$ for those non-zero $a_i f_i$'s since $multi\deg(p)$ decreases from that of f as the steps go forward till termination. □

COROLLARY 2.1. *When this division algorithm is terminated, we have*

$$LM(f) = \max(\max_{1 \leq i \leq s}(LM(a_i)LM(f_i)), LM(r)).$$

Moreover, when the above algorithm is terminated with at least one $a_i \neq 0$, then $LT(f)$ is the sum of those $LT(a_i)LT(f_i)$'s such that

$$LM(f) = LM(a_i)LM(f_i)$$

among all $LT(a_i)LT(f_i)$'s.

PROOF. We first note that the initial conditions for p was f, r was 0, and all a_i was 0. As the step goes, the multidegree of the monomial $LT(p)$ is decreased and that of some of the a_i's is increased, unless $f = r$. But still at each step of the algorithm, we force to maintain $LM(p) \leq LM(f)$, where p is a new p, Then a new quotient a_i is obtained by adding a non-zero (monomial) term $\frac{LT(p)}{LT(f_i)}$, where p is the old p, to the old a_i only when $\frac{LT(p)}{LT(f_i)} \cdot f_i$ cancels the leading term of the old p. As the above examples show, we all the time maintain

$$f - \sum a_i f_i - r = p$$

at each step of the algorithm, where a_i, r, and p are either old a_i, old r , and old p, or new a_i, new r, and new p. In particular, we only add to the old r those terms of the old p that are not divisible by any of the leading terms of $f_1, ..., f_s$, and thus this equality also implies that we always have

$$LM(a_i)LM(f_i) \leq LM(f),$$

regardless of the quotients a_i's being old or new.

Thus, particularly when the algorithm is terminated with at least one a_i being a non-constant polynomial, then $LT(f)$ must be equal to the sum of those $LT(a_i)LT(f_i)$'s such that

$$LM(f) = LM(a_i)LM(f_i)$$

among all $LT(a_i)LT(f_i)$'s. We note that this expression due includes all possible cancellations among them because of their different coefficients in k with different signs. Otherwise, $f = r$. □

REMARK 2.6. *If the divisors $f_1, ..., f_s$ are all* monomials, *then the division algorithm shrinks to a trivial one. For example, let $f = x^2y + xy^2 + y^2$ and $f_1 = y^2, f_2 = xy$. Then*

$$\begin{aligned} f &= (x+1) \cdot f_1 + x \cdot f_2 \\ &= 1 \cdot f_1 + (x+y) \cdot f_2. \end{aligned}$$

Otherwise, the result of the division usually depends on who performs it.

REMARK 2.7. *We note that in the above examples, Example 2.10, Example 2.11, and Example 2.12, each step of our division processes all satisfies the condition that* "either $r = 0$ *or* no term of the remainder r is divisible by any of the leading term of the divisors, $F = \{f_1, ..., f_s\}$."

DEFINITION 2.7. *Such a condition is called a* ***Reduction Condition (RC) modulo*** F.

We note that, as long as this condition is satisfied, a variant of the above division algorithm is essentially the same, and hence the only thing matters is which version of the division algorithm is more *time efficient* than others.

For simplicity in our description of the division process of polynomials in several variables, we introduce the following notations:

DEFINITION 2.8. *Give $f, g, r \in k[x_1, ..., x_n]$, with $g \neq 0$, we say that f is* ***reduced to*** r ***modulo*** g *in one step if $LT(g)$ divides a non-zero term x^α of f in the following sense:*

$$r = f - \frac{x^\alpha}{LT(g)} \cdot g,$$

and we denote it by

$$f \xrightarrow{g} r.$$

More generally, we have

Definition 2.9. *Given polynomials* $f, h,$ *and* $f_1, ..., f_s$ *in* $k[x_1, ..., x_n]$ *with* $f_i \neq 0$ *for* $1 \leq i \leq s$, *write* $F = \{f_1, ..., f_s\}$. *We say that* f ***is reduced to*** r ***modulo*** F *if*

$$f \xrightarrow{f_{i_1}} h_1 \xrightarrow{f_{i_2}} h_2 \xrightarrow{f_{i_3}} \cdots \xrightarrow{f_{i_{t-1}}} h_{t-1} \xrightarrow{f_{i_t}} r,$$

for some $i_1, ..., i_t \in \{1, ..., s\}$ *and* $h_1, ..., h_{t-1} \in k[x_1, ..., x_n]$, *and we denote it by*

$$f \xrightarrow{F}_+ r.$$

Definition 2.10. *A polynomial* $r \in k[x_1, ..., x_n]$ *is called* ***reduced*** *with respect to a set of non-zero polynomials* $F = \{f_1, ..., f_s\}$ *if the Reduction Condition (RC) modulo* F *is satisfied. In other words,* r *cannot be reduced modulo* F *any further.*

With these better terminologies, we have:

Definition 2.11. *If* $f \xrightarrow{F}_+ r$ *and* r *is reduced with respect to* F, *the* r *is called a* ***remainder of*** f ***on division by*** F.

We note that r is called "a" remainder, because we do not know yet when it is going to be uniquely determined under what condition.

Exercises for Section 2.2

Exercise 2.4. *Compute the remainder and the quotients on the division of*

$$f = x^5y^6 + x^4y^3 - xy + 1$$

by the ordered set $F = (x^2y - y, x^2 - y)$, *or* $F = (x^2 - y, x^2y - y)$, *respectively. Are they the same? For example, see Program 2 for a further discussion on this matter.*

3. Monomial Ideals and Groebner Bases

In the previous section, we have seen that the reduction process is done in terms of $LT(f_i)$'s. It turns out that ideals whose generators are monomials are very important, so we start our discussion with the definition of monomials ideals in $k[x_1, ..., x_n]$.

DEFINITION 2.12. *A polynomial ideal in* $k[x_1, .., x_n]$ *is called a* ***monomial ideal*** *if its generating set (finite or not) consists of only monomials.*

EXAMPLE 2.13. $I = (x^4y^2, x^3y^4, x^2y^5) \subset k[x, y]$.

In particular, we note that $k[x_1, ..., x_n]$ itself is a monomial ideal generated by all monomials in $k[x_1, ..., x_n]$. Then we recall that the set of all monomials in $k[x_1, ..., x_n]$ is a k-basis of the infinite k-dimensional vector space $k[x_1, ..., x_n]$. In particular, this says

LEMMA 2.8. *Every monomial ideal is a* k*-vector subspace of* $k[x_1, ..., x_n]$.

PROOF. Its basis consists of all the monomials in it. □

Thus, it is monomial ideals which allow us to have a connection with k-vector space structure of $k[x_1, ..., x_n]$.

LEMMA 2.9. *Let* $I = (x^\alpha : \alpha \in A$, *where* A *is a subset of* $\mathbb{Z}^n_{\geq 0})$ *be a monomial ideal. Then a monomial* x^β *is an element of* I *if and only if it is divisible by some of its generating monomial* x^α.

PROOF. Write x^β in terms of finite number of generators x^α's:

$$x^\beta = \sum_{finite} h_i x^{\alpha_i},$$

where each $h_i \in k[x_1, ..., x_n]$; then write each of these polynomials h_i as a k-linear combination of monomials in $k[x_1, ..., x_n]$, to have a nontrivial linear relation among monomials in $k[x_1, ..., x_n]$. Then since x^β in the left- hand side of the equality is a basis element, there must be cancellation among the terms in the right-hand side of the equality, and moreover x^β itself must appear in the right-hand side of the equality, so that it must be divisible by some x^{α_i}. All this is because the set of all monomials in $k[x_1, ..., x_n]$ is a k-basis of the infinite k-dimensional vector space $k[x_1, ..., x_n]$. □

LEMMA 2.10. *Let* I *be a monomial ideal in* $k[x_1, ..., x_n]$, *and let* f *be a polynomial* $\in k[x_1, ..., x_n]$. *Then the following statements are equivalent to each other:*

(1) $f \in I$.
(2) *Every term of f is an element of I.*
(3) *f is a k-linear combination of the monomials in I.*

PROOF. Part 3 implies part 2 is easy since each monomial in I is, by Lemma 2.9, a monomial multiple of one of the monomials in the generating set for I. Then Part 2 implies part 1 is obvious. Part 1 implies part 3 is also obvious. □

COROLLARY 2.2. *Two monomial ideals are the same if and only if they contain the same monomials.*

PROOF. Because part 3 of the lemma guarantees the $\Longleftarrow$ direction. □

3.1. Dickson's Lemma. We recall that we have proved Hilbert's Basis Theorem (Theorem 1.5).

THEOREM 2.1. *(Dickson's Lemma) Every monomial ideal I is finitely generated by a finite number of generators from its generating set of monomials as a monomial ideal. In particular, every monomial ideal is finitely generated.*

PROOF. Let $I = (x^\alpha : \alpha \in A$, where A is a subset of $\mathbb{Z}^n_{\geq 0})$ be a monomial ideal. Then since it is an ideal in $k[x_1, ..., x_n]$, it is finitely generated by some polynomials in $k[x_1, ..., x_n]$ by Hilbert Basis Theorem (Theorem 1.5):

$$I = (f_1, ..., f_s).$$

Then since each $f_i \in I$, by Lemma 2.10 each f_i is a k-linear combination of the monomials x^β in I; and then by Lemma 2.9 each monomial x^β in this expression is in turn a monomial multiple of some monomial x^α that is in the generating set of I as a monomial ideal; hence replacing each f_i with all but still finite number of those mentioned monomials x^α in the generating set of I, we finally see that I is contained in an ideal J that is finitely generated by those monomials in the generating set of I as a monomial ideal in $k[x_1, ..., x_n]$:

$$I = (f_1, ..., f_s) \subseteq J \subseteq I,$$

where J is the monomial ideal generated by those above-mentioned monomials x^α in the generating set of I as a monomial ideal. □

COROLLARY 2.3. *Let I be a monomial ideal generated by monomials $x^{\alpha_1}, ..., x^{\alpha_s}$ in $k[x_1, ..., x_n]$, and let $f \in k[x_1, ..., x_n]$. Then $f \in I$ if and only if the remainder of f on division by $x^{\alpha_1}, ..., x^{\alpha_s}$ is zero.*

PROOF. The $\Longleftarrow$ direction is obvious, and the $\Longrightarrow$ direction holds by Lemma 2.10. □

The following corollary says something on the well-ordering property of a monomial ordering on $k[x_1, ..., x_n]$.

COROLLARY 2.4. *Let $>$ be a relation on $\mathbb{Z}^n_{\geq 0}$ such that*

(1) *it is a total ordering.*

(2) *$\alpha > \beta$ implies $\alpha + \gamma > \beta + \gamma$ for any $\gamma \in \mathbb{Z}^n_{\geq 0}$.*

Then $\alpha \geq 0$ for all $\alpha \in \mathbb{Z}^n_{\geq 0}$ if and only if $>$ is a well-ordering.

PROOF. Suppose that $>$ is a well-ordering. Let α be the smallest element of $\mathbb{Z}^n_{\geq 0}$. If this $0 > \alpha$, then by adding infinitely many time the same α on both sides of this inequality, we would have an infinite decreasing sequence,

$$0 > \alpha > 2\alpha > \cdots$$

but this is impossible. Conversely, suppose that $\alpha \geq 0$ for all $\alpha \in \mathbb{Z}^n_{\geq 0}$. Let A be a non-empty subset of $\mathbb{Z}^n_{\geq 0}$. Consider the monomial ideal I generated by A. By Dickson's Lemma (Theorem 2.1) we know that I is generated by finitely many elements $\alpha_1, ..., \alpha_s$ of A. We may assume that $\alpha_1 < \alpha_2 < \cdots < \alpha_s$. We are going to prove that α_1 is the smallest element of A. For this, let $\alpha \in A$. Then $x^\alpha \in I$, so by Lemma 2.9, it is a monomial multiple of some x^{α_i}; hence $\alpha = \alpha_i + \gamma$ for some $\gamma \in \mathbb{Z}^n_{\geq 0}$. Then since $\gamma \geq 0$, so

$$\alpha = \alpha_i + \gamma \geq \alpha_i \geq \alpha_1.$$

□

REMARK 2.8. *We may compare this result with the statement of Definition 2.2.*

3.2. Axiomatic Definition of Groebner Bases. We have already seen that leading terms of polynomials play an important role in the division algorithm.

DEFINITION 2.13. *Let I be a non-zero polynomial ideal in $k[x_1, ..., x_n]$.*

1. *We denote by $LT(I)$ the set of leading terms $LT(f)$ of elements f of I:*

$$LT(I) := \{cx^\alpha : \text{ there exists } f \in I \text{ such that } LT(f) = cx^\alpha\}.$$

2. *We denote by $(LT(I))$ the ideal generated by the elements of $LT(I)$.*

LEMMA 2.11. *Let I be an ideal in $k[x_1, ..., x_n]$.*

1. $(LT(I))$ is a monomial ideal.
2. There exist polynomials $g_1, ..., g_s \in I$ such that

$$(LT(I)) = (LT(g_1), ..., LT(g_s)).$$

PROOF. $(LT(I)) = (LM(g) : g \in I - (0))$, so the first statement is trivial. Then, by Dickson's Lemma (Theorem 2.1), $(LT(I)) = (x^{\alpha_1}, ..., x^{\alpha_s})$, where each x^{α_i} is the leading term $LT(g_i)$ of a polynomial $g_i \in I$; hence the second statement is true. □

Thus, given a non-zero polynomial ideal I, $(LT(I))$ is the associated monomial ideal.

Now suppose we are given a polynomial ideal I in $k[x_1, ..., x_n]$. Then by Hilbert, it is finitely generated, so

$$I = (g_1, ..., g_s).$$

Then its associated monomial ideal clearly contains:

$$(LT(I)) \supseteq (LT(g_1), ..., LT(g_s)).$$

We note that if I is a monomial ideal generated by finitely many monomials $x^{\alpha_1}, ..., x^{\alpha_s}$, then we always have

$$(LT(I)) = I = (x^{\alpha_1}, ..., x^{\alpha_s}) = (LT(x^{\alpha_1}), ..., LT(x^{\alpha_s})).$$

But the following example shows that the equality does not hold in general.

EXAMPLE 2.14. *let* $I = (x^3 - 2xy, x^2y - 2y^2 + x)$. *Then under the grlex ordering with* $x > y$, *we have a polynomial identity, called*

$$S\text{-}polynomial((x^3 - 2xy + x, x^2y - 2y^2)$$

which is the resulting polynomial after cancellation of the leading terms of the two polynomials:

$$x \cdot (x^2y - 2y^2 + x) - y \cdot (x^3 - 2xy) = x^2.$$

Thus, $x^2 \in I$; *hence it is an element of* $(LT(I))$. *But, it is not divisible by either* $LT(x^3 - 2xy) = x^3$ *or* $LT(x^2y - 2y^2 + x) = x^2y$, *so it is not an element of* (x^3, x^2y).

Now, we are going to prove Hilbert Basis Theorem from Dickson's Lemma. This not only establishes that these two important results are equivalent but also its proof tells a very important thing: Existence of not-yet defined basis for a given polynomial ideal, called a Groebner basis.

THEOREM 2.2. *(Hilbert Basis Theorem) Every polynomial ideal* I *in* $k[x_1, ..., x_n]$ *is finitely generated.*

PROOF. If $I = (0)$, there is nothing to prove. Suppose that $I \neq (0)$. Then By Dickson's Lemma (Theorem 2.1), we know that its associated monomial ideal $(LT(I))$ is finitely generated by monomials in it:

$$(LT(I)) = (LT(g_1), ..., LT(g_s)),$$

for some polynomials $g_1, ..., g_s \in I$. (See Lemma 2.11). In what follows, we are going to prove that $I = (g_1, ..., g_s)$.

Clearly, $(g_1, ..., g_s) \subseteq I$. For the reverse inclusion, let $f \in I$. We simply apply the division.algorithm, Algorithm 5, under a fixed monomial order on $k[x_1, ..., x_n]$:

$$f = a_1 g_1 + \cdots a_s g_s + r,$$

where either $r \equiv 0$ or none of the terms of the remainder polynomial r is divisible by any $LT(g_i)$. But if $r \neq 0$, then

$$r = f - a_1 g_1 + \cdots a_s g_s \in I,$$

and hence $LT(r) \in (LT(I)) = (LT(g_1), ..., LT(g_s))$; hence by Lemma 2.9, it is divisible by some $LT(g_i)$, which is impossible. Thus, we must have $r \equiv 0$, and this just means that $I = (g_1, ..., g_s)$. □

Therefore, we have proved that

COROLLARY 2.5. *Hilbert's Basis Theorem and Dickson's Lemma are equivalent.* □

COROLLARY 2.6. *For a fixed monomial order on $k[x_1, ..., x_n]$, every non-zero polynomial ideal I in $k[x_1, ..., x_n]$ has a basis, i.e., a generating set, $\{g_1, ..., g_s\}$ such that*

$$(LT(I)) = (LT(g_1), ..., LT(g_s)). \quad \square$$

DEFINITION 2.14. *Fix a monomial order on $k[x_1, ..., x_n]$. A finite subset $G = \{g_1, ..., g_s\}$ of a polynomial ideal I in $k[x_1, ..., x_n]$ is called a* ***Groebner basis*** *(or* ***Standard basis****)if*

$$(LT(I)) = (LT(g_1), ..., LT(g_s)).$$

Thus the above corollary equivalently says that

COROLLARY 2.7. *Every non-zero ideal I in $k[x_1, ..., x_n]$ has a Groebner basis.* □

COROLLARY 2.8. *Any Groebner basis for a polynomial ideal I is a basis of I, but not conversely.*

PROOF. If $(LT(I)) = (LT(g_1), ..., LT(g_s))$ holds, then as the proof of the above Hilbert Basis Theorem shows, $I = (g_1, ..., g_s)$. □

In other words,

COROLLARY 2.9. *Any Groebner basis of a polynomial ideal I is a Hilbert basis of I, but not conversely.* □

We have thus proved existence of a Groebner basis for a non-zero polynomial ideal $I \subset k[x_1, ..., x_n]$ under a fixed monomial ordering on $k[x_1, ..., x_n]$.

3.3. Groebner Bases Plus the Division Algorithm. We now describe commonly used statements which are equivalent to the statement that G is a Groebner basis of I.

THEOREM 2.3. *Let I be a non-zero ideal in $k[x_1, ..., x_n]$ and $G = \{g_1, ..., g_s\}$ be a subset of I. Then the following statements are equivalent to each other:*

(1) *For each non-zero polynomial $f \in I$, some $LM(g_i)$ divides $LM(f)$.*
(2) *$f \in I$ if and only if $f \xrightarrow{G}_+ 0$.*
(3) *$f \in I$ if and only if $f = \sum_{i=1}^{s} a_i g_i$ with*

$$LM(f) = \max_{1 \leq i \leq s}(LM(a_i)LM(g_i)).$$

(4) *G is a Groebner basis for I, i.e., $(LT(g_1), ..., LT(g_s)) = (LT(I))$.*

PROOF. (4) $\Longrightarrow$ (1): Let $f \in I$. Then since

$$LT(f) \in (LT(I)) = (LT(g_1), ..., LT(g_s)),$$

we have

$$LT(f) = \sum_{i=1}^{s} h_i LT(g_i),$$

for some polynomials $h_i \in k[x_1, ..., x_n]$. We note that this is a k-linear equation of monomials in $k[x_1, ..., x_n]$. Since monomials in $k[x_1, ..., x_n]$ form a k-basis for the k-vector space $k[x_1, ..., x_n]$, the summation in the right side should contain $LT(f)$ and there should be a cancellation resulting to zero among other terms in the right side; hence in particular $LT(f)$ is divisible by some $LT(g_i)$.

(1) $\Longrightarrow$ (2): Let $f \in k[x_1, ..., x_n]$. By Division Algorithm (Algorithm 5), we know that there exists $r \in k[x_1, ..., x_n]$ such that $f \xrightarrow{G}_+ r$ where r is reduced modulo G. Thus, $f - r \in I$. This simply means that $f \in I$ if and only if $r \in I$. Thus if we assume that $r = 0$, then we clearly have $f \in I$; this proves the $\Longleftarrow$ direction in (2). For the converse, we now assume that $f \in I$. If $r \neq 0$ under this assumption, then we would have $r = f - a_1 g_1 + \cdots + a_s g_s \in I$, so that $LT(r) \in LT(I)$; hence it would have to be divisible by some $LT(g_i)$ by hypothesis (1). But this is a contradiction to the fact that r is reduced modulo G; this proves the $\Longleftarrow$ direction in (2).

(2) $\Longrightarrow$ (3) holds by Algorithm 5.

(3) $\Longrightarrow$ (4): By hypothesis, $r = 0$. Here, we only have to show that $(LT(I)) \subseteq (LT(g_1),,,, LT(g_s))$. For this, let $f \in I$. Then $LT(f) \in (LT(I))$. We note that since $r = 0$ in the hypothesis, at least one $a_i \neq 0$ in the hypothesis; hence particularly when the division algorithm terminates, we have, by Corollary 2.1,

$$LT(f) = \sum_i LT(a_i) LT(g_i),$$

where the sum is taken over all those i such that

$$LM(f) = LM(a_i) LM(g_i),$$

as we have seen before. This proves the inclusion, by induction. □

LEMMA 2.12. *Let $G = \{g_1, ..., g_s\}$ be a set of non-zero polynomials in $k[x_1, ..., x_n]$ such that for all $f \in k[x_1, ..., x_n]$, the remainder of the division of f by G is unique. Let $c \neq 0 \in k$ and x^α a monomial in $k[x_1, ...x_n]$. If a polynomial $h \in k[x_1, ...x_n]$ is reduced to r modulo G with a reduced r, that is,*

$$h \xrightarrow{G}_+ r, \ r \text{ reduced},$$

then $h - cx^\alpha g_i \xrightarrow{G}_+ r$ for each $g_i \in G$.

PROOF. We note first that the leading coefficient of $cx^\alpha g_i = cLC(g_i)$ and h may have a term $x^\alpha LM(g_i)$ with a non-zero constant. Thus, let

$dLC(g_i)$ be a coefficient of a possible term $x^\alpha LM(g_i)$ of h. The we have a few cases:

Case 1: If there is no such a term in h, i.e., $d = 0$, then the coefficient of $x^\alpha LM(g_i)$ in the difference $h - cx^\alpha g_i$ is $-cLC(g_i) \neq 0$; hence we have

$$h - cx^\alpha g_i \xrightarrow{g_i} h \xrightarrow{G}_+ r.$$

Case 2: If there is such a term in h, i.e., $d \neq 0$, then we have two cases, either $d = c$ or $d \neq c$.

Case 2 − 1: $0 \neq d = c$. Assume that $h - cx^\alpha g_i \xrightarrow{G}_+ r_1$ with r_1 reduced. (We want to show that $r_1 = r$.) Since $d = c \neq 0$, we have by composition,

$$h \xrightarrow{g_i} h - cx^\alpha g_i \xrightarrow{G}_+ r_1.$$

Then since $h \xrightarrow{G}_+ r$, with reduced r, is unique by hypothesis, $r = r_1$.

Case 2 − 2: $0 \neq d \neq c$. Put $u = h - dx^\alpha g_i$. Then the coefficient of $x^\alpha LM(g_i)$ in u is zero. Then since $d \neq 0$, we have

$$h \xrightarrow{g_i} u.$$

Also, since $d \neq c$ we have

$$h - cx^\alpha g_i \xrightarrow{g_i} u.$$

Thus, if $u \xrightarrow{G}_+ r_2$ is any reduction modulo G with a reduced r_2, we get by composition,

$$h \xrightarrow{g_i} u \xrightarrow{G}_+ r_2,$$

and hence $r_2 = r$ by hypothesis. Therefore,

$$h - cx^\alpha g_i \xrightarrow{g_i} u \xrightarrow{G}_+ r.$$

□

THEOREM 2.4. *Let $G = \{g_1, ..., g_s\}$ be a set of non-zero polynomials in $k[x_1, ..., x_n]$. Then G is a Groebner basis if and only if, for each polynomial $f \in k[x_1, ..., x_n]$, the remainder of the division of f by G is unique.*

PROOF. The $\Longrightarrow$ direction. Let $f \xrightarrow{G}_+ r_1$ and $f \xrightarrow{G}_+ r_2$ be reduction modulo G with both r_1 and r_1 reduced with respect to G. Since both $f - r_1$

and $f - r_2$ are in the ideal (G) generated by G, their difference $r_1 - r_2$ is also in (G). Since $r_1 - r_2$ is automatically reduced too, by (2) of Theorem 2.3, $r_1 - r_2 = 0$.

The $\Longleftarrow$ direction. To finish the proof, we are going to prove (2) of Theorem 2.3. For this, let $f \in (G)$, an ideal generated by $\{g_1, ..., g_s\}$. Suppose that $f \xrightarrow{G}_+ r$ with r reduced. We have to show that $r = 0$.

Since $f \in (G)$, $f = \sum h_i g_i$ for some $h_i \in k[x_1, ..., x_n]$; and hence by expressing each h_i as a k-linear combination of monomials in $k[x_1, ..., x_n]$, we have

$$f = \sum_{j=1}^{m} c_j x_j^\alpha g_{i_j},$$

where x_j^α is a monomial in $k[x_1, ..., x_n]$, c_j is non-zero scalar in k, and each $i_j \in \{1, ..., s\}$. Then by Lemma 2.12 with $h = f$ there, we see that

$$f - c_1 x_1^\alpha g_{i_1} \xrightarrow{G}_+ r$$

holds. Now we apply Lemma 2.12 to $h = f - c_1 x_1^\alpha g_{i_1}$ to see that

$$f - c_1 x_1^\alpha g_{i_1} - c_2 x_2^\alpha g_{i_2} \xrightarrow{G}_+ r.$$

By repeatedly applying Lemma 2.12 in this way, we see that

$$0 = f - \sum_{j=1}^{m} c_j x_j^\alpha g_{i_j} \xrightarrow{G}_+ r.$$

Then $0 \xrightarrow{G}_+ r$ immediately implies that $r = 0$. $\square$

REMARK 2.9. *We put a few remarks in order.*

(1) *So far, we have defined a Groebner basis of a given polynomial ideal I with respect to a given monomial order on $k[x_1, ..., x_n]$. In the following section, we will learn how to actually compute a Groebner basis of I.*
(2) *We note that although the remainder is uniquely determined for division by a Groebner basis, the quotients are not necessarily uniquely determined.*

(3) *We also note that a Groebner basis with respect to one monomial order is not usually a Groebner basis with respect to a different monomial order.*

(4) *The reality is that, depending on a particular monomial order, computation of Groebner basis usually takes different amount of time. Thus, we sometimes compute a Groebner basis with respect to a monomial order which produces the Groebner basis which take less time than other monomial orders, and then we try to "transform" the computed Groebner basis to another Groebner basis with respect to a different monomial order of our choice. As long as such an indirect approach takes less time than directly computing a Groebner basis with respect to a monomial order of our choice, it is perfectly OK.*

Exercises for Section 2.3

EXERCISE 2.5. *Let* $I = (x^3 - 2xy, x^2y - 2y^2 + x)$. *Show that the set*

$$\{x^3 - 2xy, x^2y - 2y^2 + x\}$$

is not a Groebner basis for the ideal I, *say, under the lex order.*

EXERCISE 2.6. *Let* I *be an ideal of the polynomial ring* $k[x_1, ..., x_n]$. *Prove that the quotient ring* $k[x_1, ..., x_n]/LT(I)$ *is a* k*-vector space spanned by* $\{LM(f) : f \in I\}$.

EXERCISE 2.7. *Let* h *be a non-zero polynomial in* $k[x_1, ..., x_n]$, *and let* $I = (g_1, ..., g_s)$ *be an ideal of* $k[x_1, ..., x_n]$. *Prove that* $\{g_1, ..., g_s\}$ *is a Groebner basis of* I *if and only if so is* $\{hg_1, ..., hg_s\}$.

EXERCISE 2.8. *Let* G *be a Groebner basis for an ideal* I *of* $k[x_1, ..., x_n]$ *and let* K *be a field extension of* k. *Prove that* G *is a Groebner basis for the ideal* J *of* $K[x_1, ..., x_n]$ *generated by* I *over* K.

4. Buchberger's Algorithm for Groebner Bases

ACKNOWLEDGEMENT 1. *This section is dedicated to Prof. Bruno Buchberger for his discovery in 1965 of an algorithm for computing Groebner bases.*

We note that Groebner basis is in fact not far away from us. We live with them without recognizing it. The following two examples are about this comment:

EXAMPLE 2.15. *In the case of a system of m linear equations in n unknowns,*

$$AX = B,$$

the (uniquely determined) row echelon form of the augmented matrix

$$\left[\begin{array}{c|c} A & B \end{array}\right]$$

in fact gives us a groebner basis of the original system of linear equations.

EXAMPLE 2.16. *In the case of a system of polynomial equations in one unknown,*

$$\begin{cases} f_1(x) & = & 0 \\ f_2(x) & = & 0 \\ \cdots & & \\ f_s(x) & = & 0 \end{cases}$$

The principal ideal $(\gcd(f_1(x), ..., f_s(x)))$ *is in fact the Groebner basis for the ideal generated by* $f_i(x)$*'s.*

But, we need a strong computing tool for other cases, i.e., when we are given a general polynomial ideal in more than one variable.

4.1. Buchberger's Criterion for Groebner Basis.

Let

$$I = (f_1, ..., f_s)$$

be a polynomial ideal in $k[x_1, ..., x_n]$, and let $F = \{f_1, ..., f_s\}$, where each $f_i \neq 0$, as usual. Then by part (2) of Theorem 2.3, we know that F is a Groebner basis if and only if for each $f \neq 0$ in I, some $LM(f_i)$ divides

$LM(f)$. Thus if there is a polynomial f such that none of $LM(f_i)$ divides $LM(f)$, then F is not a Groebner basis. Let S be a collection of such polynomials in I:

$$S := \{f \in I : LM(f) \text{ is not divisible by any of } LM(f_i)\}.$$

On the other hand, $f \in I$ just means that

$$f = \sum_{i=1}^{s} h_i f_i, \tag{2.3}$$

for some $h_i \in k[x_1, ..., x_n]$. Now, if we express both sides of this equality as k-linear combinations of monomials, we have

$$\begin{aligned} LC(f)LM(f) + \text{ the rest } &= \sum_{i=1}^{m} LC(h_i)LM(h_i) \cdot LC(f_i)LM(f_i) \\ &\quad + \text{other terms.} \end{aligned}$$

Now, if there is no such cancellation among the terms in the summation in the right side of the equality, then certainly, $LM(f)$ must be one of these products, $LM(h_i) \cdot LM(f_i)$, because we have a k-linear combination of monomials in $k[x_1, ..., x_n]$ and the monomials are a k-basis for the k-vector space $k[x_1, ..., x_n]$; hence this $LM(f_i)$ divides $LM(f)$. Otherwise, i.e., if there exists a cancellation among those products $LC(h_i)LM(h_i) \cdot LC(f_i)LM(f_i)$ such that

$$\sum_{i=1}^{m} LC(h_i)LM(h_i) \cdot LC(f_i)LM(f_i) = 0,$$

then it *could happen* that $LM(f) < LM(f_i)$ for each i, and hence none of $LM(f_i)$ will divide $LM(f)$.

We note that this is the kind of situation we would like not to have at all, as long as we are concerned with figuring out a Groebner basis from the given F.

The argument so far we have done also means that S is a subset of another subset U of I consisting of those f having such cancellation,

$$\sum_{i=1}^{m} LC(h_i)LM(h_i) \cdot LC(f_i)LM(f_i) = 0,$$

where f is expressed as in (2.3).

That is, if we define

$$U := \{f \in I : \sum_{i=1}^{m} LC(h_i)LM(h_i) \cdot LC(f_i)LM(f_i) = 0\},$$

then

$$S \subseteq U.$$

Thus, if F turns out not to be a Groebner basis, our strategy should be to "do something" in such a way that "something" makes the new subset U to be an empty subset, then the new S will be automatically an empty subset.

Thus, we would like to know more about how the cancellation among the product of leading terms $h_i \cdot f_i$ in $\sum_{i=1}^{s} h_i f_i$ happens so that $LM(f)$ is never divisible by some of $LM(f_i)$. If any two products of $h_i \cdot f_i$ and $h_j \cdot f_j$ have the same leading monomial and the same leading coefficient but with different signs, then the leading term of their difference is obviously less than $LM(f)$. On the other hand, if their difference happens to have

$$h_1 \cdot f - h_2 \cdot g = \frac{L}{LT(f)} \cdot f - \frac{L}{LT(g)} \cdot g,$$

where $L = \operatorname{lcm}(LM(f), LM(g))$, then the leading term of the resulting difference $h_1 f - h_2 g$ will be less than that of $LM(h_1 f) = LM(h_2 g)$. More generally,

LEMMA 2.13. *Given two non-zero polynomials f and g in $k[x_1, ..., x_n]$,*

$$\frac{L}{LT(f)} \cdot f - \frac{L}{LT(g)} \cdot g,$$

where $L = \operatorname{lcm}(LM(f), LM(g))$ is their least common multiple, cancels out the leading terms of f and g. $\square$

DEFINITION 2.15. *Such a difference is called the S-**polynomial of** f **and** g, and denoted by*

$$S(f,g) = \frac{L}{LT(f)} \cdot f - \frac{L}{LT(g)} \cdot g.$$

Now in order to see more about such cancellations, we first collect these products $h_i f_i$ into several different groups such that the leading monomial of $h_i f_i$'s in the same group is the same. In the following example each new f_i is the one $h_i f_i / LC(h_i f_i)$ with the same leading monomial:

LEMMA 2.14. *Let $f_1, ..., f_s \in k[x_1, ..., x_n]$ all have the same non-zero leading monomial $X = x^\alpha$. If the leading monomial $LM(f)$ of*

$$f = \sum_{i=1}^{s} c_i f_i,$$

where $c_i \in k$, is less than X, then f is a k-linear combination of $S(f_i, f_j)$, for $1 \le i < j \le s$.

PROOF. We write each $f_i = a_i X +$ lower terms, where $a_i \in k$. The by hypothesis, we see that

$$\sum_{i=1}^{s} c_i a_i = 0.$$

Now, for each pair (i,j) we simply have

$$S(f_i, f_j) = \frac{1}{a_i} \cdot f_i - \frac{1}{a_j} \cdot f_j,$$

since $LM(f_i) = LM(f_j) = X$, i.e., the same. Thus, we have through the method of telescoping:

$$\begin{aligned} f &= c_1 f_1 + \cdots c_s f_s \\ &= c_1 a_1 (\frac{1}{a_1} f_1) + \cdots + c_s a_s (\frac{1}{a_s} f_s) \\ &= c_1 a_1 (\frac{1}{a_1} f_1 - \frac{1}{a_2} f_2) + (c_1 a_1 + c_2 a_2)(\frac{1}{a_2} f_2 - \frac{1}{a_3} f_3) + \cdots \\ &\quad + (c_1 a_1 + \cdots + c_{s-1} a_{s-1})(\frac{1}{a_{s-1}} f_{s-1} - \frac{1}{a_s} f_s) \\ &\quad + (c_1 a_1 + \cdots + c_s a_s) \frac{1}{a_s} f_s \\ &= c_1 a_1 S(f_1, f_2) + (c_1 a_1 + c_2 a_2) S(f_2, f_3) + \cdots \\ &\quad + (c_1 a_1 + \cdots + c_{s-1} a_{s-1}) S(f_{s-1}, f_s) + 0, \end{aligned}$$

because $c_1 a_1 + \cdots + c_s a_s = 0$. □

REMARK 2.10. *Thus, this lemma exactly says that all cancellations in the product $\sum_{i=1}^{s} h_i f_i$ can be expressed as k-linear combinations of appropriate S-polynomials.*

The following is a very important criterion originally discovered by Bruno Buchberger in 1965.

THEOREM 2.5. ***(B. Buchberger's Criterion for Groebner Basis)*** *Let $G = \{g_1, ..., g_s\}$ be a set of non-zero polynomials in $k[x_1, ..., x_n]$. Then G is a groebner basis for the ideal I it generates if and only if for all $i \neq j$,*

$$S(g_i, g_j) \xrightarrow{G}_+ 0.$$

PROOF. We note that $S(g_i, g_j) \in I$. Thus by part (2) of Theorem 2.3, the $\Longrightarrow$ direction holds. For the converse, we are going to use part (3) of Theorem 2.3. Let $f \in I$. Then since $g_1, ..., g_s$ are generators of I, we can write, by the division algorithm, Algorithm 5,

$$f = \sum_{i=1}^{s} h_i g_i,$$

for some polynomials $h_i \in k[x_1, ..., x_n]$. We note that f can be written in many different ways like this. But since we are working under a monomial ordering on $k[x_1, ..., x_n]$, it is a well-ordering, and hence among all kinds of expressions for f, the non-empty set S defined by

$$S := \{\max_{1 \le i \le s}(LM(h_i)LM(g_i) : f = \sum_{i=1}^{s} h_i g_i\} \tag{2.4}$$

has *the least element*, say $X = x^\alpha$, and we particularly choose *that* expression: $f = \sum_{i=1}^{s} h_i g_i$, with the least element $X = \max_{1 \le i \le s}(LM(h_i)LM(g_i)$ of S.

Now, if $LM(f) = X$, we are done by part (3) of Theorem 2.3.

Otherwise, we will end up with a contradiction, and it goes as follows: So suppose $LM(f) < X$. With respect to *that* expression for f, let

$$L = \{i \in [1, s] : X = LM(h_i)LM(g_i)\}.$$

For each $i \in L$, write $h_i = c_i X_i +$ lower terms, and define

$$g := \sum_{i \in L} c_i X_i g_i.$$

We note that

$$f = g + \text{others},$$

where "others" are still polynomial multiples of some g_i and their leading terms are all less than that of g. Here, the leading monomial of each $c_i X_i g_i$ is clearly the same as $X = x^\alpha$, but $LM(g) < X$ should hold by hypothesis. Thus, by Lemma 2.14, there exist $d_{ij} \in k$ such that

$$g = \sum_{\substack{i,j \in L \\ i \neq j}} d_{ij} S(X_i g_i, X_j g_j).$$

Then since $\text{lcm}(X_ig_i, X_jg_j) = X$, we can express each $S(X_ig_i, X_jg_j)$:

$$\begin{aligned}
S(X_ig_i, X_jg_j) &= \frac{X}{LT(X_ig_i)} \cdot X_ig_i - \frac{X}{LT(X_jg_j)} \cdot X_jg_j \\
&= \frac{X}{LT(g_i)} \cdot g_i - \frac{X}{LT(g_j)} \cdot g_j \\
&= \frac{X}{X_{ij}} S(g_i, g_j),
\end{aligned}$$

where $X_{ij} = \text{lcm}(g_i, g_j)$. Thus, by hypothesis we know that

$$S(g_i, g_j) \xrightarrow{G}_+ 0,$$

and hence

$$S(X_ig_i, X_jg_j) \xrightarrow{G}_+ \frac{X}{X_{ij}} \cdot 0 = 0.$$

Thus, we can write

$$S(X_ig_i, X_jg_j) = \sum_{p=1}^{s} h_{ijp}g_p,$$

for some $h_{ijp} \in k[x_1, ..., x_n]$ where, by Division Algorithm (Algorithm 5) (and its corollary, Corollary 2.1), we have

$$\begin{aligned}
\max_{i \le p \le s}(LM(h_{ijp})LM(g_p)) &= LM(S(X_ig_i, X_jg_j)) \\
&< \max_{\substack{i,j \in L \\ i \neq j}}(LM(X_ig_i), LM(X_jg_j)) = X.
\end{aligned}$$

Thus, we have another expression for f:

$$\begin{aligned}
f &= g + \text{others} \\
&= \sum_{\substack{i,j \in L \\ i \neq j}} d_{ij}S(X_ig_i, X_jg_j) + \text{others} \\
&= \sum_{i=1}^{s} h'_ig_i,
\end{aligned}$$

for some polynomials $h'_i \in k[x_1, ..., x_n]$ with $\max_{1 \le i \le s}(LM(h'_i)LM(g_i)) < X$. But it contradicts to our choice of the expression for f with the least element X of S in Eq. (2.4). □

By Corollary 2.1, the immediate corollary of this theorem is:

COROLLARY 2.10. *Let $G = \{g_1, ..., g_s\}$ be a subset in $k[x_1, ..., x_n]$ consisting of non-zero polynomials. Then G is a Groebner basis if and only if for all $i \neq j, 1 \leq i, j \leq s$, we have*

$$S(g_i, g_j) = \sum_{i=1}^{s} h_{ijp} g_p,$$

where $LM(S(g_i, g_j)) = \max_{1 \leq p \leq s}(LM(h_{ijp})LM(g_p))$. □

Buchberger's Criterion for Groebner basis is extremely useful, because it allows to check computationally whether a given set of polynomials is a Groebner basis or not.

EXAMPLE 2.17. *A revisit to Example 2.12. Let $G = \{x - y, y^2 - 1\}$. Then $S(x - y, y^2 - 1) \xrightarrow{G}_+ 0$ since*

$$\begin{aligned} S(x - y, y^2 - 1) &= y^2(x - y) - x(y^2 - 1) \\ &= -y^3 + x \\ &= 1 \cdot (x - y) - y \cdot (y^2 - 1) + 0. \end{aligned}$$

Thus, G is a Groebner basis for the ideal it generates. As we have seen in the previous Example, the remainder is uniquely determined by Theorem 2.4.

REMARK 2.11. *In fact, the above theorem says a lot more than this. It actually provides how we computationally obtain a Groebner basis from a non-Groebner basis.*

4.2. Buchberger's Algorithm for Groebner Bases. Let

$$G = \{g_1, ..., g_s\}$$

be a subset of polynomials in $k[x_1, ..., x_n]$. For each pair (g_i, g_j) taken from G, we note that

$$S(g_i, g_j) = \frac{L}{LT(g_i)} \cdot g_i - \frac{L}{LT(g_j)} \cdot g_j,$$

where $L = \text{lcm}(LM(g_i), LM(g_j))$, is an element of the ideal I generated by the set G; hence if there is a pair (g_i, g_j) such that $S(g_i, g_j) \xrightarrow{G}_+ r \neq 0$, i.e., if it is not reduced to the zero polynomial, then adding it to G does not change the ideal they generate, i.e., we still have

$$I = (g_1, ..., g_s) = (g_1, ..., g_s, g_{s+1} = S(g_i, g_j)).$$

Then with respect to a new set $G = \{g_1, ..., g_s, g_{s+1} = S(g_i, g_j)\}$, we obviously have $S(g_i, g_j) \xrightarrow{G}_+ 0$ for the same pair (g_i, g_j). Thus if there is another pair (g_i, g_j) taken from the new G such that $S(g_i, g_j) \xrightarrow{G}_+ r \neq 0$, then we add that $S(g_i, g_j)$ to G to obtain a new set G, and so on, until we add enough S-polynomials, which are computed from elements of the original G and from the intermediate G, to the original G in order to obtain a final G such that, for every pair (g_i, g_j) taken from the final G, $S(g_i, g_j) \xrightarrow{G}_+ 0$.

Now, the question remained in this process is whether this adding of $S(g_i, g_j)$ will terminate after finitely many steps. Although adding non-zero $S(g_i, g_j)$ to G repeatedly does not increase the involved ideals, those monomial ideals $(LT(G))$ generated by all monomials in G are strictly increasing each time whenever we add a non-zero $S(g_i, g_j)$ to the latest G,

$$(LT(g_1, ..., g_s)) \subsetneqq (LT(g_1, ..., g_s, g_{s+1} = S(g_i, g_j))) \subsetneqq \cdots.$$

Thus, if our process does not stop after finitely many steps, we are going to have a strictly increasing chain of monomial ideals, which is impossible since $k[x_1, ..., x_n]$ is a Noetherian ring.

In this way, we have proved and established the following algorithm:

ALGORITHM 6. ***(B. Buchberger 1965)*** *Let $I = (f_1, ..., f_s)$ be a non-zero polynomial ideal in $k[x_1, ..., x_n]$ with a monomial order on it. Then the following procedure constructs a Groebner basis for I in a finite number of steps:*

Input: *$F = \{f_1, ..., f_s\}$, a list of the f_i's*

Output: *a Groebner basis $G = \{g_1, ..., g_t\}$, a list of g_i's for I, with $F \subseteq G$*

Initialization: $G := F, D := \{\{f_i, f_j\} : f_i \neq f_j \in F\}$
WHILE $D \neq \varnothing$ ***DO***
Choose any $\{f, g\}$ *out of* D
$D := D - \{\{f, g\}\}$
$S(f, g) \xrightarrow{G}_+ h$, *where* h *is reduced with respect to* G
IF $h \neq 0$ ***THEN***
$G := G \cup \{h\}$
$D := D \cup \{\{u, h\} : u \in G\}$ □

Before we move into the next section, we would like to have two examples.

EXAMPLE 2.18. *We consider the ideal generated by two polynomials in* $\mathbb{Q}[x, y]$*:*

$$g_1 = x - y, g_2 = y^2 - 1.$$

We have seen in Example 2.12 that when we divide out a given polynomial $f = x^2y + xy^2 + y^2$ *by these two polynomials in 4 different procedures, we have obtained the same remainder, but the quotients are not the same. Thus, this suggests that the set* $\{g_1, g_2\}$ *could by a Groebner basis for the ideal they generate.*

In this example, we are going to apply Buchberger's algorithm to actually verify that it is indeed the case. Here is the result:

——————

Initialization: $G := \{g_1, g_2\}, D = \{\{g_1, g_2\}\}$.
First pass through the WHILE loop
$D := \varnothing$
$S(g_1, g_2) = -x + y^3 = -1 \cdot g_1 + y \cdot g_2$, *so that*
$S(g_1, g_2) \xrightarrow{G}_+ 0 = h$ *(reduced with respect to* G*)*
The WHILE loop stops since $D = \varnothing$.

——————

Thus, by Theorem 2.5, we conclude that $\{g_1, g_2\}$ *is a Groebner basis for the ideal,* (g_1, g_2)*, they generate.*

REMARK 2.12. *In fact, this example is prepared using REDUCE under the lex order with $x > y$:*[1]

———————

```
off nat$
load_package groebner$
g1:=x-y$
g2:=y^2-1$
gspoly(g1,g2);
- x + y**3$
preduce(gspoly(g1,g2),{g1,g2},{x,y});
0$
```

———————

EXAMPLE 2.19. *We now consider the ideal generated by two polynomials in $\mathbb{Q}[x, y]$:*

$$f_1 = y^2 - 1, f_2 = xy - 1.$$

We have seen in Example 2.10 and Example 2.11 that when we divide out a given polynomial $f = x^2y + xy^2 + y^2$ by these two polynomials in 4 different procedures, we have neither obtained the same remainder nor the quotients. Thus, this suggests that the set $\{f_1, f_2\}$ is not a Groebner basis for the ideal (f_1, f_2) they generate.

In this example, we are going to apply Buchberger's Algorithm to actually compute a Groebner basis for the ideal (f_1, f_2). As we have already seen in Example 2.12, we must be able to obtain $\{g_1, g_2\}$ as its Groebner basis, because we have already verified that

$$(f_1, f_2) = (g_1, g_2),$$

in Example 2.12.

Here is the result:

———————

Initialization: $G := \{f_1, f_2\}, D = \{\{f_1, f_2\}\}$.

[1] A working program for REDUCE for this example is also available from: http://ch.catholic.or.kr/pundang/4/soh/2006_aa_i/Example_2_18.htm

First pass through the WHILE loop
$D := D - \{\{f_1, f_2\}\} = \varnothing$
$S(f_1, f_2) = x - y$
$S(f_1, f_2) \xrightarrow{G}_+ x - y = h$ *(reduced with respect to G)*
Since $h \neq 0$, *let* $f_3 := x - y$
$G := \{f_1, f_2\} \cup \{f_3\}$
$D := D \cup \{\{f_1, f_3\}, \{f_2, f_3\}\}$
Second pass through the WHILE loop
$D := D - \{\{f_1, f_3\}\} = \{\{f_2, f_3\}\}$
$S(f_1, f_3) = x - y^3$
$S(f_1, f_3) \xrightarrow{G}_+ 0 = h$ *(reduced with respect to G)*
Third pass through WHILE loop
$G := G$
$D := D - \{\{f_2, f_3\}\} = \varnothing$
$S(f_2, f_3) = -y^2 + 1$
$S(f_2, f_3) \xrightarrow{G}_+ 0$
The WHILE loop stops since $D = \varnothing$.

Thus, by Theorem 2.5, we conclude that $\{f_1, f_2, f_3 = x - y\}$ *is a Groebner basis for the ideal,* (f_1, f_2).

Moreover, since

$$f_2 = y \cdot f_1 + 1 \cdot f_3,$$

$\{f_1, f_2, f_3\} - \{f_2\} = (f_1, f_3\}$ *still generates the ideal* (f_1, f_2). *Then the previous example says that the set* $\{f_1, f_3\}$, *which is equal to the set* $\{g_2, g_1\}$, *is a Groebner basis for the ideal* (f_1, f_2). *Thus, this additional consideration reduces the size of a Groebner basis. In fact, we can have a better criterion for checking which polynomials can be deleted (Lemma 2.15) below) from a Groebner basis constructed by Buchberger's Algorithm, and so we will discuss this matter in the next section, Reduced Groebner Basis, below.*

REMARK 2.13. *Like before, this example is prepared using REDUCE under the lex order:*[2]

————————

```
off nat$
load_package groebner$
f1:=y^2-1$
f2:=x*y-1$
gspoly(f1,f2);
x - y$
preduce(gspoly(f1,f2),{f1,f2},{x,y});
x - y$
f3:=x -y$
gspoly(f1,f3);
x - y**3$
gspoly(f2,f3);
-y**2+1$
preduce( gspoly(f1,f3),{f1,f2,f3},{x,y});
0$
preduce( gspoly(f2,f3),{f1,f2,f3},{x,y});
0$
```

————————

EXAMPLE 2.20. *Here is another example which is a little bit complicated for us to compute only using our hand.*

Let $f_1 = x^2 - xy + 1, f_2 = x^2y - x^2 + x$.

Using Computer Algebra System, REDUCE under the lex order with $x > y$*, we have:*[3]

————————

```
off nat$
load_package groebner$
```

[2]A working program for REDUCE for this example is also available from: http://ch.catholic.or.kr/pundang/4/soh/2006_aa_i/Example_2_19.htm

[3]A working program for REDUCE for this example is also available from: http://ch.catholic.or.kr/pundang/4/soh/2006_aa_i/Example_2_20.htm

```
f1:=x^2-x*y+1;
f2:=x^2*y-x^2+x;
preduce(gspoly(f1,f2),{f1,f2},{x,y});
x*y**2 - x*y + x - y + 1$
f3:=ws$
preduce(gspoly(f1,f3),{f1,f2,f3},{x,y});
- x*y + x - 1$
f4:=ws$
preduce(gspoly(f1,f4),{f1,f2,f3,f4},{x,y});
0$
preduce(gspoly(f2,f4),{f1,f2,f3,f4},{x,y});
0$
preduce(gspoly(f3,f4),{f1,f2,f3,f4},{x,y});
x - 2*y + 1$
f5:=ws$
preduce(gspoly(f1,f5),{f1,f2,f3,f4,f5},{x,y});
0$
preduce(gspoly(f2,f5),{f1,f2,f3,f4,f5},{x,y});
0$
preduce(gspoly(f3,f5),{f1,f2,f3,f4,f5},{x,y});
2*y**3 - y**2 - y + 2$
f6:=ws$
preduce(gspoly(f1,f6),{f1,f2,f3,f4,f5,f6},{x,y});
0$
preduce(gspoly(f2,f6),{f1,f2,f3,f4,f5,f6},{x,y});
0$
preduce(gspoly(f3,f6),{f1,f2,f3,f4,f5,f6},{x,y});
- 2*y**2 + 3*y - 2$
f7:=ws$
```

————————

Thus, we have $G = \{f_1, f_2, f_3, f_4, f_5, f_6, f_7\}$, *where*

$$\begin{aligned} f_1 &= x^2 - xy + 1, \\ f_2 &= x^2y - x^2 + x, \\ f_3 &= xy^2 - xy + x - y + 1, \\ f_4 &= -xy + x - 1, \\ f_5 &= x - 2y + 1, \\ f_6 &= 2y^3 - y^2 - y + 2 \\ f_7 &= -2y^2 + 3y - 2. \end{aligned}$$

After this, we can routinely check with REDUCE that for each pair (i, j) *with* $i \neq j$ *and* $1 \leq i < j \leq 7$,

$$S(f_i, f_j) \xrightarrow{G} 0,$$

and hence by Theorem 2.5 this G *is indeed a Groebner basis for the ideal* (f_1, f_2).

EXAMPLE 2.21. *The following verifies what it says at the end of the above Example:*
Using Computer Algebra System, REDUCE under the lex order with $x > y$, *we have:*[4]

———————————

```
off echo$
off nat$
load_package groebner$
torder({x,y},lex)$
f1:=x^2-x*y+1$
f2:=x^2*y-x^2+x$
f3:=x*y^2-x*y+x-y+1$
f4:=-x*y+x-1$
```

[4]A working program for REDUCE for this example is also available from: http://ch.catholic.or.kr/pundang/4/soh/2006_aa_i/Example_2_21.htm

```
f5:=x-2*y+1$
f6:=2*y^3-y^2-y+2$
f7:=-2*y^2+3*y-2$
kmax:=7$
k:=1$
jmax:=7$
j:=k+1$
while k <= kmax do
<<while j<=jmax do
<<a:=mkid(f,k)$
b:=mkid(f,j)$
result1:=gspoly(a,b)$
write "S-poly(",mkid(f_,k),",",mkid(f_,j),") := ",result1;
result2:=preduce(gspoly(a,b), {f1,f2,f3,f4,f5,f6,f7}, {x,y})$
write "reduction of S-poly(",mkid(f_,k),",",mkid(f_,j),") := ", result2;
j:=j+1$
>>$
k:=k+1$
j:=k+1$
>>$
```

Here are the computed results:

```
S-poly(f_1,f_2) := - x**2 + x*y**2 + x - y$
reduction of S-poly(f_1,f_2) := 0$
S-poly(f_1,f_3) := - x**2*y + x**2 + x*y**3 - x*y + x - y**2$
reduction of S-poly(f_1,f_3) := 0$
S-poly(f_1,f_4) := x**2 - x*y**2 - x + y$
reduction of S-poly(f_1,f_4) := 0$
S-poly(f_1,f_5) := - x*y + x - 1$
reduction of S-poly(f_1,f_5) := 0$
S-poly(f_1,f_6) := - x**2*y**2 - x**2*y + 2*x**2 + 2*x*y**4 - 2*y**3$
```

```
reduction of S-poly(f_1,f_6) := 0$
S-poly(f_1,f_7) := 3*x**2*y - 2*x**2 - 2*x*y**3 + 2*y**2$
reduction of S-poly(f_1,f_7) := 0$
S-poly(f_2,f_3) := x*(x - 2*y + 1)$
reduction of S-poly(f_2,f_3) := 0$
S-poly(f_2,f_4) := 0$
reduction of S-poly(f_2,f_4) := 0$
S-poly(f_2,f_5) := x*(x - 2*y**2 + y - 1)$
reduction of S-poly(f_2,f_5) := 0$
S-poly(f_2,f_6) := x*(x*y**2 - x*y + 2*x - 2*y**2)$
reduction of S-poly(f_2,f_6) := 0$
S-poly(f_2,f_7) := x*(x*y - 2*x + 2*y)$
reduction of S-poly(f_2,f_7) := 0$
S-poly(f_3,f_4) := x - 2*y + 1$
reduction of S-poly(f_3,f_4) := 0$
S-poly(f_3,f_5) := x*y - x - 2*y**3 + y**2 + y - 1$
reduction of S-poly(f_3,f_5) := 0$
S-poly(f_3,f_6) := x*y**2 - 3*x*y + 2*x + 2*y**2 - 2*y$
reduction of S-poly(f_3,f_6) := 0$
S-poly(f_3,f_7) := x*y - 2*y + 2$
reduction of S-poly(f_3,f_7) := 0$
S-poly(f_4,f_5) := - x + 2*y**2 - y + 1$
reduction of S-poly(f_4,f_5) := 0$
S-poly(f_4,f_6) := - x*y**2 + x*y - 2*x + 2*y**2$
reduction of S-poly(f_4,f_6) := 0$
S-poly(f_4,f_7) := - x*y + 2*x - 2*y$
reduction of S-poly(f_4,f_7) := 0$
S-poly(f_5,f_6) := - x*y**2 - x*y + 2*x + 4*y**4 - 2*y**3$
reduction of S-poly(f_5,f_6) := 0$
S-poly(f_5,f_7) := 3*x*y - 2*x - 4*y**3 + 2*y**2$
reduction of S-poly(f_5,f_7) := 0$
```

*S-poly(f_6,f_7) := 2*y**2 - 3*y + 2$*
reduction of S-poly(f_6,f_7) := 0$

REMARK 2.14. *We have so far, in our examples, used the field $\mathbb{Q}$. But the theory developed so far is valid over any computable field k. Moreover, we also note that Buchberger's algorithm is independent of other parts of mathematics. This in particular means that we can learn and use it quite easily.*

Exercises for Section 2.4

EXERCISE 2.9. *Let*

$$\begin{aligned} f_1 &= x - y^2w, \\ f_2 &= y - zw, \\ f_3 &= z - w^3, \\ f_4 &= w^3 - w \end{aligned}$$

be polynomials over $\mathbb{Q}$. Show that they form a Groebner basis for the ideal they generate over $\mathbb{Q}$ with respect to the lex order with $x > y > z > w$, but they are not a Groebner basis with respect to the lex order with $w > x > y > z$.

5. Reduced Groebner Bases

As we have seen in Example 2.19, Groebner bases computed using Buchberger's Algorithm (Algorithm 6) contain unnecessary polynomials. The following lemma allows to eliminate those unneeded generators.

LEMMA 2.15. *Let G be a Groebner basis for the polynomial ideal I. Let $p \in G$ be a polynomial such tat*

$$LT(p) \in (LT(G - \{p\})).$$

Then $G - \{p\}$ is also a Groebner basis for I.

PROOF. Write $G = \{g_1, ...g_s\}$. Suppose that $LT(p) \in (LT(G - \{p\}))$ holds. Then by definition of a monomial ideal, any polynomial $f(x)$ in the monomial ideal $(LT(G))$ is a finite sum of polynomial multiples of the various $LT(g_i)$'s, so if $LT(p)$ is in the ideal $(LT(G - \{p\}))$ this means that $LT(p)$ is a multiple of other monomial $LT(g_i)$ by Lemma 2.9; hence, $f(x)$ becomes an element of $(LT(G - \{p\}))$. This establishes

$$(LT(G - \{p\})) = (LT(G)) = (LT(I)).$$

Therefore, $G - \{p)$ is also a Groebner basis for I. by the second part of the proof of Theorem 2.2. □

Thus, using this lemma and adjusting constants to make all leading coefficients 1 and removing any p with $LT(p) \in LT(G - \{p\})$, we can obtain what we call a minimall Groebner basis:

DEFINITION 2.16. *A Groebner basis G for a polynomial ideal I is called a* ***minimal Groebner basis*** *for I if it furthermore satisfies the following two more conditions:*

(1) $LC(p) = 1$ *for each* $p \in G$.
(2) *For each* $p \in G, LT(p) \notin (LT(G - \{p\}))$,

where $(LT(G - \{p\}))$ is the monomial ideal generated by the leading terms of elements of the set $G - \{p\}$.

EXAMPLE 2.22. *In the construction of a Groebner basis $G = (f_1 = y^2 - 1, f_2 = xy - 1, f_3 = x - y)$ for the ideal $(f_1 = y^2 - 1, f_2 = xy - 1)$ in Example 2.19, we note that $LT(f_2) = xy$ is divisible by $LT(f_3) = x$, and hence we delete f_2 from G to obtain a minimal Groebner basis (f_1, f_3) for the ideal (f_1, f_3).*

EXAMPLE 2.23. *This is a continuation of Example 2.20. There, we started with two polynomials*

$$\begin{aligned} f_1 &= x^2 - xy + 1, \\ f_2 &= x^2y - x^2 + x. \end{aligned}$$

And we have computed a Groebner basis $G = \{f_1, f_2, f_3, f_4, f_5, f_6, f_7\}$ *of the ideal* (f_1, f_2), *where*

$$\begin{aligned} f_1 &= x^2 - xy + 1, \\ f_2 &= x^2y - x^2 + x, \\ f_3 &= xy^2 - xy + x - y + 1, \\ f_4 &= -xy + x - 1, \\ f_5 &= x - 2y + 1, \\ f_6 &= 2y^3 - y^2 - y + 2 \\ f_7 &= -2y^2 + 3y - 2. \end{aligned}$$

Now, if we examine these polynomials, we quickly find that $LT(f_5)$ *divides all the leading term of other* f_i*'s with* $i < 5$, *so we can delete these* $f_i, i < 5$ *from* G. *Similarly, since* $LT(f_7)$ *divides* $LT(f_6)$ *we delete* f_6 *out of* G. *Therefore, the resulting*

$$\begin{aligned} G &= \{f_5, \frac{-1}{2} \cdot f_7\} \\ &= \{x - 2y + 1, y^2 - \frac{3}{2}y + 1\} \end{aligned}$$

is a minimal Groebner basis for the ideal (f_1, f_2).

COROLLARY 2.11. *If* $F = \{f_1, ..., f_s\}$ *and* $G = \{g_1, ..., g_t\}$ *are minimal Groebner bases for the same polynomial ideal* I, *then* $s = t$ *and after renumbering if necessary,* $LT(f_i) = LT(g_i)$ *for each* $i = 1, ..., s$.

PROOF. We note that both F and G are subsets of the ideal I. Since $f_1 \in F$ is in I and since G is a Groebner basis of I, there exists i such that $LM(g_i)$ divides $LM(f_1)$. After renumbering if necessary, we may assume that $i = 1$. Then since $g_1 \in G$ is in I and since F is a Groebner basis of I, there exists j such that $LM(f_j)$ divides $LM(g_1)$. Thus, $LM(f_j)$ divides $LM(f_1)$; and hence $j = 1$ because F is a minimal Groebner basis of I. Thus, $LM(f_1) = LM(g_1)$. Then by induction on the index set, we conclude that $s = t$ and $LM(f_i) = LM(g_i)$ for each $i = 1, ..., s$. □

On the other hand, a minimal Groebner basis of a polynomial ideal I is not uniquely determined in general.

EXAMPLE 2.24. *Let* $f_1 = x^2 + xy + y^2, f_2 = x + y, f_3 = x$. *Then we can verify that the following polynomials are a Groebner basis for the ideal* $I = (f_1, f_2, f_3)$ *generated by* f_1, f_2, *and* f_3 *under the lex order with* $x > y$:

$$\begin{aligned} f_1 &= x^2 + xy + y^2, \\ f_2 &= x + y, \\ f_3 &= x, \\ f_4 &= y^2, \\ f_5 &= y. \end{aligned}$$

Thus, both $\{f_2, f_5\}$ *and* $\{f_3, f_5\}$ *are minimal Groebner bases for* I.

LEMMA 2.16. *Let* $G = \{g_1, ...g_t\}$ *be a minimal Groebner basis for a polynomial ideal* I *under a monomial order on* $k[x_1, ..., x_n]$. *Then* $H = \{h_1, ..., h_t\}$ *is also a minimal Groebner basis for* I, *where each* h_i *is the polynomial obtained from* g_i *successively*

$$g_i \xrightarrow{H_i} h_i, \qquad i = 1, 2, ..., t.$$

by reduction modulo $H_i = \{h_1, ..., h_{i-1}, g_{i+1}, ..., g_t\}$, *where we ignore those* h_j *with* $j \leq 0$.

PROOF. Since G is a subset of I so is H. Since G is a minimal Groebner basis for I, each h_i satisfies

$$LM(h_i) = LM(g_i),$$

and hence H is again a minimal Groebner basis for I by the first part of Theorem 2.3.

In particular, we note that since any term in each h_i which is divisible by some leading polynomial $LM(g_j) = LM(h_j)$ is taken out by this process, none of the non-zero terms in each $h_i \in H$ except the leading term $LT(h_i)$ is divisible by any $LM(h_j)$ for $j \neq i$. □

DEFINITION 2.17. *A Groebner basis $G = \{g_1, ..., g_t\}$ for a polynomial ideal I is called a* **reduced Groebner basis** *if it satisfies the following two conditions:*

(1) *$LC(p) = 1$ for each $p \in G$.*
(2) *None of the non-zero terms of each $p_i \in G$ is divisible by some $LM(p_j)$ of $p_j \in G$ for $j \neq i$.*

So far by Lemma 2.16, we have proved:

COROLLARY 2.12. *Any polynomial ideal I of $k[x_1, ..., x_n]$ has a reduced Groebner basis under a given monomial order on $k[x_1, ..., x_n]$.* □

Then the following result further claims that a reduced Groebner basis is uniquely determined.

THEOREM 2.6. *(B. Buchberger) Given a monomial order on a polynomial ring $k[x_1, ..., x_n]$, every non-zero polynomial ideal of $k[x_1, ..., x_n]$ has a* unique *reduced Groebner basis.*

PROOF. Since any polynomial ideal I of $k[x_1, ..., x_n]$ has a reduced Groebner basis under a given monomial ordering on $k[x_1, ..., x_n]$, let $G = \{g_1, ..., g_t\}$ and $H = \{h_1, ..., h_t\}$ be two reduced Groebner bases for I. Since they are automatically minimal Groebner bases, they have the same cardinality and each $LM(g_i) = LM(h_i)$ for $i = 1, ..., t$. If there were i such that $g_i \neq h_i$ then, since their difference $g_i - h_i$ would be an element of $I, LM(g_i - h_i)$ in the monomial ideal $(LT(H))$ would be divisible by some $LM(h_j) = LM(g_j)$ by Lemma 2.9, but this is impossible because $LM(g_i - h_i) < LM(h_j)$ for any j. □

REMARK 2.15. *Therefore, whoever computes a reduced Groebner basis for a polynomial ideal I under a given monomial order, we must have the same result. Because of this reason, each Computer Algebra System computes "the" reduced Groebner basis for I. Or, a Groebner basis satisfying the second condition of the definition for the reduced Groebner basis for I, so that one can get the reduced Groebner basis immediately. For example, The Computer Algebra System, REDUCE is the case.*

On the other hand, if we consider the amount of time needed for computing "the" reduced Groebner basis for a polynomial ideal I, it is not a good idea to compute a Groebner basis first, a minimal Groebner basis from the computed Groebner basis, and then finally "the" reduced Groebner basis from the computed minimal Groebner basis. Fortunately, another theorem of B. Buchberger, which we do not study in this book, guarantees that we can detect unnecessary polynomials quite effectively while computing a Groebner basis as a first phase, and it turns out this usually reduces the whole computation time quite a lot.

Although it is completely another matter, there is still something in common in two subjects, "the" Taylor series expansion and the reduced Groebner basis: one reason why the Taylor series expansion of a real-valued function, if exists, is very important is that, it does not matter which methods are used for what different manners. As long as one obtains the Taylor series expansion of a real-valued function, it is uniquely determined. That is, it is independent of the path for obtaining it, and this is the kind of computation we would like to appreciate most of the time. We can apply the same arguments to "the" reduced Groebner basis for a polynomial ideal I under a given monomial order on $k[x_1, ..., x_n]$.

Exercises for Section 2.5

EXERCISE 2.10. *Using a Computer Algebra System, for instance, REDUCE, verify that the polynomials $f_1, ..., f_5$ are a Groebner basis for the ideal $I \in Q[x, y]$ generated by $f_1 = x^2 + xy + y^2, f_2 = x + y$, $f_3 = x$, $f_4 = y^2$, $f_5 = y$.*

EXERCISE 2.11. *What is going to be the reduced Groebner basis for the ideal of $k[x_1, ..., x_n]$ generated by a finite number of linear polynomials over $\mathbb{Q}$ with respect to the lex order with $x_1 > ... > x_n$?*

CHAPTER 3

Immediate Consequences of Groebner Bases

1. Solving a System of Polynomial Equations

Suppose that we are given a system of polynomial equations,

$$\begin{cases} f_1(x_1, ..., x_n) &= 0, \\ f_2(x_1, ..., x_n) &= 0, \\ \quad \cdots\cdot & \\ f_s(x_1, ..., x_n) &= 0 \end{cases}$$

with coefficients in an algebraically closed field k, where $n \geq 1$.

In Subsection 4.2 of Chapter 1, we have reviewed some common notions on solution sets of systems of polynomial equations. In particular, if we are given a system of polynomial equations in one variable, i.e., $n = 1$ case, then Section 3.3 of Chapter 1 explains in detail why Gauss' Fundamental Theorem of Algebra provides the final answer. There, we have seen the usefulness of Euclid's Algorithm for computing the greatest common divisor, $\gcd(f(x), g(x))$ of two polynomials (Algorithm 4), in reducing the given system of polynomial equation to a single polynomial equation, and the underlying idea in doing so was the invariance of the solution set we are looking for.

Otherwise, namely when $n > 1$, then Lemma 1.7 in Section 5 of Chapter 1 states that

$$V(I) = V(f_1, ..., f_s).$$

That is, the solution set of infinitely many polynomials equations in the left side is the same as the solution set of finitely many $f_1, ..., f_s$ polynomials in the right side, provided that the ideal I is generated by these polynomials $f_1, ..., f_s$.

Thus, the study on how to solve systems of finite number of polynomial equations becomes that of ideals generated by finitely many polynomials $f_1, ..., f_s$, by Hilbert's Basis Theorem (Theorem 1.4) in Subsection 5.1 of Chapter 1. One might think that this change of the subjects only increases the difficulty, but it turns out that it is not the case at all. We will explain why it is the case in this section.

We now have a computational means which converts the given Hilbert basis for a polynomial ideal I to a Groebner basis for the same ideal I, thanks to Bruno Buchberger's Theorem and Algorithm (Theorem 2.5, Algorithm 6 in Section 4 of Chapter 2. Then Corollary 1.10 or Theorem 1.10 becomes

COROLLARY 3.1. *$V(I)$ is not an empty set if and only if the Groebner basis of I contains a non-constant polynomial.* □

Why then Buchberger's discovery is really amazing?

1.1. The Role of the Dictionary Ordering. The purpose of this section is to provide a good and solid explanation on the significance of Buchberger's discovery. Let's start with a system of linear equations in several variables:

$$\begin{cases} 3 - 6y - 2z &= 0, \\ 2x - 4y - 4w &= 0, \\ x - 2y - z - w &= 0. \end{cases}$$

In matrix notation, we have

$$AX = 0.$$

When we solve such a system of linear equations, we silently assume that

$$x > y > z > w,$$

for instance, and when we obtain the row echelon form of the matrix A,

$$B = \begin{bmatrix} 1 & -2 & 0 & 2 \\ 0 & 0 & 1 & 3 \\ 0 & 0 & 0 & 0 \end{bmatrix}$$

of the matrix A:

$$BX = 0.$$

This B is very special in the sense that, in this case, we know from the "third" row of B that w can be any number, i.e., w is a parameter, and then we solve for the unknown z using the "second" row of B in terms of the parameter w, and then finally we solve for the unknown x, using the "first" row of B, again putting that y is a parameter. Thus, x will be expressed in terms of two parameters w and y. This way of solving a system of linear equations is usually called Gaussian elimination or **Gaussian backward substitutions**, etc. In this case, the solution set consists of those column vectors

$$(2s - 2t, s, -3t, t)^t,$$

with t and s are parameters.

Here, we should note that listing the variables x, y, z and w, and considering that x is bigger than y is bigger than z is is bigger than w is the dictionary ordering, namely, the one used in the ordinary dictionary for listing words.

Then among (infinitely) many monomial orderings on $k[x_1, ..., x_n]$, which one do you think will provide us with such a backward substitution process at all?

The answer is the lexicographic order or the dictionary order, and under this monomial ordering, the Groebner basis, especially the (reduced) Groebner basis G for an ideal I generated by polynomials $f_1, ..., f_t$, shines over any other monomial orderings, because, it will allows us to solve the given system of polynomial equations, by first solving the last equation in the only variable x_n, and then solving for x_{n-1} the one just before the last variable x_n, etc., and finally by solving for the first variable x_1.

Of course, to solve each of these at most n equations for one variable each time, we will definitely need Gauss's Fundamental Theorem of Algebra and, as a last resort, we may also have to apply certain numerical method of solving the equations in one variable at each intermediate step if solving them exactly is not possible at all.

REMARK 3.1. *There is one practical problem related to the dictionary ordering, which has relatively more time consuming nature of computing a Groebner basis than other monomial ordering, in general. Because of this reason, it is sometimes wise to compute a Groebner basis, first with respect to a monomial ordering which can be computed spending less time than computing a Groebner basis with respect to the dictionary ordering, and then try to convert it to a Groebner basis with respect to a dictionary ordering.*

Since Algebraic Geometry is for the study of the properties of solution sets of polynomial equations in several variables, the dictionary ordering is more needed in the study of those subjects which are directly or closely related to subjects in Algebraic Geometry. On the other hand, those subjects requiring only heavy amount of computations do not usually require computation of a Groebner basis with respect to the dictionary ordering.

EXAMPLE 3.1. *Suppose we want to decide whether the following system of polynomial equations has a solution or not, and we want to find all solutions of it, if it has a solution:*

$$\begin{cases} x^2 + y^2 + z^2 - 1 &= 0, \\ x^2 + y^2 + z^2 - 2x &= 0, \\ 2x + 3y - z &= 0. \end{cases}$$

To answer these two questions, we let $f_1 = x^2+y^2+z^2-1,\ f_2 = x^2+y^2+z^2-2x,\ f_3 = 2x+3y-z$. *Then we easily see that the system has a solution if and only if the Groebner basis of the polynomial ideal* I *generated by* f_1, f_2, *and* f_3 *is not generated by* 1.

We compute the followings, using the Computer Algebra System REDUCE:[1]

[1]A working program for REDUCE for this example is also available from: http://ch.catholic.or.kr/pundang/4/soh/2006_aa_i/Example_3_01.htm

```
off echo$
off nat$
load_package groebner$
torder({x,y,z},lex)$
f1:=x^2+y^2+z^2-1$
f2:=x^2+y^2+z^2-2*x$
f3:=2*x+3*y-z$
groebner({f1,f2,f3});
{2*x - 1, 3*y - z + 1, 40*z**2 - 8*z - 23}$
solve({f1,f2,f3},{x,y,z});
{{x=1/2, y=(sqrt(26) - 6)/20, z=(3*sqrt(26) + 2)/20},
{x=1/2, y=( - sqrt(26) - 6)/20, z=( - 3*sqrt(26) + 2)/20}}$
```

Thus, the solution set of the system is non-empty, because the Groebner basis

$$2x - 1,$$
$$3y - z + 1,$$
$$40z^2 - 8z - 23,$$

computed contains a non-constant polynomial. If you examine these polynomials, the last one is a polynomial in one variable z only, so that we can find its solution set. And then by solving the second equation, we can solve for y. Then, we have explicitly computed all of its solutions.

1.2. Program for the Method of Lagrange Multipliers. Here is such a program written using REDUCE commands.

PROGRAM 1. *(Program for Lagrange Multipliers)*

```
off echo$
off nat$
on fullroots$
```

```
write "comment The following program computes minimum and maximum of a real-valued polynomial function f in 3 variables under a polynomial constraint g=0. It is a prototype example of Lagrange's method of multipliers, lamda";
% inputs:
f:=x^3+2*x*y*z-z^2$
g:=x^2+y^2+z^2-1$
% If you choose inputs different from the above ones,
% then you may have to modify the following lines:
f1:=df(f,x)-lamda*df(g,x)$
f2:=df(f,y)-lamda*df(g,y)$
f3:=df(f,z)-lamda*df(g,z)$
result:=solve({f1,f2,f3,g},{lamda,x,y,z})$
%% You don't have to modify the rest of lines.
write "critical points := ",result;
write "comment Now mimimum and maximum of f under the constraint g=0 are determined as:";
   minimum:=sub(part(result,1),f)$
   ctps_min:=part(result,1)cons{}$
   maximum:=sub(part(result,1),f)$
   ctps_max:=part(result,1)cons{}$
j:=2$
while j leq length(result) do
   begin a:=sub(part(result,j),f)$
     if a geq maximum then
        begin if a greaterp maximum then
           begin maximum:=a$
              ctps_max:=part(result,j)cons{}$
           end
        else if a leq minimum then
           begin
              ctps_min:=part(result,j).ctps_min$
```

```
            end
         else
            begin
               ctps_max:=part(result,j).ctps_max$
            end$
      end
   else if a leq minimum then
      begin if a lessp minimum then
         begin minimum:=a$
            ctps_min:=part(result,j)cons{}$
         end
      else
         begin
            ctps_min:=part(result,j).ctps_min$
         end$
      end
   else begin end$
  j:=j+1$
  end$
  write "The maximum value of f under g=0 is, ",maximum,", and it
occurs at the critical points ",ctps_max;
  write "The minimum value of f under g=0 is, ",minimum,", and it
occurs at the critical points ",ctps_min;
  write "comment Check: Here are the values of f at all the computed
critical points";
  j:=1$
  while j leq length(result) do
   begin a:=sub(part(result,j),f)$
      write "The value of f at the critical point ",part(result,j)," is: ",
a;
      j:=j+1$
   end$
```

——————————

EXAMPLE 3.2. *When the inputs are*

$$f = x^3 + 2xyz - z^2,$$
$$g = x^2 + y^2 + z^2 - 1,$$

we obtain the following results:

——————————

critical points := {{x=(- 2)/3,z=2/3,y=1/3,lamda=(- 4)/3},
{x=(- 2)/3,z=(- 2)/3,y=(- 1)/3,lamda=(- 4)/3},
*{x=(- 3)/8, z=sqrt(11)/(8*sqrt(2)), y=(- 3*sqrt(11))/(8*sqrt(2)), lamda=1/8},*
*{x=(- 3)/8, z=(- sqrt(11))/(8*sqrt(2)), y=(3*sqrt(11))/(8*sqrt(2)), lamda=1/8},*
{x=1,z=0,y=0,lamda=3/2}, {x=0,z=1,y=0,lamda=-1},
{x=0,z=0,y=1,lamda=0},
{x=0,z=0,y=-1,lamda=0},
{x=0,z=-1,y=0,lamda=-1},
{x=-1,z=0,y=0,lamda=(- 3)/2}}$

comment Now mimimum and maximum of f under the constraint g=0 are determined as:$

The maximum value of f under g=0 is, 1, and it occurs at the critical points

{{x=1,z=0,y=0,lamda=3/2}}$

The minimum value of f under g=0 is, (- 28)/27, and it occurs at the critical points

{{x=(- 2)/3,z=(- 2)/3,y=(- 1)/3,lamda=(- 4)/3},
{x=(- 2)/3,z=2/3,y=1/3,lamda=(- 4)/3}}$

comment Check: Here are the values of f at all the computed critical points$

The value of f at the critical point {x=(- 2)/3,z=2/3,y=1/3,lamda=(- 4)/3}

is: (- 28)/27$

The value of f at the critical point {x=(- 2)/3,z=(- 2)/3,y=(- 1)/3, lamda=(- 4)/3} is: (- 28)/27$
*The value of f at the critical point {x=(- 3)/8, z=sqrt(11)/(8*sqrt(2)), y=(- 3*sqrt(11))/(8*sqrt(2)), lamda=1/8} is: 7/128$*
*The value of f at the critical point {x=(- 3)/8, z=(- sqrt(11))/(8*sqrt(2)), y=(3*sqrt(11))/(8*sqrt(2)), lamda=1/8} is: 7/128$*
The value of f at the critical point {x=1,z=0,y=0,lamda=3/2} is: 1$
The value of f at the critical point {x=0,z=1,y=0,lamda=-1} is: -1$
The value of f at the critical point {x=0,z=0,y=1,lamda=0} is: 0$
The value of f at the critical point {x=0,z=0,y=-1,lamda=0} is: 0$
The value of f at the critical point {x=0,z=-1,y=0,lamda=-1} is: -1$
The value of f at the critical point {x=-1,z=0,y=0,lamda=(- 3)/2} is: -1$
comment The following is the timex report of the system in seconds:
real 1.70
user 0.40
sys 0.13
$

Exercises for Section 3.1

EXERCISE 3.1. *Compute solutions of*

$$\begin{cases} x^2y - z^3 - 3 &= 0, \\ xy - 4z - 1 &= 0, \\ z - x^2 &= 0, \end{cases}$$

by first computing a Groebner basis of the polynomials. [Hint: You need to use two commands, on rounded$ and solve(f1,f2,f3)$ for a numerical computation of solutions.]

2. The Ideal Membership Problem

The following short historical remark is prepared for those who do not know about what's really happened in the development of the theory of Groebner basis in its beginning stage, which I myself happed to know only in the middle of 1980.

Suppose that

(1) you are given a polynomial $f(x_1, ..., x_n)$,
(2) a polynomial ideal I of the polynomial ring $k[x_1, ..., x_n]$ in n variables with $n \geq 1$, and
(3) you want to know whether $f = f(x_1, ..., x_n)$ is a member of I or not.

The problem is that, in most cases, the ideal I will have infinitely many polynomials in it, so that there is no easy way of deciding its membership to I, because even if you are allowed to apply the division algorithm, Algorithm 5 directly, sometimes it will tell you that the remainder is zero so that f belongs to I. But, since the resulting remainder is really dependent on your own path of your actual division process, you will not be able to get the zero polynomial most of the time as a remainder of your division process, even though f is in fact a member of the ideal I.

If you do not mind whether you are able to compute something out completely or not, like the posed question, it may not be a problem to you at all. Otherwise, this problem is a really serious one, because if mathematicians refuse to computational business who else could figure out the underlying reason in computational phenomena around us? In the mean while the history of mathematics tells us that ever since Euclid's Algorithm for computing the greatest common divisor of two numbers, there was no comprehensible generalization of Euclid's algorithm for more than, say, 2500 years. But at the year 1965, the situation has suddenly been changed completely.

First of all, it should be quite the case that such an incomplete division algorithm, Algorithm 5, was not very well-known to us or even not well-formulated as it is, because itself alone is not computationally useful at all, although someone says that we have some traces of such a division process in the works of Gauss.

Secondly, in his famous Ph.D. thesis in 1964 prepared under the guidance of Oscar Zariski, a Japanese mathematician Heisuke Hironaka (1931 -) claimed the existence of the so-called "standard bases" in the domain of formal power series and proved its existence by a method of contradiction, so he never actually tried to construct such a basis explicitly.

But, after Bruno Buchberger (1942 -) has computationally constructed his very famous and very important algorithm, Algorithm 6 for Groebner bases in his Ph.D. thesis in the year 1965 without knowing what Hironaka has claimed its existence in his thesis of the year 1964, such an incomplete algorithm, Algorithm 6 somewhat began to remain as the current formulation that *no terms of the remainder r should be divisible by any of the leading terms of the divisors*, because it suddenly becomes to have a significant meaning of computing a remainder in a *unique* manner, *provided that the divisors are a Groebner basis for the given ideal I.*

Now going back to the posted problem, we know by Hilbert Basis Theorem (Theorem 1.5 or Theorem 2.2) a finite number of generating elements of I are going to be given to you.

If $n = 1$, then it is very easy because $k[x]$ is a *PID*, so the single generator of I is given to you, and hence you simply apply the division algorithm, Algorithm 3, to see if f is a pure multiple of the single generator of I.

Otherwise, let $f_1, ..., f_s$ be generators of the ideal I. Then under a monomial order on $k[x_1, ..., x_n]$ of your own choice, you compute the (reduced) Groebner basis $G = \{g_1, ..., g_t\}$ for I from these generators, using Buchberger's Algorithm, Algorithm 6 or an improved version of this algorithm for computing the (reduced) Groebner basis. After that, you apply the division algorithm, Algorithm 5 to the given f in order to see whether the remainder r is the zero polynomial or not. If it turns out that r is the zero polynomial, this equivalently means that f is indeed an element of the ideal I. Otherwise, since the remainder is uniquely determined by a Groebner basis, f is not a member of the ideal I.

Nowadays, most Computer Algebra Systems, say, REDUCE, MAPLE, etc., since 1970s, do provide a command for computing the (reduced) Groebner basis which satisfies at least the second condition of the definition for the reduced Groebner basis. Once it satisfies the second condition, it is very easy to normalize the leading term of each of them, if it is ever possible. As long as we are working over a field, it is OK. Otherwise, it is not.

Some Computer Algebra System, for instance, (not the "preduce" command but) the "greduce" command in REDUCE, does provide a command which computes the unique remainder via the (reduced) Groebner basis.

EXAMPLE 3.3. *We want to computationally decide whether*

$$\begin{aligned} f &= x^4y - 2x^5 + 2x^2y^2 - 2x^3y - 2x^4 \\ &\quad -2y^3 + 4xy^2 - 3x^2y + 2x^3 - y + 2x, \end{aligned}$$

is a member of the ideal I generated by the following two polynomials:

$$\begin{aligned} f_1 &: = x^2y - y + x, \\ f_2 &: = xy^2 - x. \end{aligned}$$

For example, we can use the REDUCE command

$$greduce(f, \{f_1, ..., f_t\})$$

to see if the remainder is zero or not. By Theorem 2.4, f is a member of I if and only if the remainder is zero.

The main purpose of this example is to demonstrate the importance of B. Buchberger's Algorithm for computing a Groebner basis for a polynomial ideal.

———————————

Here is the computed result:[2]

```
off echo$
off nat$
load_package groebner$
f:=x^4*y - 2*x^5 + 2*x^2*y^2 - 2*x^3*y - 2*x^4 - 2*y^3 + 4*x*y^2
- 3*x^2*y + 2*x^3 - y + 2*x$
f1:=x^2*y-y+x$
f2:=x*y^2-x$
greduce(f,{f1,f2});
0$
```

———————

Thus, f is indeed a member of I generated by f_1 and f_2.

2.1. A Program for the Quotients of the Division Algorithm, Algorithm 5. Here we provide our program in REDUCE commands which decides whether a given polynomial f is a member of a given ideal I generated by $f_1, ..., f_s$. Moreover, this program also computes the quotients of f with respect to the computed (reduced) Groebner basis for I, and verify whether these quotients are indeed correct or not. In this program, we use a few REDUCE commands in groebner package. Among them, "groebnerprot" and "groebnert" computes the Groebner basis plus their relations to the given polynomials, and "groebprot́" is a switch which, if it is set to on, allows us to trace back the quotients with respect to the computed Groebner basis.

PROGRAM 2. *(Sun T. Soh, 2000) (Program for the Quotients).*

———————

```
off echo$
off nat$
load_package groebner$
%%%% Beginning of Inputs. %%%%%
torder({y,x}, gradlex)$ % You need to change this line.
```

[2]A working program for REDUCE for this example is also available from: http://ch.catholic.or.kr/pundang/4/soh/2006_aa_i/Example_3_03.htm

```
f:=x^4*y - 2*x^5 + 2*x^2*y^2 - 2*x^3*y - 2*x^4 - 2*y^3 + 4*x*y^2 - 3*x^2*y + 2*x^3 - y + 2*x$
write "f := ",f;
write "comment Is the given f a member of the ideal I generated by the following polynomials?";
% You need to change the following a few lines.
f1:=x^2*y-y+x$
f2:=x*y^2-x$
write "f1 := ",f1;
write "f2 := ",f2;
write "comment For this, we first compute the reduced Groebner basis for the ideal I.";
gbt:=groebnert({given_f1=f1, given_f2=f2})$ % You need to change this line.
gb:=for k:=1:length(gbt) collect lhs(part(gbt,k))$
% gb:=groebner({f1,f2})$
%%%%% End of Inputs. You don't have to change any lines below this line. %%%%%
write "gb := ",gb;
on groebprot$
write "comment Here is the remainder of f modulo the computed groebner basis:";
write "comment The following result is zero iff f is a member of the ideal I.";
preduce(f,gb)$
write "preduce(f,gb) := ",preduce(f,gb);
write "comment Now we would like to compute the quotients of f modulo the computed groebner basis.";
groebprotfile;
saveas groebprotfile_out$
write "groebprotfile_out := ",groebprotfile_out;
```

```
write "comment Thus we rewrite these relations for f in terms of the
computed Groebner basis elements:";
j:=length(gb)+2$
candidate:=given_f$
while j leq length(groebprotfile_out) do
  begin
    candidate:=rhs(part(groebprotfile_out,j))$
    write "candidate := ",candidate;
    j:=j+1$
  end$
result:=(preduce(f,gb)-(candidate-given_f))$
write "comment Here is the result:";
write "given_f := ",result;
write "comment Now we would like to verify that this result is indeed
correct.";
pre_sub:= for k:=2:length(gb)+1 collect part(groebprotfile_out,k)$
check:=sub(pre_sub,result);
write "comment The following line should be zero.";
write "comment Otherwise, the quotients we computed above are not
correct ones.";
write "check - f := ",check-f;
pre_sub4final_result:=for k:=1:length(gb) collect
mkid(poly,k)=rhs(part(gbt,k))$
write "pre_sub4final_result := ",pre_sub4final_result;
write "comment Finally, we would like to express the given f in terms
of the given polynomials.";
final_result:=sub(pre_sub4final_result,result)$
write "given_f := ",final_result;
```

———————

REMARK 3.2. *Even if the Computer Algebra System, REDUCE, is not available to you, you can still compute these quotients if you are willing to use our InetCompu Service.*

For instance, if you send an E-mail whose main body consists of the above program in the following format,

input:

the exact copy of Program 2

end input:

to the E-mail address:

reduce@trinitas.mju.ac.kr

then our Middleware, called InetCompu (Copyrighted by the author), will handle your request for such a computation for your own sake, and the computed results will be sent back to you immediately, even with the system's timex stamp for the time spent.

Note 1: An E-mail in plain text mode is highly appreciated.

Note 2: Because of the spam-mail filtering Software, there should be no "$<$", "$>$", "$<<$", or "$>>$" symbols in your REDUCE program if you are willing to use our InetCompu Service. This is because these symbols are also used in HTML text format, and the spam-mail filtering Software usually recognizes these symbols as the beginning or the end of HTML tags.

EXAMPLE 3.4. *In this example, we apply Program 2 to compute the quotients modulo the groebner basis G for I which is generated by the following two polynomials*

$$\begin{aligned} f_1 &:= x^2y - y + x, \\ f_2 &:= xy^2 - x. \end{aligned}$$

of a given polynomial f:

$$\begin{aligned} f &= x^4y - 2x^5 + 2x^2y^2 - 2x^3y - 2x^4 \\ &\quad -2y^3 + 4xy^2 - 3x^2y + 2x^3 - y + 2x. \end{aligned}$$

The monomial order we use is gradlex with $y > x$

$$torder(\{y, x\}, gradlex)\$$$

We note that the quotients are not uniquely determined.

We use the InetCompu Service of the above Remark to obtain the following results:[3]

———————————

```
Here are computed results.
f := - 2*x**5 + x**4*y - 2*x**4 - 2*x**3*y + 2*x**3 + 2*x**2*y**2
- 3*x**2*y + 4*x*y**2 + 2*x - 2*
y**3 - y$
comment Is the given f a member of the ideal I generated by the fol-
lowing polynomials?$
f1 := x**2*y + x - y$
f2 := x*(y**2 - 1)$
comment For this, we first compute the reduced Groebner basis for the
ideal I.$
gb := {x**2*y + x - y,
x**3 - 2*x + y,
- x**2 - x*y + y**2}$
comment Here is the remainder of f modulo the computed groebner
basis:$
comment The following result is zero iff f is a member of the ideal I.$
preduce(f,gb) := 0$
comment Now we would like to compute the quotients of f modulo the
computed groebner basis.$
groebprotfile_out := {candidate= - 2*x**5 + x**4*y - 2*x**4 - 2*x**3*y
+ 2*x**3 + 2*x**2*y**2 - 3*
x**2*y + 4*x*y**2 + 2*x - 2*y**3 - y,
poly1=x**2*y + x - y,
poly2=x**3 - 2*x + y,
poly3= - x**2 - x*y + y**2,
candidate=candidate - poly1*x**2,
candidate=candidate + 2*poly2*x**2,
```

[3] A working program for REDUCE for this example is available from: http://ch.catholic.or.kr/pundang/4/soh/2006_aa_i/Example_3_04.htm

```
candidate=candidate - 2*poly1*y,
candidate=candidate + 2*poly1*x,
candidate=candidate + 2*poly2*x,
candidate=candidate + 2*poly3*y,
candidate=candidate - 2*poly3*x,
candidate=candidate + poly2,
candidate=candidate - 2*poly3}$
comment Now we rewrite these relations for f in terms of the computed
Groebner basis elements:$
candidate := given_f - poly1*x**2$
candidate := given_f - poly1*x**2 + 2*poly2*x**2$
candidate := given_f - poly1*x**2 - 2*poly1*y + 2*poly2*x**2$
candidate := given_f - poly1*x**2 + 2*poly1*x - 2*poly1*y
+ 2*poly2*x**2$
candidate := given_f - poly1*x**2 + 2*poly1*x
- 2*poly1*y + 2*poly2*x**2 + 2*poly2*x$
candidate := given_f - poly1*x**2 + 2*poly1*x
- 2*poly1*y + 2*poly2*x**2 + 2*poly2*x + 2*poly3*y$
candidate := given_f - poly1*x**2 + 2*poly1*x
- 2*poly1*y + 2*poly2*x**2 + 2*poly2*x - 2*poly3*x + 2*poly3*y$
candidate := given_f - poly1*x**2 + 2*poly1*x
- 2*poly1*y + 2*poly2*x**2 + 2*poly2*x + poly2
- 2*poly3*x + 2*poly3*y$
candidate := given_f - poly1*x**2 + 2*poly1*x
- 2*poly1*y + 2*poly2*x**2 + 2*poly2*x + poly2
- 2*poly3*x + 2*poly3*y - 2*poly3$
comment Here is the result:$
given_f := poly1*x**2 - 2*poly1*x + 2*poly1*y
- 2*poly2*x**2 - 2*poly2*x - poly2 + 2*poly3*x
- 2*poly3*y + 2*poly3$
comment Now we would like to see verify that this result is indeed
correct.$
```

*check := - 2*x**5 + x**4*y - 2*x**4 - 2*x**3*y*
*+ 2*x**3 + 2*x**2*y**2 - 3*x**2*y + 4*x*y**2*
*+ 2*x - 2*y**3 - y$*

comment The following line should be zero.$

comment Otherwise, the quotients we computed above are not correct ones.$

check - f := 0$

pre_sub4final_result := {poly1=given_f1,
*poly2=given_f1*x*y - given_f1 - given_f2*x**2 + given_f2,*
*poly3= - given_f1*y + given_f2*x}$*

comment Finally, we would like to express the given f in terms of the given polynomials.$

given_f := - 2*given_f1*x**3*y - 2*given_f1*x**2*y
+ 3*given_f1*x**2 - 3*given_f1*x*y + 2*given_f1*y**2
+ given_f1 + 2*given_f2*x**4 + 2*given_f2*x**3
+given_f2*x**2 - 2*given_f2*x*y - given_f2$

comment The following is the timex report of the system for the above computation (in seconds):

real 0.23

user 0.16

sys 0.07

$

————————

Thus,

(1) *we have verified that f is indeed a member of the ideal I generated by f_1 and f_2, and*
(2) *we have also verified that the computed quotients of f with respect to the computed Groebner basis are indeed correct.*
(3) *We even expressed the given f in terms of the given generating polynomials of I.*

EXAMPLE 3.5. *We use the same program in Program 2 to see if the given polynomial*

$$f = 5y^2 + 2x^2y + 5/2xy + 3/2y + 8x^2 + 3/2x - 9/2$$

is a member of the ideal generated by

$$\begin{aligned} f_1 &= 3x^2y + 2xy + y + 9x^2 + 5x - 3, \\ f_2 &= 2x^3y - xy - y + 6x^3 - 2x^2 - 3x + 3, \\ f_3 &= x^3y + x^2y + 3x^3 + 2x^2, \end{aligned}$$

under the lex order with $x > y$

$$torder(\{x, y\}, lex)\$$$

Our program also computes the quotients of f with respect to the computed Groebner basis for I, and verify whether these quotients are indeed correct or not.[4]

———————

Here are computed results.

*f := (4*x**2*y + 16*x**2 + 5*x*y + 3*x + 10*y**2 + 3*y - 9)/2$*

comment Is the give f a member of the ideal I generated by the following polynomials?$

*f1 := 3*x**2*y + 9*x**2 + 2*x*y + 5*x + y - 3$*

*f2 := 2*x**3*y + 6*x**3 - 2*x**2 - x*y - 3*x - y + 3$*

*f3 := x**2*(x*y + 3*x + y + 2)$*

comment For this, we first compute the reduced Groebner basis for the ideal I.$

*gb := {8*x - 2*y**2 + 5*y + 3,2*y**3 - 3*y**2 - 16*y + 21}$*

comment Here is the remainder of f modulo the computed groebner basis:$

comment The following result is zero iff f is a member of the ideal I.$

preduce(f,gb) := 5*y**2$

[4]A working program for REDUCE for this example is available from: http://ch.catholic.or.kr/pundang/4/soh/2006_aa_i/Example_3_05.htm

```
comment Now we would like to compute the quotients of f modulo the
computed groebner basis.$
groebprotfile_out := {candidate=(4*x**2*y + 16*x**2 + 5*x*y + 3*x
+ 10*y**2 + 3*y - 9)/2,
poly1=8*x - 2*y**2 + 5*y + 3,
poly2=2*y**3 - 3*y**2 - 16*y + 21,
candidate=(4*candidate - poly1*x*y)/4,
candidate=candidate - poly1*x,
candidate=(16*candidate - poly1*y**3)/16,
candidate=(32*candidate - 3*poly1*y**2)/32,
candidate=(32*candidate + 13*poly1*y)/32,
candidate=(16*candidate + 3*poly1)/16,
candidate=(16*candidate - poly2*y**2)/16,
candidate=(32*candidate - poly2*y)/32,
candidate=(16*candidate + 3*poly2)/16}$
comment Now we rewrite these relations for f in terms of the computed
Groebner basis elements:$
candidate := (4*given_f - poly1*x*y)/4$
candidate := (4*given_f - poly1*x*y - 4*poly1*x)/4$
candidate := (16*given_f - 4*poly1*x*y - 16*poly1*x
- poly1*y**3)/16$
candidate := (32*given_f - 8*poly1*x*y - 32*poly1*x
- 2*poly1*y**3 - 3*poly1*y**2)/32$
candidate := (32*given_f - 8*poly1*x*y - 32*poly1*x
- 2*poly1*y**3 - 3*poly1*y**2 + 13*poly1*y)/32$
candidate := (32*given_f - 8*poly1*x*y - 32*poly1*x
- 2*poly1*y**3 - 3*poly1*y**2 + 13*poly1*y + 6*
poly1)/32$
candidate := (32*given_f - 8*poly1*x*y - 32*poly1*x
- 2*poly1*y**3 - 3*poly1*y**2 + 13*poly1*y + 6*
poly1 - 2*poly2*y**2)/32$
candidate := (32*given_f - 8*poly1*x*y - 32*poly1*x
```

```
- 2*poly1*y**3 - 3*poly1*y**2 + 13*poly1*y + 6*
poly1 - 2*poly2*y**2 - poly2*y)/32$
candidate := (32*given_f - 8*poly1*x*y - 32*poly1*x
- 2*poly1*y**3 - 3*poly1*y**2 + 13*poly1*y + 6*
poly1 - 2*poly2*y**2 - poly2*y + 6*poly2)/32$
comment Here is the result:$
given_f := (8*poly1*x*y + 32*poly1*x + 2*poly1*y**3
+ 3*poly1*y**2 - 13*poly1*y - 6*poly1 + 2*poly2*y**2
+ poly2*y - 6*poly2 + 160*y**2)/32$
comment Now we would like to see verify that this result is indeed correct.$
check := (4*x**2*y + 16*x**2 + 5*x*y + 3*x + 10*y**2 + 3*y - 9)/2$
comment The following line should be zero.$
comment Otherwise, the quotients we computed above are not correct ones.$
check - f := 0$
pre_sub4final_result := {poly1=5*given_f1*x**2*y
+ 15*given_f1*x**2 + 4*given_f1 *x*y + 7*given_f1*x
- 2*given_f1*y + 16*given_f1 + 17*given_f2
- 15*given_f3*x*y- 45*given_f3*x - 7*given_f3*y
- 50*given_f3,
poly2= - 5*given_f1*x**2*y**2 - 37*given_f1*x**2*y
- 66*given_f1*x**2 - 4*given_f1*x*y**2 - 28*given_f1*x*y
- 41*given_f1*x + 2*given_f1*y**2 + 3*given_f1*y - 51*given_f1
- 44*given_f2 + 15*given_f3*x*y**2 + 111*given_f3*x*y
+ 198*given_f3*x + 7*given_f3*y**2 + 57*given_f3*y
+ 189*given_f3}$
comment Finally, we would like to express the given f in terms of the given polynomials.$
given_f := (40*given_f1*x**3*y**2
+ 280*given_f1*x**3*y+ 480*given_f1*x**3
```

```
- 34*given_f1*x**2*y**3
- 127*given_f1*x**2*y**2 + 115*given_f1*x**2*y
+ 530*given_f1*x**2 - 34*given_f1*x*y**3
- 133*given_f1*x*y**2 + 76*given_f1*x*y
+ 716*given_f1*x + 34*given_f1*y**3
- 37*given_f1*y**2 - 265*given_f1*y + 210*given_f1
+ 136*given_f2*x*y + 544*given_f2*x
+ 34*given_f2*y**3 - 37*given_f2*y**2
- 265*given_f2*y + 162*given_f2
- 120*given_f3*x**2*y**2 - 840*given_f3*x**2*y
- 1440*given_f3*x**2 + 102*given_f3*x*y**3
+ 421*given_f3*x*y**2 - 417*given_f3*x*y
- 2518*given_f3*x + 334*given_f3*y**2
+ 539*given_f3*y - 834*given_f3 + 160*y**2)/32$
comment The following is the timex report of the system for the above computation (in seconds):
real 0.23
user 0.17
sys 0.07
$
```

———————

Thus,

(1) *we have verified that f is not a member of the ideal I generated by f_1, f_2, and f_3, and*
(2) *we have also verified that the computed quotients of f with respect to the computed Groebner basis are indeed correct.*
(3) *We even expressed the given f in terms of the given generating polynomials of I.*

Exercises for Section 3.2

EXERCISE 3.2. *In Program 2, show that the one line*

```
pre_sub:= for k:=2:length(gb)+1 collect
```

part(groebprotfile_ out,k)$

can be replaced with the following lines:

```
on nat$
out "temp.out"$
write "pre_sub:={";
% write "part({given_f=f},1),";
j:=1$
while j leq length(gb) do
  begin
     out "temp.out"$
     if j leq length(gb) then
        begin
           write "part(groebprotfile_ out,",j+1,"),";
        end
     else
        begin
           write "part(groebprotfile_ out,",j+1,")";
        end$
   j:=j+1$
end$
write "}$";
shut "temp.out"$
off nat$
in "temp.out"$
```

The main point of this exercise is about how to create a list consisting of elements taken from more than one list, using REDUCE commands.

EXERCISE 3.3. *Let* $f = x + y^2 + z^3$ *and* $I = (x^2y - z^3 - 3, xy - 4z - 1, z - x^2)$.

(1) *Compute the remainder of the division of* f *by the given generators of the ideal* I.

(2) *Compute the remainder of the division of f by a Groebner basis for I.*

3. A Criterion for Polynomial Ideals To Be The Same

We now know by Theorem 2.6 that given a monomial order on a polynomial ring $k[x_1, ..., x_n]$ the reduced Groebner basis of a polynomial ideal I is uniquely determined, and hence if $I = (f_1, ..., f_s)$ and $J = (g_1, ..., g_r)$ are two polynomial ideals, then we just compute their reduced Groebner bases *under a given monomial order* to see if they are exactly the same. If their reduced Groebner bases are exactly the same, I and J are the same ideal, and otherwise, not. Thus, Theorem 2.6 provides the right answer to this question.

Exercises for Section 3.3

EXERCISE 3.4. *Let $I = (x^2y - z^3 - 3, xy - 4z - 1, z - x^2)$ and $J = (x^2y - z^3 - 2, xy - 4z - 3, z - x^2)$. Decide whether they are the same ideal or not.*

4. Computations in a Quotient Ring $k[x_1, ..., x_n]/I$

Historically, the subjects we are going to study in this section were the first applications of Groebner basis of Bruno Buchberger: He originally tried to develop a computational means for finding "standard sets of representatives" for the classes in quotient rings.

We recall that the quotient ring $k[x_1, ..., x_n]/I$ consists of the cosets $f(x_1, ..., x_n)+I$ for various $f(x_1, ..., x_n) \in k[x_1, ..., x_n]$. If $G = \{g_1, ..., g_t\}$ is a Groebner basis for I, then by Theorem 2.4 we know that the remainder $r(x_1, ..., x_n)$ of $f(x_1, ..., x_n)$ by G is uniquely determined.

DEFINITION 3.1. *This polynomial $r(x_1, ..., x_n)$ is called the **standard representative** or the **normal form** of $f(x_1, ..., x_n)$ with respect to G, and is denoted by*

$$N_G(f).$$

PROPOSITION 3.1. *Under this setting, let f, g be two polynomials in $k[x_1, ..., x_n]$. Then we have*

(1) *The map N_G: $k[x_1, ..., x_n] \longrightarrow k[x_1, ..., x_n]$ is a k-linear map.*

(2)
$$f \equiv g \pmod{I} \Longleftrightarrow N_G(f) = N_G(g).$$

(3) *The collection $\{N_G(f) : f \in k[x_1, ..., x_n]\}$ is a set of coset representatives for the quotient ring $k[x_1, ..., x_n]/I$.*

(4)
$$N_G(f \cdot g) = N_G(N_G(f) \cdot N_G(g)).$$

PROOF. Part (3) is a consequence of part (2).

Part (1). We first note that $f = q + N_G(f)$ for some $q \in k[x_1, ..., x_n]$ by Division Algorithm, Algorithm 5; hence $f - N_G(f) = q \in I = (g_1, ..., g_t)$ if and only if $f + I = N_G(f) + I$. Moreover, the same Division Algorithm also tells us that no monomial component of $N_G(f)$ is divisible by any $LM(g_i)$. Thus, if we compute $N_G(f)$ and $N_G(g)$ separately, i.e., $f = q_1 + N_G(f)$ and $g = q_2 + N_G(g)$ for some $q_1, q_2 \in I$ by the same Division Algorithm, we see that they are uniquely determined and none of $LM(g_i)$ divides any monomial component of both $N_G(f)$ and $N_G(g)$, and hence their sum is also uniquely determined and none of $LM(g_i)$ divides any monomial component of their sum, $N_G(f) + N_G(g)$. That is, we have:

$$f + g = (q_1 + q_2) + N_G(f) + N_G(g),$$

where none of $LM(g_i)$ divides any monomial component of $N_G(f)+N_G(g)$. On the other hand, we have by the same Division Algorithm,

$$f + g = q' + N_G(f + g),$$

for some $q' \in I$. Then since the remainder of $f + g$ with respect to a Groebner basis G is *uniquely* determined by Theorem 2.4, we have

$$N_G(f) + N_G(g) = N_G(f + g).$$

Similarly, if $f = q_1 + N_G(f)$ for some $q_1 \in I$, then for any scalar $\alpha \in k$, we have by the same Division Algorithm,

$$\alpha f = \alpha q_1 + \alpha N_G(f),$$

where we obviously see that $N_G(f)$ is uniquely determined and no monomial component of $\alpha N_G(f)$ is possibly divisible by any $LM(g_i)$. On the other hand, we have by the same Division Algorithm,

$$\alpha f = q' + N_G(\alpha f).$$

Since the remainder of αf with respect to a Groebner basis is *uniquely* determined again by Theorem 2.4, we have

$$\alpha N_G(f) = N_G(\alpha f).$$

This proves that N_G: $k[x_1, ..., x_n] \longrightarrow k[x_1, ..., x_n]$ is indeed a k-linear map.

Part (2). We know that $f \equiv g \pmod{I}$ if and only if $f = q + g$ for some $q \in I$, thus $f \equiv g$ mod(I) implies that $N_G(f) = N_G(q + g) = N_G(q) + N_G(g) = 0 + N_G(g) = N_G(g)$. Conversely, if $N_G(f) = N_G(g)$, then $f - g = (f - N_G(f)) - (g - N_G(g)) \in I$, that is, $f \equiv g \pmod{I}$

Part (4). We first note that

$$N_G(N_G(f)) = N_G(f).$$

Moreover, we have

$$\begin{aligned} N_G(fg) + I &= fg + I = (f + I)(g + I) \\ &= (N_G(f) + I)(N_G(g) + I) \\ &= N_G(f)N_G(g) + I. \end{aligned}$$

That is, we have $N_G(f)N_G(g) \equiv N_G(fg) \pmod{I}$. Thus, by part (2) we equivalently have

$$N_G(fg) = N_G(N_G(fg)) = N_G(N_G(f)N_G(g)).$$

□

We note that $k[x_1, ..., x_n]$ is a vector space over k with a basis consisting of the monomials of various total degree and hence the quotient ring $k[x_1, ..., x_n]/I$ is a vector space over k spanned by the cosets of the monomials of $k[x_1, ..., x_n]$ under the canonical projection

$$\pi : k[x_1, ..., x_n] \longrightarrow k[x_1, ..., x_n]/I.$$

PROPOSITION 3.2. *A basis for the k-vector space $k[x_1, ..., x_n]/I$ consists of the cosets of all those monomials $X_j = x_1^{\alpha_1} \cdots x_n^{\alpha_n}$ not divisible by any $LM(g_i)$, for $i = 1, ..., t$, where $G = \{g_1, ..., g_t\}$ is a Groebner basis of I.*

PROOF. By Division Algorithm, Algorithm 5, the unique remainder $N_G(f)$ of a polynomial f is a k-linear combination of those monomials which are not divisible by any of the $LM(g_i)$, $i = 1, ..., t$.

Suppose that we have a linear equation among those cosets of monomials X_j not divisible by any of the $LM(g_i)$, $i = 1, 2, ..., t$:

$$\beta_1(X_1 + I) + \cdots + \beta_u(X_u + I) = 0 + I,$$

in $k[x_1, ..., x_n]/I$. This equivalently means that we have

$$\beta_1 X_1 + \cdots + \beta_u X_u \in I.$$

On the other hand, for each $j = 1, ..., u$, we have $N_G(X_j) = X_j$. Thus we have by Proposition 3.1,

$$\begin{aligned} 0 &= N_G(\beta_1 X_1 + \cdots + \beta_u X_u) \\ &= N_G(\beta_1 X_1) + \cdots + N_G(\beta_u X_u) \\ &= \beta_1 N_G(X_1) + \cdots + \beta_u N_G(X_u) \\ &= \beta_1 X_1 + \cdots + \beta_u X_u, \end{aligned}$$

in $k[x_1, ..., x_n]$. Thus, each $\beta_j = 0$ for $j = 1, ..., u$, because these monomials in $k[x_1, ..., x_n]$ are linearly independent over k. □

REMARK 3.3. *Thus, via the standard representatives $N_G(f)$, these two propositions enable us to do an explicit computations such as sum and product operations in the quotient ring $k[x_1, ..., x_n]/I$.*

EXAMPLE 3.6. *This is a continuation of Example 3.4, where* $G = \{x^2y + x - y, x^3 - 2x + y, y^2 - x^2 - xy\}$ *is a Groebner basis for the ideal* $I = (x^2y - y + x, xy^2 - x)$ *under gradlex with* $y > x$. *Thus, the* $\mathbb{Q}$*-basis for the quotient ring* $\mathbb{Q}[x, y]/I$ *consists of cosets of* $\{1, x, y, xy, x^2\}$, *so that its dimension as a* $\mathbb{Q}$*-vector space over* $\mathbb{Q}$ *is* 5.

COROLLARY 3.2. *For those basis elements* $X_j + I$ *of the* k*-vector space* $k[x_1, ..., x_n]/I$ *with respect to* G, *the* $N_G(X_iX_j)$*'s for all* i, j *enable us to have a multiplication table for the standard representatives for the cosets of* $k[x_1, ..., x_n]/I$. $\square$

Therefore, Proposition 3.1 and Proposition 3.2 enable us to compute the sum and product operations in $k[x_1, ..., x_n]/I$ very explicitly in a unique manner.

EXAMPLE 3.7. *This is a continuation of Example 3.6. We have the following multiplication table for the standard representatives*

$$\{1, x, y, xy, y^2\}$$

of the cosets of $\mathbb{Q}[x, y]/I$.

$\times$	1	x	y	xy	x^2
1	1	x	y	xy	x^2
x	x	x^2	xy	$-x + y$	$2x - y$
y	y	xy	$x^2 + xy$	x	$-x + y$
xy	xy	$-x + y$	x	x^2	$xy - x^2$
x^2	x^2	$2x - y$	$-x + y$	$xy - x^2$	$-xy + 2x^2$

For instance, the product $(2x^2 + y + I)(3xy - 5 + I) = 6xy - 16x^2 - 5y + 3x + I$*, because we have*

$$\begin{aligned} N_G((2x^2+y)(3xy-5)) &= N_G(6x^3y - 10x^2 + 3xy^2 - 5y) \\ &= 6N_G(N_G(xy) \cdot N_G(x^2)) - 10x^2 + 3xy^2 - 5y \\ &= 6N_G((xy - x^2)) - 10x^2 + 3xy^2 - 5y \\ &= 6(xy - x^2) - 10x^2 + 3xy^2 - 5y \\ &= 6xy - 16x^2 - 5y + 3x. \end{aligned}$$

PROPOSITION 3.3. *Let* $I = (f_1, ..., f_s)$ *be a polynomial ideal of* $k[x_1, ..., x_n]$*. Then a coset* $f + I$ *in* $k[x_1, ..., x_n]/I$ *has the multiplicative inverse* $g + I$ *if and only if a Groebner basis G for the ideal* $J = (f_1, ..., f_s, f)$ *is* $\{1\}$*.*

PROOF. We note that a coset $f+I$ in $k[x_1, ..., x_n]/I$ has the multiplicative inverse $g+I$ if and only if $fg-1 \in I$ if and only if $1 \in (f_1, ..., f_s, f)$. □

This gives us an algorithm.

ALGORITHM 7. *(Sun T. Soh, 2000) To compute the multiplicative inverse of a coset* $f + I$ *in* $k[x_1, ..., x_n]/(f_1, ..., f_s)$*, we do the followings:*

(1) *Compute a Groebner basis G of the ideal* $(f_1, ..., f_s, f)$*.*
(2) *If* $G = \{1\}$*, then* $f + I$ *has the inverse* $g + I$*. To compute* $g + I$*, modify Program 2 appropriately.*
(3) *If* $G \neq \{1\}$*, then* $f+I$ *is not invertible in* $k[x_1, ..., x_n]/(f_1, ..., f_s)$*.* □

EXAMPLE 3.8. *We are going to use a modified version of Program 2 in order to compute the inverse of* $x + y + 1 + I$ *in* $\mathbb{Q}[x, y]/I$*, where* $I = (x^2y - y + x, xy^2 - x)$ *under gradlex with* $y > x$*. Here are modified lines of Program 2 for this purpose:*

—————————

PROGRAM 3. *(Sun T. Soh, 2000) (Program for computing multiplicative inverses in a quotient ring)*

```
off echo$
```

```
off nat$
load_package groebner$
%%%% Beginning of Inputs. %%%%%
torder({y,x}, gradlex)$ % You need to change this line.
f:=1$
write "f := ",f;
write "comment Is the given f a member of the ideal I generated by the following polynomials?";
% You need to change the following a few lines.
f1:=x^2*y-y+x$
f2:=x*y^2-x$
f3:=x+y+1$    % We want to compute an inverse of f3+I, where I is the ideal generated by f1 and f2.
write "f1 := ",f1;
write "f2 := ",f2;
write "f3 := ",f3;
write "comment For this, we first compute the reduced Groebner basis for the ideal I.";
gbt:=groebnert({given_f1=f1, given_f2=f2, given_f3=f3})$ % You need to change this line.
gb:=for k:=1:length(gbt) collect lhs(part(gbt,k))$
% gb:=groebner({f1,f2})$
%%%%% End of Inputs. You don't have to change any lines below this line. %%%%%
```

——————————

Here is the computed results:

```
gb := {1}$
poly1= - given_f1*x**2 - 2*given_f1*x + given_f1 -
given_f2*x**2 - given_f2*x + 3*given_f2 + given_f3*x**3*y
+ given_f3*x**2*y - 3*given_f3*x*y + given_f3*x + given_f3
```

——————————

Here, poly1=1. Thus, the multiplicative inverse of $x+y+1+I$ *is*

$$\begin{aligned} x^3y+x^2y-3xy+x+1+I &= xy-x^2+y-x-3xy+x+1+I \\ &= -x^2-2xy+y+1+I, \end{aligned}$$

where for the final equality, we have used the multiplicative table in Example 3.7.

To verify that this result is indeed correct, we use Program 2 again with

$$\begin{aligned} f &= (x+y+1)(x^3y+x^2y-3xy+x+1), \text{ or} \\ f &= (x+y+1)(-x^2-2xy+y+1), \end{aligned}$$

to see if f *is reduced to* 1 *with respect to the computed Groebner basis* G *for the ideal* I*, which can be easily checked using the REDUCE command preduce with respect to* G *(Program 3_1).*

Exercises for Section 3.4

EXERCISE 3.5. *Let* $I=(x^2y-y+x, xy^2-x)$*. Is* $x+y+2+I$ *invertible in* $\mathbb{Q}[x,y]/I$*?*

5. The Implicitization Problem

Around 180 B.C., Diocles wrote the book *On Burning-Glasses*, and one of the curves he considers was the **cissoid** . He described the cissoid using the following geometric construction using a ruler and compass:

(1) Given a line L_1, draw a circle C of radius a (which we will take as centered at the origin).
(2) Let A and B be the two points they meet, respectively. Then the magnitude, i.e., the distance, from the center of the circle C to the point A is a.
(3) Pick arbitrary point x between a and $-a$, and draw two vertical lines passing through x and $-x$, respectively. Let L_2 be the vertical line passing through the point, $-x$, so that its magnitude from A is always bigger than that from B.

(4) Let P be one of the two intersection points of the circle C and the vertical line L_2 passing through $-x$.

(5) Then draw another straight line L_3 passing through the point P and the point A.

(6) Let Q be the intersection point of the line L_3 and the vertical line at the point x.

(7) Repeat Steps $4-6$ using another intersection point P' of the circle C and the vertical line L_2 passing through $-x$.

Thus, when we move the point P (and P') on the circumference of the circle C, the point Q will move accordingly and this constructs a curve, and by definition, this locus of Q is called the **cissoid**. Thus, this curve cissoid is a parametrized curve. This is just one example of how the Greek knew how to draw plane curves, parametrically.

Let's introduce the XY-coordinates into the plane and let the center of the circle C be the origin of the coordinate. Then the equation for the line L_1 is just the X-axis, and the equation for the line L_3 is:

$$Y = \frac{-\sqrt{a^2 - x^2}}{a + x}(X - a).$$

Here, we use the famous Theorem of Pythagoras for the rectangular triangles. Now, let (x, y) denote the point Q. Then x and y certainly satisfy the equation:

$$y = \frac{-\sqrt{a^2 - x^2}}{a + x}(x - a).$$

By squaring both sides of this equality, we easily obtain the defining equation for the cissoid:

$$y^2(a + x) = (a - x)^3.$$

In other words, we have a system of *non-polynomial* functions in t,

$$\begin{cases} x &= & t, \\ y &= & \frac{-\sqrt{a^2-t^2}}{a+t}(t - a), \end{cases}$$

but by replacing t in the second equation with x, and squaring it, we obtain the above defining equation for the cissiod. In this business, we note that, by squaring both sides, we have gotten rid of the intermediate "geometric" parameter, the slope

$$\frac{-\sqrt{a^2 - t^2}}{a + t},$$

which is a function of the free variable, i.e., a parameter, t, and obtained an implicit equation for the point on the cissoid, in terms of the two coordinates, x and y.

DEFINITION 3.2. *Such a process is called the "**implicitization**" of a given* "parametrized" *geometric object, and the reverse process of implicitization is called the **parametrization** of the* "implicitized" *geometric object.*

EXAMPLE 3.9. *Here is the graph of the cissiod with* $a = 2$.

$$y^2(2 + x) = (2 - x)^3$$

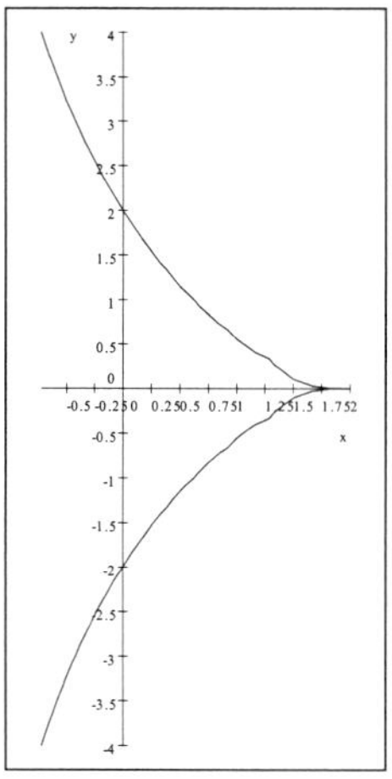

Historically, this curve has a very interesting aspect: If we draw another line L_4 which passes through the two points $(-a, 0)$ and $(0, a/2)$, then L_4 certainly hit a point (x, y) on the cissoid. Thus, if we further examine this point (x, y) by solving

$$\begin{cases} y^2(a + x) &= (a - x)^3, \\ y &= (x + a)/2, \end{cases}$$

then the second equations says $x + a = 2y$, so that the first equation becomes, $2y^3 = (a - x)^3$, i.e.,

$$2 = \left(\frac{a - x}{y}\right)^3,$$

where (x, y) is a point of the cissoid. This means the following: if we are allowed to use the cissoid, then we can construct the cube root of 2. In other words, the "*duplication of the cube*" is possible if we use a ruler and compass, as well as the plane curve, called the cissoid.

On the other hand, we recall that it is impossible to construct the cube root of 2 only using the ruler and compass, because there is no positive integer r such that

$$\left[\mathbb{Q}(\sqrt[3]{2}) : \mathbb{Q}\right] = 3 = 2^r,$$

where $\mathbb{Q}(\sqrt[3]{2})$ is the extension field of the field $\mathbb{Q}$ of rational numbers.

The curves of the Greeks are basically all parameterized curves, but we have seen that some of them are parametrized in terms of *non-polynomial* functions, like the cissoid.

Unlike the cissoid of Diocles, suppose that we are given a geometric object represented by a system of parametric expressions,

$$\begin{aligned} x_1 &= f_1(t_1, .., t_m), \\ x_2 &= f_2(t_1, .., t_m), \\ &\ldots.. \\ x_n &= f_n(t_1, .., t_m), \end{aligned}$$

where the f_i's are either *polynomial functions* or *rational functions* in the parameters $t_1, ..., t_m$. We then would like to obtain an implicit expression for the geometric object by getting rid of all the parameters t_i.

EXAMPLE 3.10. *Consider a space curve, i.e., a curve C in the 3-space, $\mathbb{R}^3$, parametrized by t:*

$$\begin{aligned} x &= t^4, \\ y &= t^3, \\ z &= t^2. \end{aligned}$$

Then we easily see that this curve is irreducible i.e. it is not a union of two curves, because it is a parametrized curve, and that the three coordinates, x, y, and z, satisfy the following system of polynomial equations:

$$\begin{cases} x - z^2 &= 0, \\ y^2 - z^3 &= 0. \end{cases}$$

Thus, any point on the curve C is in the intersection of two surfaces in $\mathbb{R}^3$ whose defining equations are

$$x = z^2,$$
$$y^2 = z^3,$$

respectively. But, it is still not clear whether the curve C fills up all the solutions of the system, that is, it is not clear whether the intersection is the smallest geometric object containing the curve C or not, and moreover, even if it is the case, we do not yet know how to argue on whether it is exactly the same as C or not.

Since this example is so simple, we have reached this much without any help or any difficulties. But, the next example really requires a strong computational tool to reach up to such a level of partial understandings, as we have seen in this example.

EXAMPLE 3.11. *Let's consider the twisted cubic C in $\mathbb{R}^3$, parametrically defined by:*

$$\begin{aligned} x &= t, \\ y &= t^2, \\ z &= t^3. \end{aligned}$$

It is obviously the intersection of two surfaces defined by $x^3 = z, y^3 = z^2$, *respectively. Since the twisted cubic is smooth at each point on the curve, there exists a surface* S *in* $\mathbb{R}^3$ *which is tangent to* C *at each point of the curve* C. *This surface* S *is called the tangent surface of the twisted cubic.*

Let (x, y, z) *be a point in* S, *and let* (x_0, y_0, z_0) *be point in* C. *Then we have an equation relating for the tangent line at* (x_0, y_0, z_0), *which passes through a point* (x, y, z) *of the tangent surface* S*:*

$$(x, y, z) = (x_0, y_0, z_0) + (x - x_0, y - y_0, z - z_0).$$

Then since the line segment $(x - x_0, y - y_0, z - z_0)$ *of the tangent line at the point* $(x_0, y_0 z_0) = (t, t^2, t^3)$ *is*

$$(x - x_0, y - y_0, z - z_0) = (1, 2t, 3t^2)u,$$

for some parameter u, *where*

$$(1, 2t, 3t^2) = (t', (t^2)', (t^3)')$$

is the slope of the tangent line at the point $(x_0, y_0 z_0) = (t, t^2, t^3)$, *we have a parametric representation of the tangent surface* S *of the twisted cubic* C*:*

$$\begin{aligned} x &= t + u, \\ y &= t^2 + 2tu, \\ z &= t^3 + 3t^2 u. \end{aligned}$$

We first note that the right sides of these equalities are polynomials in parameters t *and* u.

Now, the immediate question we can issue is: Is it ever possible to express the tangent surface S *in terms of the coordinates,* $x, y,$ *and* z *only? Is it moreover possible to figure out the smallest algebraic variety containing the tangent surface* S*?*

Here comes exactly the place for the application of Buchberger's algorithm for computing Groebner basis under the dictionary monomial ordering, or, the lex ordering, with $t > u > x > y > z$. Under this monomial ordering, we in fact have a system of polynomial equations.

We use a Computer Algebra System, REDUCE, to go further:[5]

———————

```
off echo$
off nat$
load_package groebner$
torder({t,u,x,y,z},lex)$
f1:=x - (t+u);
f2:=y - (t^2+2*t*u);
f3:=z - (t^3+3*t^2*u);
gb:=groebner({f1,f2,f3});
gb := {t + u - x,
u**2 - x**2 + y,
2*u*x**2 - 2*u*y - 2*x**3 + 3*x*y - z,
u*x*y - u*z - x**2*y - x*z + 2*y**2,
2*u*x*z - 2*u*y**2 + 2*x**2*z - x*y**2 - y*z,
2*u*y**3 - 2*u*z**2 - 4*x**2*y*z + x*y**3 - 2*x*z**2 + 5*y**2*z,
4*x**3*z - 3*x**2*y**2 - 6*x*y*z + 4*y**3 + z**2}$
```

———————

[5]A working program for REDUCE for this example is also available from: http://ch.catholic.or.kr/pundang/4/soh/2006_aa_i/Example_3_11.htm

Thus, if we examine the computed Groebner basis, gb, very carefully,

$$t + u - x,$$

$$u^2 - x^2 + y,$$

$$2ux^2 - 2uy - 2x^2 + 3xy - z,$$

$$uxy - uz - x^2y - xz + 2y^2,$$

$$2uxz - 2uy^2 + 2x^2z - xy^2 - yz,$$

$$2uy^3 - 2uz^2 - 4x^2yz + xy^3 - 2xz^2 + 5y^2z,$$

$$4x^3z - 3x^2y^2 - 6xyz + 4y^3 + z^2.$$

we just find that the last polynomial does not have any terms involving the parameters t and u. Therefore, the tangent surface S is also the solution set of the system of these polynomial equations, and hence it is contained in the surface S' defined by a single equation:

$$4x^3z - 3x^2y^2 - 6xyz + 4y^3 + z^2 = 0.$$

Although we do not yet know whether $S = S'$ or not, we would not be able to reach up to this much, unless Buchberger's algorithm for computing Groebner bases is available to us.

In a later Chapter, we will be able to conclude that $S = S'$, once we learn the so-called the Extension Theorem, but its proof requires the theory of resultants. (See Theorem 6.1 and Section 4 of Chapter 6).

Exercises for Section 3.5

EXERCISE 3.6. *You the same computation as in Example 3.11 for a space curve C in $\mathbb{R}^3$ defined by*

$$x = t^2, y = t^3, z = t^4.$$

CHAPTER 4

Theory of Invariants

Throughout the history of invariant theory, computational methods have always been at the center of attention.

1. Symmetric Polynomials

Let's start with a definition.

DEFINITION 4.1. *A polynomial $f = f(x_1, ..., x_n)$ in n variables $x_1, ..., x_n$ is called* ***symmetric*** *if*

$$f(x_{i_1}, ..., x_{i_n}) = f(x_1, ..., x_n)$$

for each permutation $x_{i_1}, ..., x_{i_n}$ of the n variables $x_1, ..., x_n$.

EXAMPLE 4.1. *Let $f = x^3 + ax^2 + bx + c$ be a polynomial over a computable field F, having 3 distinct roots $\alpha_1, \alpha_2, \alpha_3$ in its splitting field K. Then, by expanding the right side of the identity*

$$x^3 + ax^2 + bx + c = (x - \alpha_1)(x - \alpha_2)(x - \alpha_3),$$

we obviously see that

$$\begin{aligned} a &= -(\alpha_1 + \alpha_2 + \alpha_3), \\ b &= \alpha_1\alpha_2 + \alpha_1\alpha_3 + \alpha_2\alpha_3, \\ c &= -\alpha_1\alpha_2\alpha_3, \end{aligned}$$

and hence the coefficients a, b, c are the same under any permutation of the three roots. Moreover, the same argument can be applied to conclude that any polynomial having n distinct roots is a symmetric polynomial.

EXAMPLE 4.2. *The polynomial $x^2 + y^2 + z^2$ is obviously symmetric.*

DEFINITION 4.2. *Given n variables $x_1, ..., x_n$, we define $\sigma_1, ..., \sigma_n$ by the formula*

$$\begin{aligned}
\sigma_1 &= x_1 + x_2 + \cdots + x_n, \\
\sigma_2 &= x_1x_2 + x_1x_3 + \cdots + x_1x_n \\
&\quad +x_2x_3 + \cdots + x_2x_n \\
&\quad \ldots \\
&\quad +x_{n-1}x_n, \\
\sigma_3 &= x_1x_2x_3 + x_1x_2x_4 + \cdots + x_1x_2x_n \\
&\quad +x_2x_3x_4 + x_2x_3x_5 + \cdots + x_2x_3x_n \\
&\quad \ldots \\
&\quad +x_{n-2}x_{n-1}x_n \\
&\quad . \\
&\quad . \\
\sigma_n &= x_1x_2 \cdots x_n,
\end{aligned}$$

and they are called ***elementary symmetric functions****.*

That is, for each $r = 1, 2, ..., n$, σ_r is the sum of all monomials that are products of r distinct variables:

$$\sigma_i = \sum_{i_1<i_2<\cdots<i_r} x_{i_1}x_{i_2}\cdots x_{i_r}.$$

COROLLARY 4.1. *The σ_i are symmetric polynomials in the n variables $x_1, ..., x_n$.*

PROOF. Just expand the following polynomial f with respect to a new variable x in a non-increasing order to obtain them as coefficients $(-1)^i\sigma_i$ of f:

$$f = (x - x_1)(x - x_2)\cdots(x - x_n).$$

Thus, as we have seen in the above example, the $(-1)^i\sigma_i$ are all symmetric and so are the σ_i. □

The following example shows how we can get the symmetric functions σ_i using a Computer Algebra System.

EXAMPLE 4.3. *Let* $f = (x - x_1)(x - x_2)(x - x_3)(x - x_4)$. *Then the REDUCE command,*

$$coeffn(f, x, 4 - j),$$

will compute the σ_i, *for* $i = 1, 2, 3, 4$.[1]

```
off echo$
off nat$
f:=(x-x1)*(x-x2)*(x-x3)*(x-x4);
j:=1$
while j leq 4 do
begin
 write "sigma",j," := ",coeffn(f,x,4-j);
 j:=j+1$
end$
```

Here is the results:

```
sigma1:= - (x1 + x2 + x3 + x4)$
sigma2:=x1*x2 + x1*x3 + x1*x4 + x2*x3 + x2*x4 + x3*x4$
sigma3:= - (x1*x2*x3 + x1*x2*x4 + x1*x3*x4 + x2*x3*x4)$
sigma4:=x1*x2*x3*x4$
```

We then have an easy corollary:

COROLLARY 4.2. *A polynomial expression in these* σ_i *is a symmetric function.* □

The following theorem says that the converse of this corollary is also true, which is not quite obvious.

[1] A working program for REDUCE for this example is also available from: http://ch.catholic.or.kr/pundang/4/soh/2006_aa_i/Example_4_03.htm

THEOREM 4.1. *(The Fundamental Theorem of Symmetric Polynomials) Every symmetric polynomial $f \in R[x_1, ..., x_n]$ in n variables over a commutative ring R can be written* uniquely *as a polynomial in the elementary symmetric functions $\sigma_1, ..., \sigma_n$. That is, for each symmetric polynomial $f(x_1, ..., x_n)$, there exists a unique polynomial $\varphi(z_1, ..., z_n) \in R[z_1, ..., z_n]$ such that*

$$f(x_1, ..., x_n) = \varphi(\sigma_1, ..., \sigma_n).$$

THE FIRST PROOF OF THEOREM 4.1. If $n = 1$, there is nothing to prove. Suppose that the theorem is proved for $n - 1$ variables. Let $f = f(x_1, ..., x_n)$ be a given symmetric polynomial in $x_1, ..., x_n$. Let f_0 be the polynomial obtained by putting $x_n = 0$ in f:

$$f_0 = f(x_1, ...x_{n-1}, 0).$$

Then f_0 is a symmetric polynomial in $x_1, ..., x_{n-1}$, and hence by induction hypothesis, there exists a polynomial $g_1 = g_1(x_1, ..., x_{n-1})$ such that

$$f_0 = g_1(s_1, ..., s_{n-1}),$$

where the s_i are elementary symmetric functions in $x_1, ..., x_{n-1}$.

We then note that, for each $i = 1, 2, ..., n - 1$,

$$s_i = \sigma_i(x_1, ..., x_{n-1}, 0),$$

where the σ_i are elementary symmetric functions in $x_1, ..., x_n$.

Now we consider the following polynomial,

$$p(x_1, ..., x_n) = f(x_1, ..., x_n) - g_1(\sigma_1, ..., \sigma_{n-1}),$$

as a polynomial in $x_1, ..., x_n$. If p is a polynomial in $n - 1$ variables $x_1, ..., x_{n-1}$, then we are done by induction hypothesis. Otherwise, since it is a difference of two symmetric polynomials, p is again a symmetric polynomial and it satisfies

$$p(x_1, ..., x_{n-1}, 0) = 0.$$

Thus, each monomial of p is divisible by x_n, and hence by symmetry, each monomial of p is divisible by any x_j, $j = 1, 2, .., n$; consequently, p is divisible by σ_n. Thus, we can write

$$f(x_1, ..., x_n) = g_1(\sigma_1, ..., \sigma_{n-1}) + \sigma_n h_1(x_1, ..., x_n),$$

for some symmetric polynomial $h_1 = h_1(x_1, ..., x_n)$.

Now if h_1 does not have a term with x_n, then we are done by induction hypothesis. Otherwise, we apply the same argument to $h_1(x_1, ..., x_n)$ to have

$$h_1(x_1, ..., x_n) = g_2(\sigma_1, ..., \sigma_{n-1}) + \sigma_n h_2(x_1, ..., x_n),$$

so that

$$\begin{aligned} f(x_1, ..., x_n) &= g_1 + \sigma_n\{g_2 + \sigma_n h_2(x_1, ..., x_n)\} \\ &= g_1 + \sigma_n g_2 + \sigma_n^2 h_2(x_1, ..., x_n), \end{aligned}$$

and we repeat this process if necessary.

In this process, we note, at each step of this process, that either the degree of the last variable x_n is reduced by 1 or the last variable x_n does not appear in the intermediate polynomial h_i; hence, after finitely many steps, f is expressed as

$$f(x_1, ..., x_n) = \varphi(\sigma_1, ..., \sigma_n),$$

for some polynomial $\varphi(z_1, ..., z_n) \in R[z_1, ..., z_n]$.

Now, we have to prove that such a φ is uniquely determined. For this, we consider the kernel of the substitution homomorphism

$$\Phi\colon R[z_1, ..., z_n] \longrightarrow R[x_1, ..., x_n],$$

which sends each $z_i \mapsto \sigma_i$. If we can show that the kernel of Φ is zero, then we are done. So, we suppose that

$$\Phi(\varphi(z_1, ..., z_n)) = \varphi(\sigma_1, ..., \sigma_n) = 0$$

holds for some $\varphi(z_1, ..., z_n) \in R[z_1, ..., z_n]$. Since each σ_i is a polynomial in $x_1, ..., x_n$, by setting $x_n = 0$ in this expression, we still get zero:

$$\varphi(s_1, ..., s_{n-1}, 0) = 0.$$

Thus, we have $\varphi(z_1, .., z_{n-1}, 0) = 0$ by induction hypothesis on n. It follows that z_n divides $\varphi(z_1, ..., z_n)$, so we write

$$\varphi(z_1, ..., z_n) = z_n \psi(z_1, ..., z_n),$$

for some $\psi(z_1, ..., z_n) \in R[z_1, ..., z_n]$. Thus, we have

$$0 = \varphi(\sigma_1, ..., \sigma_n) = \sigma_n \psi(\sigma_1, ..., \sigma_n) = x_1 \cdots x_n \psi(\sigma_1, ..., \sigma_n).$$

We note that the monomial $x_1 \cdots x_n$ is not a zero-divisor in the polynomial ring $R[z_1, ..., z_n]$, and thus, from this expression, we conclude that

$$\psi(\sigma_1, ..., \sigma_n) = 0.$$

Then since $\psi(z_1, ..., z_n)$ has lower total degree in $z_1, ..., z_n$ than $\varphi(z_1, ..., z_n)$, we thus conclude by induction on the total degree that $\psi(z_1, ..., z_n) = 0$. Therefore, $\varphi(z_1, ..., z_n) = 0$. □

REMARK 4.1. *The problem with this proof is that we do not know how to construct the intermediate polynomials g_i, and this proof is not very useful in actually expressing f as a polynomial in the n elementary symmetric functions σ_i.*

THE SECOND PROOF OF THEOREM 4.1. We use the dictionary order with $x_1 > x_2 > ... > x_n$. We first prove an existence of such an expression. Given a symmetric polynomial $f(x_1, ..., x_n)$, we consider first its leading term

$$LT(f) = ax^\alpha,$$

where α is the exponent vector $(\alpha_1, ..., \alpha_n)$ of x^α. If there exists j such that $\alpha_j < \alpha_{j+1}$, then we let

$$\beta = (..., \alpha_{j+1}, \alpha_j, ...)$$

be the exponent vector obtained from α by switching α_j and α_{j+1}. Then since ax^α is a term of f, ax^β is the term of

$$f(..., x_{j+1}, x_j, ...).$$

Thus, since f is symmetric,

$$f(..., x_{j+1}, x_j, ...) = f,$$

and hence it follows that ax^β is a term of f. But, this is impossible because the monomial term ax^α is the leading term of f under the lex order and $x^\beta > x^\alpha$. Therefore, the exponent $\alpha = (\alpha_1, ..., \alpha_n)$ of the leading term $LT(f) = ax^\alpha$ should satisfy

$$\alpha_1 \geq \alpha_2 \geq \cdots \geq \alpha_n.$$

Having said this, we now consider a symmetric polynomial:

$$h = \sigma_1^{\alpha_1-\alpha_2} \cdot \sigma_2^{\alpha_2-\alpha_3} \cdots \sigma_{n-1}^{\alpha_{n-1}-\alpha_n} \cdot \sigma_n^{\alpha_n}.$$

Then

$$\begin{aligned} (4.1) \qquad LM(h) &= LM(\sigma_1^{\alpha_1-\alpha_2} \cdot \sigma_2^{\alpha_2-\alpha_3} \cdots \sigma_{n-1}^{\alpha_{n-1}-\alpha_n} \cdot \sigma_n^{\alpha_n}) \\ &= LM(\sigma_1)^{\alpha_1-\alpha_2} \cdot LM(\sigma_2)^{\alpha_2-\alpha_3} \cdots LM(\sigma_n)^{\alpha_n} \\ &= x_1^{\alpha_1-\alpha_2}(x_1x_2)^{\alpha_2-\alpha_3} \cdots (x_1 \cdots x_n)^{\alpha_n} \\ &= x_1^{\alpha_1}x_2^{\alpha_2} \cdots x_n^{\alpha_n} \\ &= x^\alpha. \end{aligned}$$

Thus, we have

$$LM(f) > LM(f - ah),$$

whenever $f - ah \neq 0$. If $f = ah$, then we are done. Otherwise, $f_1 = f - ah$ is again a symmetric polynomial, so that we can apply the same argument to f_1 to obtain $f_2 = f_1 - a_1h_1$, where a_1 is a constant and h_2 is a polynomial in σ_i's and we have

$$LM(f) > LM(f_1) > LM(f_2).$$

Continuing this way, we obtain a sequence of symmetric polynomials

$$f, f_1, f_2, \dots$$

such that their leading monomials $LM(f_i)$ satisfies a strictly decreasing sequence

$$LM(f) > LM(f_1) > LM(f_2) > \dots.$$

But, since the lex ordering is a well-ordering, the sequence must be finite, and hence $f_{q+1} = 0$ for some q. Thus, putting these results together, we obtain

$$f = ah + a_1h_1 + \cdots a_qh_q,$$

showing that f is a polynomial in the n elementary symmetric functions σ_i.

To finish the proof, we need to show that such an expression is unique. Suppose that we have two such expressions for the same symmetric polynomial f:

$$\begin{aligned} f &= g_1(\sigma_1, \dots, \sigma_n), \\ f &= g_2(\sigma_1, \dots, \sigma_n), \end{aligned}$$

where $g_1 = g_1(z_1, \dots, z_n)$ and $g_2 = g_2(z_1, \dots, z_n)$ are two polynomials in $R[z_1, \dots, z_n]$. Thus, we have

$$g_1(\sigma_1, \dots \sigma_n) - g_2(\sigma_1, \dots, \sigma_n) = 0.$$

But, if

$$g = g(z_1, \dots, z_n) = g_1 - g_2 \not\equiv 0$$

as polynomials, then we write

$$g = \sum_\beta a_\beta z_1^{\beta_1} \cdots z_n^{\beta_n}, \tag{4.2}$$

in a unique way, i.e., the $\beta = (\beta_1, \dots, \beta_n)$'s are distinct, and hence

$$g(\sigma_1, \dots, \sigma_n) = \sum_\beta a_\beta \sigma_1^{\beta_1} \cdots \sigma_n^{\beta_n}, \tag{4.3}$$

where each σ_i is an elementary symmetric functions so that it is a polynomial in $x_1, ..., x_n$. Only for simplicity, we write

$$g_\beta = a_\beta \sigma_1^{\beta_1} \cdots \sigma_n^{\beta_n},$$

for each β, and consider their leading monomials $LM(g_\beta)$. First of all, we have, for each β,

$$\begin{aligned} LM(g_\beta) &= LM(\sigma_1^{\beta_1} \cdots \sigma_n^{\beta_n}) \\ &= LM(\sigma_1)^{\beta_1} \cdots LM(\sigma_n)^{\beta_n} \\ &= x_1^{\beta_1} \cdot (x_1 x_2)^{\beta_2} \cdots (x_1 x_2 \cdots x_n)^{\beta_n} \\ &= x_1^{\beta_1 + \cdots + \beta_n} x_2^{\beta_2 + \cdots + \beta_n} \cdots x_n^{\beta_n}. \end{aligned}$$

Then since the mapping sending

$$(\beta_1, ..., \beta_n) \mapsto (\beta_1 + \cdots + \beta_n, \beta_2 + \cdots + \beta_n, ..., \beta_n)$$

is obviously injective, the distinct exponents of (4.2) are sent to distinct exponents of (4.3), and hence if we choose the largest one, $LM(g_\beta)$, among these leading monomials, this one will not be cancelled out by other leading terms of (4.3), and consequently, $g(\sigma_1, ..., \sigma_n) \neq 0$, which is a contradiction. □

REMARK 4.2. *We note that the second proof is due to Gauss (1777 - 1855) , who needed the properties of symmetric polynomials for his second proof of his very famous proof (dated 1816) of the Fundamental Theorem of Algebra, Theorem 1.3. In his proof, he also mentioned that he uses the dictionary ordering explicitly, and it is known as the earliest explicit statement of the lex order.*

REMARK 4.3. *We also note that the above proof provides us with an algorithm for expressing a given symmetric polynomial in terms of the elementary symmetric functions $\sigma_1, ..., \sigma_n$.*

EXAMPLE 4.4. *Obviously, when* $\sigma_1 = x_1 + x_2 + x_3, \sigma_2 = x_1x_2 + x_1x_3 + x_2x_3, \sigma_3 = x_1x_2x_3$,

$$\begin{aligned} \sigma_1^2 - \sigma_1\sigma_3 &= -x_1^2x_2x_3 + x_1^2 - x_1x_2^2x_3 - x_1x_2x_3^2 \\ &\quad +2x_1x_2 + 2x_1x_3 + x_2^2 + 2x_2x_3 + x_3^2, \end{aligned}$$

is a symmetric polynomial. We want to apply the algorithm described in the above proof as follows:

Write

$$f = -x_1^2x_2x_3 + x_1^2 - x_1x_2^2x_3 - x_1x_2x_3^2 + 2x_1x_2 + 2x_1x_3 + x_2^2 + 2x_2x_3 + x_3^2.$$

Since $LM(f) = x_1^2x_2x_3 = LM(\sigma_1\sigma_3)$, *we have*

$$\begin{aligned} f_1 &= f + \sigma_1\sigma_3 \\ &= x_1^2 + 2x_1x_2 + 2x_1x_3 + x_2^2 + 2x_2x_3 + x_3^2. \end{aligned}$$

Then since $LM(f_1) = LM(\sigma_1^2)$, *we have*

$$\begin{aligned} f_2 &= f_1 - \sigma_1^2 \\ &= 0. \end{aligned}$$

Thus,

$$f = f_1 - \sigma_1\sigma_3 = \sigma_1^2 - \sigma_1\sigma_3.$$

REMARK 4.4. *The problem with this algorithm based on Gauss's proof of Theorem 4.1 is that, we do not know whether the given polynomial* f *is symmetric or not yet.*

PROPOSITION 4.1. *Let* $G = \{g_1, ..., g_t\}$ *be a Groebner basis for the ideal* $(\sigma_1 - y_1, ..., \sigma_n - y_n)$ *of the polynomial ring* $k[x_1, ..., x_n, y_1, ..., y_n]$ *under a monomial order satisfying that any monomial involving one of the* x_i*'s is bigger than all monomials in* $k[y_1, ..., y_n]$ *(Such a monomial order is called an* ***elimination order*** *(Definition 5.2 in Chapter 5), where the* σ_i*'s are the elementary symmetric functions in the* x_i*'s. Given* $f \in k[x_1, ..., x_n]$, *let* $g = \bar{f}^G$ *be the remainder of* f *on division by* G. *That is,*

$$f \xrightarrow{G}_+ g.$$

Then

(1) *f is symmetric if and only if $g \in k[y_1, ..., y_n]$.*
(2) *If f is symmetric, then $f = g(\sigma_1, ..., \sigma_n)$ is the unique polynomial expression of f as a polynomial in the elementary symmetric polynomials $\sigma_1, ..., \sigma_n$.*

One of such monomial ordering is the dictionary ordering, i.e., the lex order on $k[x_1, ..., x_n, y_1, ..., y_n]$ with

$$x_1 > x_2 > ... > x_n > y_1 > ... > y_n.$$

PROOF. Part (2) follows immediately from part (1).

Part (1). By Division Algorithm (Algorithm 5), we have

$$f = a_1 g_1 + \cdots + a_t g_t + g$$

for some non-zero polynomials $a_i \in k[x_1, ..., x_n, y_1, ..., y_n]$. We know that the ideal $I = (\sigma_1 - y_1, ..., \sigma_n - y_n)$ is also generated by G, so that each

$$g_i = h_1(\sigma_1 - y_1) + \cdots + h_n(\sigma_n - y_n),$$

for some $h_i \in k[x_1, ..., x_n, y_1, ..., y_n]$. Now, we suppose that the remainder $g \in k[y_1, ..., y_n]$. If we substitute each y_i with σ_i in both sides of this expression, each g_i becomes zero, but the $f \in k[x_1, ..., x_n]$ in the left side is the same, because there is no place for substitution. Thus,

$$f = g(\sigma_1, ..., \sigma_n).$$

Conversely, suppose that $f \in k[x_1, ..., x_n]$ is symmetric. By Proposition 4.1, we know that there exists $g \in k[y_1, ..., y_n]$ such that

$$f = g(\sigma_1, ..., \sigma_n).$$

We want to show that g is the unique remainder of f on division by G. For this, we go as follows: Since f is a polynomial in the σ_i's, it is a k-linear combination

$$f = \sum_{\alpha} k_\alpha \sigma_1^{\alpha_1} \sigma_2^{\alpha_2} \cdots \sigma_n^{\alpha_n}$$

of the various monomials

$$\sigma_1^{\alpha_1}\sigma_2^{\alpha_2}\cdots\sigma_n^{\alpha_n}.$$

Now we need to rewrite each of these monomials

$$\begin{aligned}\sigma_1^{\alpha_1}\sigma_2^{\alpha_2}\cdots\sigma_n^{\alpha_n} &= (y_1+(\sigma_1-y_1))^{\alpha_1}\cdots(y_n+(\sigma_n-y_n))^{\alpha_n}\\ &= y_1^{\alpha_1}\cdots y_n^{\alpha_n} + \text{ an element } h_\alpha \text{ in } (\sigma_1-y_1,...,\sigma_n-y_n).\end{aligned}$$

Thus,

$$\begin{aligned} f &= \sum_\alpha k_\alpha y_1^{\alpha_1}\cdots y_n^{\alpha_n} + \sum_\alpha k_\alpha h_\alpha \\ &= g(y_1,...,y_n) + \sum_\alpha k_\alpha h_\alpha.\end{aligned}$$

Then since $\sum_\alpha k_\alpha h_\alpha \in (\sigma_1-y_1,...,\sigma_n-y_n)$, we can express this in terms of the elements of $G=\{g_1,...,g_t\}$ with the zero remainder, and thus if we show that none of the terms of $g(y_1,...,y_n)$ is divisible by any $LT(g_i)$ then this $g(y_1,...,y_n)$ is indeed the unique remainder g (in the statement of the theorem we want to prove) of f on division by G, by Theorem 2.4.

Now, suppose that there exists $g_i \in G$ such that $LT(g_i)$ divides some term of g. Then $LT(g_i)$ involves only $y_1,...,y_n$ and hence $g_i \in k[y_1,...,y_n]$ because the variables $x_i > y_j$ for all i,j under our monomial order. Then since $g_i \in (\sigma_1-y_1,...,\sigma_n-y_n)$, $g_i(y_1,...,y_n)$ is sent to $g_i(\sigma_1,...,\sigma_n)=0$, and then by the uniqueness of Theorem 4.1, we see that $g_i(y_1,...,y_n)$ is the zero polynomial. But, this is impossible because $g_i \in G$. Therefore, no term of $g(y_1,...,g_n)$ is divisible by any $LT(g_i)$, and hence it is the unique remainder of f on division by G which is an element of $k[y_1,...,y_n]$, and this finishes our proof. □

REMARK 4.5. *In application of Proposition 4.1 we need to compute a Groebner basis for the ideal $(\sigma_1-y_1,...,\sigma_n-y_n)$ under a given monomial order described in the proposition. But, if we use the dictionary order with*

$$x_1 > x_2 > ... > x_n > y_1 > ... > y_n,$$

it turns out that it is quite easy to get a Groebner basis for

$$(\sigma_1 - y_1, ..., \sigma_n - y_n).$$

Thus, the following proposition is very useful for this purpose when n gets big.

Moreover, when n is quite small, say, $n \leq 5$, it is very easy to compute a Groebner basis under the dictionary ordering, and we will have a few computed examples in the next section, Discriminant.

DEFINITION 4.3. *Given variables $u_1, ..., u_k$, we write*

$$h_i(u_1, ..., u_n) = \sum_{|\alpha|=i} u^\alpha$$

to be the sum of all monomials of total degree i in $u_1, ..., u_n$, and we call it a ***complete symmetric polynomial****.*

Thus, $h_i = \sigma_i +$ the rest, where σ_i is the elementary symmetric function in $u_1, ..., u_n$. We then have the following result whose proof we decide not to present because it would take too long.

PROPOSITION 4.2. *Fix the lex order on $k[x_1, ..., x_n, y_1, ..., y_n]$ with $x_1 > x_2 > ... > x_n > y_1 > ... > y_n$. Then the polynomials*

$$g_k = h_k(x_k, ..., x_n) + \sum_{i=1}^{k} (-1)^i h_{k-i}(x_k, ..., x_n) y_i, \quad k = 1, ..., n,$$

where $y_0 = 1$ forms a Groebner basis for the ideal $(\sigma_1 - y_1, ..., \sigma_n - y_n)$. □

REMARK 4.6. *But, this proposition again requires us to generate various complete "symmetric polynomials", $h_k(x_k, ..., x_n)$, and this is another headache even when n is quite small, say, $n = 20$. In the next section, we present our own algorithms for computing the Groebner basis for the ideal $(\sigma_1 - y_1, ..., \sigma_n - y_n)$. There, you will find that three different methods for computing the Groebner basis for the ideal $(\sigma_1 - y_1, ..., \sigma_n - y_n)$ are discussed, and among these three, the last one called "the direct method" is the most efficient method.*

LEMMA 4.1. *For any natural number n, one has*

$$0 = \sum_{i=0}^{k} (-1)^i h_{k-i}(x_1, ..., x_n)\sigma_i(x_1, ..., x_n), \qquad 1 \leq k \leq n,$$

where $\sigma_0 = 1$ and $\sigma_i = 0$ if $i > n$ by definition.

It has two computationally interesting corollaries, which are easily deduced from the formula.

COROLLARY 4.3. *For any k such that $1 \leq k \leq n$, replacing the list $x_1, ..., x_n$ with $x_k, ..., x_n$ in the above formula, one can compute all the*

$$h_j(x_k, ..., x_n), 0 \leq j \leq k,$$

even in a recursive manner from the $\sigma_i(x_k, ..., x_n)$'s.

This is quite easy since the number n of unknowns in Lemma 4.1 is arbitrary, one is allowed to count the actual number of variables, say, not from 1 to $n - k$ but from k to n.

Its second corollary is the recursive formula for complete symmetric polynomials h_j in terms of elementary symmetric ones σ_i, or vice versa:

COROLLARY 4.4. *For any natural number n, one has*

$$h_k = \sum_{j=1}^{k-1} (-1)^j \sigma_{j+1} h_{k-1-j}, \qquad 1 \leq k \leq n,$$

where $h_0 = 1$ by definition.

Thus, it completely generates the unique polynomial expression for each complete symmetric polynomial in terms of elementary symmetric polynomials. We actually discovered the same recursive formula as the above one in a quite different situation, that is, in our own works on graded partition functions over a multiset.

Exercises for Section 4.1

EXERCISE 4.1. *Write the following polynomial as a polynomial in elementary symmetric functions,*

$$(x^2 + y^2)(x^2 + z^2)(y^2 + z^2).$$

2. Algorithms for Symmetric Polynomials

In this section, we present our own algorithms and/or programs for generating elementary symmetric functions, complete symmetric polynomials, which in turn give us an algorithm for the Groebner basis for the ideal $(\sigma_1 - y_1, ..., \sigma_n - y_n)$. These algorithms and/or programs belong to our unpublished works.

Hardware and software limitations for computed examples in this section are: the Computer Algebra System, REDUCE, under Windows 3.1 on my Personal Computer with Pentium II running at 200 MHz with 128 Mbytes of main memory.

2.1. An Algorithm for Generating Elementary Symmetric Functions.

ALGORITHM 8. *(Sun T. Soh, 1999)*

PROGRAM 4.

```
% Computed elementary symmetric functions are in the output files,
lj, where j=1,...,r.
r:=5$ %4$  % This is the only input line you need to change.
off echo$
off nat$
linelength=72$
array d(r),w(r), v(r)$
a:=0$
for i:=1:r do <<a:=a+mkid(v,i)>>$
name:=mkid(l,1)$
out "writeto.out"$ write "out ",name; write ";end$";
```

```
shut "writeto.out"$
in "writeto.out"$ write "a := ",a; write ";end$";
out "writeto.out"$ write "shut ",name; write ";end$";
shut "writeto.out"$
in "writeto.out"$
for i:=1:r do <<d(i):=mkid(v,i)>>$
n:=1$
while n<=r do
<< if n-2*(fix(n/2))=1 then
    << u:=0$a:=0$
for i:=r-n step -1 until 1 do <<u:=u+d(i+1)$d(i+1):=0$
w(i):=mkid(v,i)*u$ a:=a+w(i) >>$
    >> else
    << u:=0$a:=0$
for i:=r-n step -1 until 1 do <<u:=u+w(i+1)$w(i+1):=0$
d(i):=mkid(v,i)*u$ a:=a+d(i) >>$
    >>$
name:=mkid(l,n+1)$
out "writeto.out"$ write "out ",name; write ";end$";
shut "writeto.out"$
in "writeto.out"$ write "a := ",a; write ";end$";
out "writeto.out"$ write "shut ",name; write ";end$";
shut "writeto.out"$
in "writeto.out"$
out t$ write "Next step := ",n+1;
n:=n+1$
>>$
end$
```

EXAMPLE 4.5. *Thanks to this algorithm, we can generate the elementary symmetric functions easily. The following table reports how long it takes in computing them for various number* r *of unknowns* $x_1, ..., x_r$ *which had been obtained more than several years ago using the Computer Algebra System, REDUCE, under Microsoft Windows 3.1 in my PC running at 200 MHz with 128 Mbytes.*

r	*Computation Time*	*Hard Disk Space (Mbytes)*
16	*13 Seconds*	*2.0*
17	*28*	*4.2*
18	*1 Minute*	*9.0*
19	*2*	*19.2*
20	*4*	*40.5*
21	*8*	*85.3*
22	*20*	*179.2*
23	*Failed after computing up to level 9*	
24	*8*	
25	*8*	

In particular, the following table reports the Hard Disk space consumption when $r = 16$.

Level #	*HD Space in Bytes*
L1	*99*
L2	*1,105*
L3	*7,105*
L4	*29,511*
L5	*86,271*
L6	*186,459*
L7	*306,879*
L8	*390,751*
L9	*387,833*
L10	*299,807*
L11	*178,989*
L12	*81,025*
L13	*26,917*
L14	*6,199*
L15	*891*
L16	*67*
L17	*14*

17 file(s) 1,989,922 bytes
0 dir(s) 622,428,160 bytes free

REMARK 4.7. *The algorithm in Algorithm 8 with undetermined n inputs x_i computes the elementary symmetric polynomials with time efficiency of* $O(2^n)$.

2.2. An Algorithm for Complete Symmetric Polynomials. The algorithm in Algorithm 8 is very good since its minor modification gives an algorithm for complete symmetric polynomials.

ALGORITHM 9. *(Sun T. Soh, 1999)*

PROGRAM 5.

```
% Computed complete symmetric polynomials are in the output files
h_pj, where j=1,..,r.
r:=5$   % This is the only input line you need to change.
array z(r+1),d(r+1),w(r+1)$
off echo$ off nat$ linelength=72$
z(0):=1$
a:=0$ for i:=1:r do <<a:=a+mkid(x,i)$>>$
z(1):=a$
for i:=1:r do <<d(i):=mkid(x,i)>>$
n:=1$
while n<=r-1 do
<< if n-2*(fix(n/2))=1 then
<< u:=0$a:=0$
for i:=r step -1 until 1 do <<u:=u+d(i)$d(i):=0$w(i):=mkid(x,i)*u$
a:=a+w(i) >>$
>>
else
<< u:=0$a:=0$
for i:=r step -1 until 1 do <<u:=u+w(i)$w(i):=0$d(i):=mkid(x,i)*u$
a:=a+d(i) >>$
>>$
z(n+1):=a$
n:=n+1$
>>$
clear d,w$
t:=1$
while t<=r do
<< out name$ write "x",t-1,":=0"; write ";end$"; shut name$ in
name$
name:=mkid(h_p,t)$ out "writeto.out"$ write "out ",name; write ";end$";
shut "writeto.out"$
in "writeto.out"$
```

```
for i:=0:r do <<write "h",t,i," := ",z(i);>>$
write ";end$";
out "writeto.out"$ write "shut ",name; write ";end$";
shut "writeto.out"$
in "writeto.out"$
clear name$
t:=t+1$
out t$ write "next t := ",t;
>>$
for j:=1:r do << out name$ write "clear x",j; write ";end$";
shut name$ in name$ >>$
% in "chk_h.srt"$
;end$
```

REMARK 4.8. *It directly computes the complete symmetric polynomials with time efficiency of* $O(2^n)$.

2.3. An Algorithm for the Reduced Groebner Basis for the Ideal $(\sigma_1 - y_1, \sigma_2 - y_2, ..., \sigma_n - y_n)$**.** Using a modified version of the algorithm in Algorithm 8, one can compute such a basis very easily. Since complete symmetric polynomials of the form $h_i(x_j, ..., x_n)$ for $j \leq i$, for each i with $1 \leq i \leq n$ are only needed, a slight modification of Algorithm 9 should be a solution.

ALGORITHM 10. *(Sun T. Soh, 1999)*

(1) Inputs are a natural number n (and the unknowns $x_1, ..., x_n$).
(2) Set $i = 1$. Compute the complete symmetric polynomials

$$S_i = \{h_1(x_1, ..., x_n), h_2(x_2, ..., x_n), ..., h_i(x_i, ..., x_n), ..., h_n(x_n)\},$$

recursively using the algorithm in Algorithm 9.

(3) Set $i = i + 1$. While $j \leq i$, substitute $x_j = 0$ to each elements of S_{i-1} for each step i, and obtain the complete symmetric polynomials

$$S_i = \{h_1(x_i, ..., x_n), h_2(x_{i+1}, ..., x_n), ..., h_{n-1}(x_{i+(n-1)})\}.$$

Repeat this step until $i \leq n$.

(4) Apply the formula in Proposition 4.2 to obtain the unique reduced Groebner basis for the ideal $(\sigma_1 - y_1, \sigma_2 - y_2, ..., \sigma_n - y_n)$ from the set $S = S_1 \cup \cdots \cup S_n$.

In the above algorithm, the only step consuming most time and memory for computation is the third one. The rest of the steps should be finished without any trouble at all. Here is the actual implementation of this algorithm.

PROGRAM 6. *(Sun T. Soh, 1999)*

```
% The computed Groebner basis is in the output file: grbn.out.
r:=15$    % This is the only input line you need to change.
          % But do not try r bigger than 20
          % unless your computer has enough Hard Disk space.
array z(r+1),d(r+1),w(r+1)$
off echo$ off nat$ linelength=72$
z(0):=1$
a:=0$
for i:=1:r do <<a:=a+mkid(x,i)$>>$
z(1):=a$
for i:=1:r do <<d(i):=mkid(x,i)>>$
n:=1$
while n<=r-1 do
<< if n-2*(fix(n/2))=1 then
<< u:=0$a:=0$
for i:=r step -1 until n do <<u:=u+d(i)$
d(i):=0$ w(i):=mkid(x,i)*u$ a:=a+w(i) >>$
```

```
>>
else
<< u:=0$a:=0$
for i:=r step -1 until n do <<u:=u+w(i)$
w(i):=0$d(i):=mkid(x,i)*u$ a:=a+d(i) >>$
>>$
z(n+1):=a$
n:=n+1$
>>$
clear w$
out name$ write "y",0,":=1"; write ";end$"; shut name$ in name$
out "grbn.out"$
write "comment This is the reduced Groebner basis for the
ideal (s_1-y_1, ..., s_n-y_n)."$
t:=1$
while t<=r do
<< out name$ write "x",t-1,":=0"; write ";end$"; shut name$ in
name$
name:=mkid(h_p,t)$ out "writeto.out"$ write "out ",name; write ";end$";
shut "writeto.out"$ in "writeto.out"$
for i:=0:t do <<write "h",t,i," := ",z(i);>>$ write ";end$";
out "writeto.out"$ write "shut ",name; write ";end$";
shut "writeto.out"$ in "writeto.out"$
clear name$
d(t):=for i:=0:t sum (-1)^i*z(t-i)*mkid(y,i)$ out "grbn.out"$ write
"g",t," := ",d(t);
t:=t+1$
out t$
write "next t := ",t;
>>$
out "grbn.out"$ write ";end$"; shut "grbn.out"$
```

```
for j:=1:r do << out name$ write "clear x",j; write ";end$"; shut
name$ in name$ >>$
;end$
```

2.4. A Few Remarks on Our Algorithms. Here we are going to discuss three different ways of computing the reduced Groebner basis for the ideal $(\sigma_1 - y_1, \sigma_2 - y_2, ..., \sigma_n - y_n)$ under lex order. Hardware and software limitations for computed examples in this section are: the Computer Algebra System, REDUCE, under Windows 3.1 on my Personal Computer with Pentium II running at 200 MHz with 128 Mbytes of main memory.

We can compute the reduced Groebner basis for the ideal $(\sigma_1 - y_1, \sigma_2 - y_2, ..., \sigma_n - y_n)$ under lex order utilizing Buchberger's algorithm, *provided that* the elementary symmetric polynomials $\sigma_j(x_1, ..., x_n)$ for various j are already at one's hand. We call this way of computation the **first indirect method**.

On the other hand, we can compute the complete symmetric polynomials $h_j(x_k, ..., x_n)$ for various k and j using the formula given in Lemma 4.1, particularly utilizing the recursive formula in its second corollary, *provided that* the elementary symmetric polynomials

$$\sigma_j(x_k, ..., x_n)$$

for various k and j are already known to us; consequently one can compute the reduced Groebner basis for the ideal

$$(\sigma_1 - y_1, \sigma_2 - y_2, ..., \sigma_n - y_n)$$

under lex order using Proposition 4.2. This way of computation is called the **second indirect method**.

The unavoidable problem with these two indirect methods is quite obvious. That is, whichever method is used to compute first the elementary symmetric polynomials for various number of unknowns, it additionally requires a lot more computing time for,

(1) either computing the reduced Groebner basis for the ideal

$$(\sigma_1 - y_1, \sigma_2 - y_2, ..., \sigma_n - y_n)$$

using Buchberger's algorithm for it, or

(2) computing the needed complete symmetric polynomials

$$h_j(x_k, ..., x_n)$$

for various k and j from them; then obtaining them from the formula in Proposition 4.2.

On the other hand, Algorithm 10 or Program 6 only requires us to directly compute the complete symmetric polynomials $h_j(x_j, ..., x_n)$ for $1 \leq j \leq n$, using a modified algorithm of the one in Algorithm 9, i.e., not from the elementary symmetric ones, and the rest of the steps are just for simpleminded substitutions of $x_k = 0$ sequentially. This way of computation is called the **direct method**.

EXERCISE 4.2. *Verify that the direct method among them is the most efficient one.*

When the input n is very small, any of these three different methods will compute the Groebner basis for the ideal $(\sigma_1 - y_1, ..., \sigma_n - y_n)$. But, when the input n gets big, the direct method announced above should perform more efficiently than the other two methods both in space and time, and perhaps it may be the only one properly working up to the upper limit of hardware and software capacities.

EXAMPLE 4.6. *It turned out, even when n is* 15, *that the direct method required about one Minute to finish the job, but the other two indirect methods failed.*

EXAMPLE 4.7. *When n =* 20, *the* ***direct method*** *of Algorithm 10 required about* 37 *Minutes to compute the reduced Groebner basis for the ideal*

$$(\sigma_1 - y_1, \sigma_2 - y_2, ..., \sigma_n - y_n).$$

Here are more detailed informations on computed data.

File Names	*HD sizes in bytes*		*Time*
GRBN.SRT	*2,246*	*99-08-26*	*11:50*
H_P1	*147*	*99-08-26*	*11:51*
H_P2	*1,930*	*99-08-26*	*11:51*
H_P3	*16,550*	*99-08-26*	*11:51*
H_P4	*94,094*	*99-08-26*	*11:51*
H_P5	*384,025*	*99-08-26*	*11:51*
H_P6	*1,184,843*	*99-08-26*	*11:51*
H_P7	*2,856,348*	*99-08-26*	*11:52*
H_P8	*5,497,265*	*99-08-26*	*11:53*
H_P9	*8,565,808*	*99-08-26*	*11:54*
H_P10	*10,899,694*	*99-08-26*	*11:57*
H_P11	*11,184,982*	*99-08-26*	*12:01*
H_P12	*9,396,106*	*99-08-26*	*12:11*
H_P13	*6,436,178*	*99-08-26*	*12:22*
H_P14	*3,564,919*	*99-08-26*	*12:25*
H_P15	*1,574,776*	*99-08-26*	*12:26*
H_P16	*543,095*	*99-08-26*	*12:26*
H_P17	*141,522*	*99-08-26*	*12:26*
H_P18	*26,499*	*99-08-26*	*12:27*
H_P19	*3,374*	*99-08-26*	*12:27*
H_P20	*425*	*99-08-26*	*12:27*
GRBN.OUT	*65,609,704*	*99-08-26*	*12:27*

In the above table, GRBN.OUT contains the reduced Groebner basis for the ideal under consideration. It requires more than 65.6 *Mbytes of Hard Disk space. The file, GRBN.SRT, is the program file of which output file is GRBN.OUT. The files H_Pk, for various k, contains the complete symmetric polynomials* $h_j(x_k, ..., x_n)$ *for* $j \leq k$*, for each* k *with* $1 \leq k \leq n$*. They are definitely needed since one uses the formula in Proposition 4.2.*

On the other hand, the two indirect methods simply failed to compute it.

Exercises for Section 4.2

EXERCISE 4.3. *When $n = 15$, compute the reduced Groebner basis for the ideal $(\sigma_1 - y_1, \sigma_2 - y_2, ..., \sigma_n - y_n)$ under an elimination order where the variables x are bigger than the variables y, first using Program 6, and then using any other methods of your own. If your own program turns out to be more efficient than Program 6, report your result to us immediately.*

3. The Discriminants

This section is a small application of the theory we have studied in Section 1 of this chapter to the Galois Theory on Galois extension fields of a given field and solvability question of a polynomial equation in one variable by radical expressions. Like the dimension of a vector space V over a field k, the discriminant D defined below is in fact an example of numerical invariants associated to a polynomial $f(x)$ over a field, whose computation generally requires the knowledge we have learned in Section 1 of this chapter.

In the next section, we will extend our point of view on symmetric functions σ_i by naturally connecting Geometry and Algebra in a contravariant functorial manner, not axiomatically but a very computable way.

We start with a definition

DEFINITION 4.4. *The* ***discriminant*** *D of a polynomial*

$$f(x) = (x - \alpha_1)(x - \alpha_2) \cdots (x - \alpha_n)$$

with distinct α_i is the product

$$D = \prod_{i<j}(\alpha_i - \alpha_j)^2 = (-1)^{n(n-1)/2} \prod_{i \neq j}(\alpha_i - \alpha_j).$$

Thus, the discriminant D of $f(x)$ is clearly invariant under permutations on the set of α_i's; and consequently, we can compute them using those results developed so far in this chapter.

REMARK 4.9. *It was a Jewish-English mathematician, J.J. Sylvester (1814 - 1897), who first discovered in 1851 the discriminant of a cubic equation and he first used the name "discriminant" for such expressions of quadratic equations and those of higher order. He published important papers in 1852 and 1853, namely* "On the principle of the calculus of forms" *and* "On the theory of syzygetic relations and two rational integer functions". *In particular he used matrix theory to study higher dimensional geometry. In 1877 Sylvester accepted a chair at The Johns Hopkins University and he founded in 1878 the American Journal of Mathematics, the first mathematical journal in the United States.*

We recall that the discriminant D of an irreducible polynomial $f(x)$ over a field k plays an important role in determining the Galois group of the splitting field of $f(x) \in k[x]$. In particular, the following result is crucial:

PROPOSITION 4.3. *Let K/F be a Galois extension which is the splitting field of an irreducible polynomial $f(x) \in F[x]$ of degree n. Let $\alpha_1, ...\alpha_n$ be the roots of $f(x) = 0$ in K. Then $\delta = \delta(\alpha) = \sqrt{D}$, where D is the discriminant of $f(x)$, is a non-zero element of K; moreover we have*

- **a:** *$\delta \in F$ if and only if the Galois group $G = G(K/F)$ is a subgroup of the alternating group A_n.*
- **b:** *In any case, the subgroup $G(K/F(\delta))$ of G is contained in A_n.*

REMARK 4.10. *In exact computation, or equivalently, symbolic computation, or equivalently, algebraic computation, all the symbols are considered as unknowns, unless an algebraic relation is assigned to these unknowns. Thus, computation of the discriminant of $f(x)$ is a very natural one we can perform with a Computer Algebra System.*

3.1. Quadratic Equations. If $f(x) = x^2 + bx + c \in \mathbb{Q}[x]$ is 2, then the discriminant D of $f(x)$ is

$$D = b^2 - 4c,$$

and everybody knows that D is a non-negative number if and only if it has two non-complex roots (including multiplicity of the roots), and we do not need a Computer Algebra System.

3.2. Cubic Equations. Let

$$f(x) = x^3 + a_2x^2 + a_1x + a_0$$

be an irreducible cubic polynomial over over a field F. By *completing the cube*, it is rewritten as

$$\begin{aligned} f(x) &= \left(x + \frac{a_2}{3}\right)^3 - a_2x^2 - \frac{a_2^2}{3}x - \frac{a_2^3}{27} + a_2x^2 + a_1x + a_0 \\ &= \left(x + \frac{a_2}{3}\right)^3 + (a_1 - \frac{a_2^2}{3})x + (a_0 - \frac{a_2^3}{27}). \end{aligned}$$

Thus, by replacing $x + \frac{a_2}{3}$ with a new variable x' , we may initially assume that our irreducible polynomial $f(x)$ over the field $k = F$ of characteristic $\neq 2$ or 3, is of the form:

$$f(x) = x^3 + px + q,$$

where $p, q \in F$. Equivalently, $g(x) = f(x - \frac{a_2}{3})$ is an irreducible polynomial over F. Thus if $\alpha_1, \alpha_2, \alpha_3$ are the three roots of $f(x) = 0$, then $\beta_i = \alpha_i + \frac{a_2}{3}, i = 1, 2, 3$ are clearly the three roots of $g(x) = 0$, and consequently the discriminant D of $f(x)$ and that of $g(x)$ are the same because D is defined as the product of the squares of the differences of the roots.

REMARK 4.11. *We note that our linear replacement of $x + \frac{a_2}{3}$ by a new variable x' does not change irreducibility of $f(x)$, because such a linear replacement has an obvious inverse linear replacement.*

Let K be a splitting field of $f(x)$ over F, and let $\alpha_1, \alpha_2, \alpha_3$ be the three roots of $f(x) = 0$ in K, so that we have an equality between two polynomials:

$$x^3 + px + q = (x - \alpha_1)(x - \alpha_2)(x - \alpha_3).$$

Thus, expanding the right side of this equality, we obtain three relations:

$$\begin{aligned} 0 &= \alpha_1 + \alpha_2 + \alpha_3; \\ p &= \alpha_1\alpha_2 + \alpha_2\alpha_3 + \alpha_1\alpha_3; \\ q &= -\alpha_1\alpha_2\alpha_3, \end{aligned}$$

where $p, q \in F$. Then

COROLLARY 4.5. $D = -4p^3 - 27q^2 \in F$.

PROOF. Using the above three relations, we have for $i \neq j$,

$$\begin{aligned} (\alpha_i - \alpha_j)^2 &= (\alpha_i + \alpha_j)^2 - 4\alpha_i\alpha_j \\ &= (-\alpha_k)^2 - 4\{p - \alpha_k(\alpha_i + \alpha_j)\} \\ &= \alpha_k^2 - 4\{p - \alpha_k(-\alpha_k)\} \\ &= -3\alpha_k^2 - 4p, \end{aligned}$$

where $k \neq i, j$. Thus,

$$\begin{aligned} D &= \prod_{i<j}(\alpha_i - \alpha_j)^2 \\ &= (\alpha_1 - \alpha_2)^2(\alpha_1 - \alpha_3)^2(\alpha_2 - \alpha_3)^2 \\ &= -(3\alpha_3^2 + 4p)(3\alpha_2^2 + 4p)(3\alpha_1^2 + 4p) \\ &= -\{27(\alpha_1\alpha_2\alpha_3)^2 + 36p(\alpha_1^2\alpha_2^2 + \alpha_1^2\alpha_3^2 + \alpha_2^2\alpha_3^2) \\ &\quad +48p^2(\alpha_1^2 + \alpha_2^2 + \alpha_3^2) + 64p^3\} \\ &= -\{27q^2 + 36p(\alpha_1\alpha_2 + \alpha_1\alpha_3 + \alpha_2\alpha_3)^2 - 2\alpha_1\alpha_2\alpha_3(\alpha_1 + \alpha_2 + \alpha_3) \\ &\quad +48p^2\{(\alpha_1 + \alpha_2 + \alpha_3)^2 - 2(\alpha_1\alpha_2 + \alpha_1\alpha_3 + \alpha_2\alpha_3) + 64p^3\} \\ &= -\{27q^2 + 36p^3 - 96p^3 + 64p^3\} \\ &= -4p^3 - 27q^2 \in F. \end{aligned}$$

□

COROLLARY 4.6. *The splitting field K of a cubic irreducible polynomial $f(x)$ over a field F is a Galois extension of degree either* 3 *or* 6.

(1) *If* $[K : F] = 3$, *then its Galois group is* A_3; *and hence no proper intermediate subfields of* K *which properly contains* F.

(2) *If* $[K : F] = 6$, *then its Galois group is* S_3; *and it has three intermediate subfields* $F(\alpha_1), F(\alpha_2), F(\alpha_3)$ *of degree* 3 *and one intermediate subfield* L *of degree* 2 *generated by a square root of the discriminant* D *of* $f(x)$:

$$\delta = (\alpha_1 - \alpha_2)(\alpha_1 - \alpha_3)(\alpha_2 - \alpha_3),$$

where the α_i *are the three roots in* K *of* $f(x) = 0$. □

COROLLARY 4.7. *The discriminant* D *of an irreducible* cubic *polynomial* $f(x) \in F[x]$ *is a square in* F *if and only if the degree of the splitting field is* 3.

PROOF. If D is not a square, then since $\delta^2 = D$, $\delta \notin F$; hence $[F(\delta) : F] = 2$. Then since $\delta \in K$, the splitting field of $f(x)$ over F, $[K : F]$ is divisible by 2, and hence by Corollary 4.6, $[K : F] = 6$. Conversely, if $\delta \in F$, then every element of the Galois group $G = G(K/F)$ is required to fix δ, but odd permutations change the sign of δ; hence they can not be an element of G. Thus, G is a proper non-trivial subgroup of S_3. Then again by Corollary 4.6, $[K : F] = 3$. □

EXAMPLE 4.8. *Let* $f(x) = x^3+3x+1 \in \mathbb{Q}[x]$. *Since neither* $(x+1)$ *nor* $(x-1)$ *are a factor of* $f(x)$, *it is irreducible over* $\mathbb{Z}$; *hence it is irreducible over* $\mathbb{Q}$. *Let* K *be the splitting field of* $f(x)$ *over* $\mathbb{Q}$. *We want to determine* $[K : \mathbb{Q}]$.

(The "first" method) It has only one real root α *because its derivative* $f'(x) = 3x^2+3 > 0$ *for any real* $x \in \mathbb{R}$. *Thus, other two roots are complex numbers; so* $\mathbb{Q}(\alpha)$ *is a proper subfield of the splitting field* K *of* $f(x)$ *over* $\mathbb{Q}$. *Thus, we must have* $[K : \mathbb{Q}] = 6$ *by Corollary 4.6.*

(The "second" method) Or, the discriminant D *of* $f(x)$ *is:*

$$\begin{aligned} D &= -4p^3 - 27q^2 = -4 \cdot 3^3 - 27 \cdot 1 \\ &= -135, \end{aligned}$$

and hence its square root δ is not a rational number. Thus, by Corollary 4.7, $[K : F] = 6$.

Thus by Corollary 4.6, $K/\mathbb{Q}$ is a Galois extension with the Galois group $G = S_3$ and K contains a unique intermediate field $L = \mathbb{Q}(\delta)$ of degree 2 , where δ is a square root of D.

EXAMPLE 4.9. *In this example, we are first going to construct an irreducible polynomial $f(x)$ over $\mathbb{Q}$ of degree 3, using a 9th root of unity. We note that 9 is not a prime number.*

Let $\omega = e^{2\pi i/9} = \cos(2\pi/9) + i\sin(2\pi/9)$, a complex root of $x^9 - 1 = 0$, and let

$$\begin{aligned}\alpha_1 &= \omega + \omega^{-1} = \omega + \omega^8 = 2\cos(2\pi/9),\\ \alpha_2 &= \omega^2 + \omega^{-2} = 2\cos(4\pi/9),\\ \alpha_3 &= \omega^4 + \omega^{-4} = 2\cos(8\pi/9).\end{aligned}$$

They are all real numbers. In what follows, we are going to show that they are algebraic over $\mathbb{Q}$.

We obviously have a relation between α_1 and α_2:

$$\alpha_1^2 - 2 = \alpha_2.$$

Define:

$$f(x) = (x - \alpha_1)(x - \alpha_2)(x - \alpha_3).$$

Expansion of $f(x)$ gives us,

$$f(x) = x^3 - (\alpha_1 + \alpha_2 + \alpha_3)x^2 + (\alpha_1\alpha_2 + \alpha_1\alpha_3 + \alpha_2\alpha_3)x - \alpha_1\alpha_2\alpha_3.$$

Then using Computer Algebra System(CAS) we may check that

$$\begin{aligned}\alpha_1 + \alpha_2 + \alpha_3 &= 0,\\ \alpha_1\alpha_2 + \alpha_1\alpha_3 + \alpha_2\alpha_3 &= -3,\\ \alpha_1\alpha_2\alpha_3 &= -1.\end{aligned}$$

For instance, in order to quickly derive these relations among the roots of $f(x) = 0$, we use MAPLE and the following MAPLE Commands:

*a1:=2*cos(2*Pi/9); a2:=2*cos(2*Pi*2/9); a3:=2*cos(2*Pi*4/9);*
*simplify(a1+a2+a3); simplify(a1*a2+a1*a3+a2*a3);*
This should establish the first two relations hold. And then
simplify(a1(a1^2-2)*(-a1-(a1^2-2)));*
should establish the third relation, where we use the already-verified relations: $\alpha_2 = \alpha_1^2 - 2$ *and* $\alpha_3 = -\alpha_1 - \alpha_2$.

In this way, we have explicitly constructed an integral polynomial

$$f(x) = x^3 - 3x + 1.$$

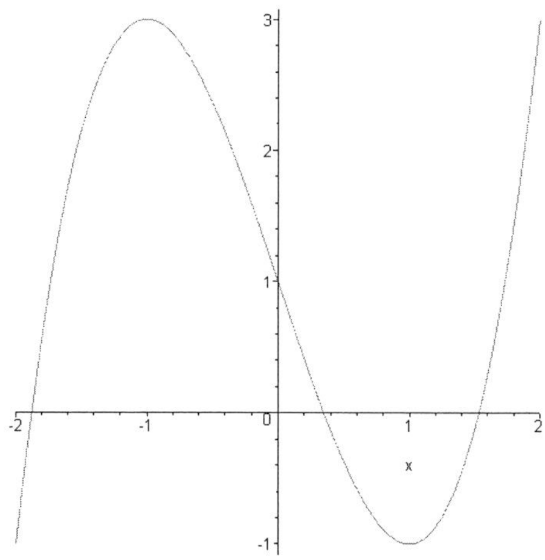

Now, since neither $(x+1)$ *nor* $(x-1)$ *is a factor of* $f(x)$, *it is irreducible over* $\mathbb{Z}$; *hence it is irreducible over* $\mathbb{Q}$. *Let* K *be the splitting field of* $f(x)$ *over* $\mathbb{Q}$. *We want to determine* $[K : \mathbb{Q}]$.

By solving $f'(x) = 0$ *for* x, *we see that* $x = -1$ *is a point where the local maximum of* $f(x)$ *occurs and* $x = 1$ *is a point where the local minimum of* $f(x)$ *occurs. Thus,* $f(x) = 0$ *has three real solutions. But, in this case, it is not easy to directly figure out these three real roots of* $f(x)$, *so that we can not try the "first" method. But, the "second" method says that*

$$\begin{aligned} D &= -4p^3 - 27q^2 \\ &= -4 \cdot (-3)^3 - 27 \cdot 1 \\ &= 81 \\ &= 9^2. \end{aligned}$$

Thus, by Corollary 4.7, we conclude that $[K : \mathbb{Q}] = 3$*, and the Galois group* $G = G(K/\mathbb{Q}) = A_3$*. Then since we have relations:*

$$\begin{aligned} \alpha_2 &= \alpha_1^2 - 2, \\ \alpha_3 &= -\alpha_1 - \alpha_2, \end{aligned}$$

the field $\mathbb{Q}(\alpha_1)$ *contains* α_2, α_3*, and consequently* $K = \mathbb{Q}(\alpha_1)$*, the splitting field of degree* 3 *of the irreducible polynomial* $f(x) = x^3 - 3x + 1$ *over* $\mathbb{Q}$*. Thus,* K *is a subfield of* $\mathbb{R}$*. Then since* 3 *is a prime number, there is no proper and non-trivial intermediate field between* $\mathbb{Q}$ *and* K*.*

3.3. Quartic Equations.

REMARK 4.12. *Unlike irreducible cubic equations, computation of the discriminant of an irreducible quartic equation by hand is quite messy. But, such a computational difficulty can be easily overcome if we apply the theory of Groebner basis.*

In what follows, we are going to see the usefulness of the Theory of Groebner Basis by explicitly computing both the discriminant of an *irreducible* quartic polynomial

$$f(x) = x^4 - \sigma_1 x^3 + \sigma_2 x^2 - \sigma_3 x + \sigma_4$$

over F, where the σ_i are the symmetric functions of the four distinct roots of $f(x) = 0$, and the associated *resolvent cubic*

$$g(x) = (x - (\alpha_1\alpha_3 + \alpha_2\alpha_4))(x - (\alpha_1\alpha_2 + \alpha_3\alpha_4))(x - (\alpha_1\alpha_4 + \alpha_2\alpha_3)).$$

of $f(x)$. Then we have a table which tells us what we need to compute to conclude the Galois group of $f(x)$:

(4.4)

	D a square in F	D not a square in F
$g(x)$ irreducible over F	$G = A_4$	$G = S_4$
$g(x)$ reducible over F	$G = V$	$G = D_4$ or C_4

3.3.1. *Computation of Discriminants.* Using Computer Algebra System REDUCE, we first compute the Groebner basis for the ideal generated by

$$\sigma_1 - y_1, \sigma_2 - y_2, \sigma_3 - y_3, \sigma_4 - y_4.$$

---

*result := {x1 + x2 + x3 + x4 - y1, x2**2 + x2*x3 + x2*x4 - x2*y1 + x3**2 + x3*x4 - x3*y1 + x4**2 - x4*y1 + y2,*

*x3**3 + x3**2*x4 - x3**2*y1 + x3*x4**2 - x3*x4*y1 + x3*y2 + x4**3 - x4**2*y1 + x4*y2 - y3,*

*x4**4 - x4**3*y1 + x4**2*y2 - x4*y3 + y4}$*

REMARK 4.13. *In Section 2, we have discussed how we can compute the Groebner basis for the ideal efficiently. The direct method introduced there is the most efficient one among the three methods discussed.*

Here, $\sigma_1 =$ *x1+x2+x3+x4,*
$\sigma_2 =$ *x1*x2+x1*x3+x1*x4+x2*x3+x2*x4+x3*x4,*
$\sigma_3 =$ *x1*x2*x3+x1*x2*x4+x2*x3*x4, and* $\sigma_4 =$ *x1*x2*x3*x4*
are the four elementary symmetric functions in the four roots x1, x2, x3, x4 of $f(x) = 0$. Then we compute:

```
result3:=preduce((x1-x2)^2*(x1-x3)^2*(x1-x4)^2*(x2-x3)^2
*(x2-x4)^2*(x3-x4)^2,
result,{x1,x2,x3,x4,y1,y2,y3,y4});
result3 := - 27*y1**4*y4**2 + 18*y1**3*y2*y3*y4 - 4*y1**3*y3**3
- 4*y1**2*y2**3*y4 + y1**2*y2**2*y3**2 + 144*y1**2*y2*y4**2
- 6*y1**2*y3**2*y4 - 80*y1*y2**2*y3*y4 + 18*y1*y2*y3**3
```

```
- 192*y1*y3*y4**2 + 16*y2**4*y4 - 4*y2**3*y3**2
- 128*y2**2*y4**2 + 144*y2*y3**2*y4 - 27*y3**4 + 256*y4**3$
```

This shows that the discriminant

$$D = (x1 - x2)^2(x1 - x3)^2(x1 - x4)^2(x2 - x3)^2(x2 - x4)^2(x3 - x4)^2$$

is indeed a symmetric polynomial by the first part of Proposition 4.1. Moreover, by the second part of Proposition 4.1, we see that

$$\begin{aligned} (4.5) \quad D &= -27\sigma_1^4\sigma_4^2 + 18\sigma_1^3\sigma_2\sigma_3\sigma_4 - 4\sigma_1^3\sigma_3^3 - 4\sigma_1^2\sigma_2^3\sigma_4 + \sigma_1^2\sigma_2^2\sigma_3^2 \\ &\quad +144\sigma_1^2\sigma_2\sigma_4^2 - 6\sigma_1^2\sigma_3^2\sigma_4 - 80\sigma_1\sigma_2^2\sigma_3\sigma_4 + 18\sigma_1\sigma_2\sigma_3^3 \\ &\quad -192\sigma_1\sigma_3\sigma_4^2 + 16\sigma_2^4\sigma_4 - 4\sigma_2^3\sigma_3^2 - 128\sigma_2^2\sigma_4^2 \\ &\quad +144\sigma_2\sigma_3^2\sigma_4 - 27\sigma_3^4 + 256\sigma_4^3 \end{aligned}$$

is the discriminant of an irreducible polynomial over F:

$$f(x) = x^4 - \sigma_1 x^3 + \sigma_2 x^2 - \sigma_3 x + \sigma_4.$$

EXAMPLE 4.10. *The discriminant D of an irreducible polynomial,*

$$f(x) = x^4 - (-r)x + s$$

over $\mathbb{Q}$ is (with $\sigma_1 = 0, \sigma_2 = 0$)

$$\begin{aligned} D &= -27\sigma_3^4 + 256\sigma_4^3 \\ &= -27(-r)^4 + 256s^3. \end{aligned}$$

Then,

```
result4:=(x-(x1*x3+x2*x4))*(x-(x1*x2+x3*x4))
 *(x-(x1*x4+x2*x3));
result4 := x**3 - x**2*x1*x2 - x**2*x1*x3 - x**2*x1*x4
 - x**2*x2*x3 - x**2*x2*x4 - x**2*x3*x4 + x*x1**2*x2*x3
 + x*x1**2*x2*x4 + x*x1**2*x3*x4 + x*x1*x2**2*x3
 +x*x1*x2**2*x4 + x*x1*x2*x3**2 + x*x1*x2*x4**2
```

```
+ x*x1*x3**2*x4 + x*x1*x3*x4**2 + x*x2**2*x3*x4
+ x*x2*x3**2*x4 + x*x2*x3*x4**2 - x1**3*x2*x3*x4
- x1**2*x2**2*x3**2 - x1**2*x2**2*x4**2
- x1**2*x3**2*x4**2 - x1*x2**3*x3*x4
- x1*x2*x3**3*x4 - x1*x2*x3*x4**3 - x2**2*x3**2*x4**2$
```

After rewriting result4 in x, we verify that each coefficient of rsult4 is a symmetric function:

```
result5:=preduce(x1*x2+x1*x3+x1*x4+x2*x3+x2*x4+x3*x4,
result,{x1,x2,x3,x4,y1,y2,y3,y4});
result5 := y2$
result6:=preduce(x1**2*x2*x3+x1**2*x2*x4+x1**2*x3*x4
+x1*x2**2*x3 +x1*x2**2*x4+x1*x2*x3**2+x1*x2*x4**2
+x1*x3**2*x4+x1*x3*x4**2 +x2**2*x3*x4+x2*x3**2*x4
+x2*x3*x4**2,result,{x1,x2,x3,x4,y1,y2,y3,y4});
result6 := y1*y3 - 4*y4$
result:=preduce(x1**3*x2*x3*x4 + x1**2*x2**2*x3**2
+ x1**2*x2**2*x4**2 + x1**2*x3**2*x4**2
+ x1*x2**3*x3*x4 + x1*x2*x3**3*x4 + x1*x2*x3*x4**3
+ x2**2*x3**2*x4**2,result,{x1,x2,x3,x4,y1,y2,y3,y4});
result := y1**2*y4 - 4*y2*y4 + y3**2$
```

Thus, the associated resolvent cubic $g(x)$ of an irreducible polynomial $f(x)$ over F is

$$g(x) = x^3 - \sigma_2 x^2 + (\sigma_1\sigma_3 - 4\sigma_4)x - (\sigma_1^2\sigma_4 - 4\sigma_2\sigma_4 + \sigma_3^2) \in F[x]. \tag{4.6}$$

In what follows, $p = \sigma_1, q = \sigma_2, r = \sigma_3$, and $s = \sigma_4$.

EXAMPLE 4.11. *Thus if* $\sigma_1 = 0$, $\sigma_2 = 0$, *then*

$$\begin{aligned} g(x) &= x^3 - 4\sigma_4 x - \sigma_3^2 \\ &= x^3 - 4sx - (-r)^2. \end{aligned}$$

EXAMPLE 4.12. *If $p = 0, q = 0, r = 8$ and $s = 12$, then by these explicitly computed results, the discriminant D of the irreducible polynomial*

$$f(x) = x^4 + 8x + 12$$

over $\mathbb{Q}$ is

$$\begin{aligned} D &= -27(-8)^4 + 256 \cdot (12)^3 \\ &= 331776 = 2^{12} \cdot 3^4, \end{aligned}$$

which is a square. Moreover, the associated resolvent cubic polynomial $g(x)$ of $f(x)$ is

$$\begin{aligned} g(x) &= x^3 - (4 \cdot 12)x - (-8)^2 \\ &= x^3 - 48x - 64, \end{aligned}$$

and we can check $g(x)$ is irreducible over $\mathbb{Q}$. Therefore, the Galois group $G = G(K/\mathbb{Q})$ of the splitting field K of the irreducible polynomial $f(x)$ over $\mathbb{Q}$ is A_4 by the table in (4.4).

EXAMPLE 4.13. *If $p = 0, q = 0, r = -4$ and $s = 1$, then by these explicitly computed results, the discriminant D of the irreducible polynomial*

$$f(x) = x^4 - 4x + 1$$

over $\mathbb{Q}$ is

$$D = -6656,$$

so D is not a square in $\mathbb{Q}$. Moreover by (4.6), the associated resolvent cubic of $f(x)$ is

$$g(x) = x^3 - 4x - 16,$$

and $g(x)$ is irreducible over $\mathbb{Q}$. Thus, the Galois group $G = G(K/\mathbb{Q})$ of the splitting field K of $f(x)$ over $\mathbb{Q}$ is S_4, by the table in (4.4).

EXAMPLE 4.14. *If $f(x) = x^4 + 6x^3 + 3x^2 + 3x + 3$, then $f(x)$ is irreducible over $\mathbb{Q}$, for instance, by Einstein, and by (4.5) we see that*

$$D = -160191,$$

and hence D is not a square in $\mathbb{Q}$. Moreover by (4.6), the associated resolvent cubic of $f(x)$ is

$$g(x) = x^3 - 3x^2 + 6x - 81$$

is irreducible over $\mathbb{Q}$, because $g(x)(\bmod 2) = x^3+x^2+1$ is irreducible over the field $F_2 = \mathbb{Z}/2\mathbb{Z}$. Thus, the Galois group $G = G(K/\mathbb{Q})$ of the splitting field K of $f(x)$ over $\mathbb{Q}$ is S_4, by the table in (4.4).

3.4. Quintic Equations. The main motivation behind Galois' work was the problem of solving fifth-degree polynomial equations in one variable over a field $\mathbb{Q}$. Here is a brief history of the solution of equations by radicals. The solution of the quadratic was known to the Hindus and in its geometric form to the Greeks. The cubic and quartic were solved by the Renaissance Italian mathematicians Scipio del Ferro (1515) and Ferrari (1545). However, not until the 19th century did Abel and Galois prove the impossibility of solving all polynomial equations of degree $n \geq 5$.

A short time earlier than Galois, Abel proved that the quintic equation

$$x^5 + a_4x^4 + a_3x^3 + a_2x^2 + a_1x + a_0 \tag{4.7}$$

with variable rational coefficients a_i could not be solved in terms of radicals, but it remained to find an explicit polynomial in one variable with rational coefficients which could not be solved by radicals. Because the problem was over 200 years old before Abel and Galois, interest in it continued. It also turned out that Galois' original ideal have turned out to be much more important than the questions which motivated them.

We start with a definition:

DEFINITION 4.5. *Let F be a subfield of the field $\mathbb{C}$ of complex numbers. We say that a complex number α is expressible by* **radicals** *over F if there exists an increasing chain of subfields*

$$F = F_0 \subset F_1 \subset \cdots \subset F_r$$

of $\mathbb{C}$such that

(1) *$\alpha \in F_r$, and*

(2) *for each* $j = 1, ..., r$, F_j *is generated over* F_{j-1} *by a radical* β_j. *That is,* $F_j = F_{j-1}(\beta_j)$ *where* $\beta_j^{n_j} \in F_{j-1}$ *for some positive integer* n_j.

We assume that F is a subfield of the field $\mathbb{C}$ of complex numbers.

THEOREM 4.2. *The roots of a quintic irreducible polynomial* $f(x)$ *over* F *whose Galois group is either* S_5 *or* A_5 *can not be expressed by radicals over* F. □

COROLLARY 4.8. *Let* $f(x)$ *be an irreducible quintic polynomial over* $\mathbb{Q}$ *with exactly three* real *roots,* $\alpha_1, \alpha_2,$ *and* α_3. *Then its Galois group is the symmetric group, and hence its roots can not be expressed by radicals.*
□

LEMMA 4.2. *There are infinitely many quintic irreducible polynomials over* $\mathbb{Q}$ *whose roots are not expressed by radicals.*

PROOF. In order first to have a polynomial with exactly three real roots, we first consider

$$f(x) = x^5 - p^4 x = x(x^2 - p^2)(x^2 + p^2),$$

where p is a prime integer. This polynomial clearly has three rational integer roots, $0, \pm p$. Then the first derivative test will allow us to add a small positive constant a satisfying

$$0 < a < p^4/\sqrt[4]{5} \simeq 0.668\,74p^4$$

to $f(x)$ (as a bias or an offset) in such a way that the resulting polynomial $f(x) + a$ will still have three distinct *real* roots.

Since we are also looking for the final polynomial to be irreducible polynomial over $\mathbb{Q}$, as well as having three distinct real roots, taking $a = p$ will always work for this purpose. That is, the polynomial

$$f(x) + p = x^5 - p^4 x + p$$

is irreducible over $\mathbb{Q}$ by Eisenstein Criterion (Proposition 1.1) and has three distinct *real* roots. □

EXAMPLE 4.15. *For instance, we may choose $p = 2$ to obtain an irreducible polynomial over $\mathbb{Q}$:*

$$x^5 - 16x + 2.$$

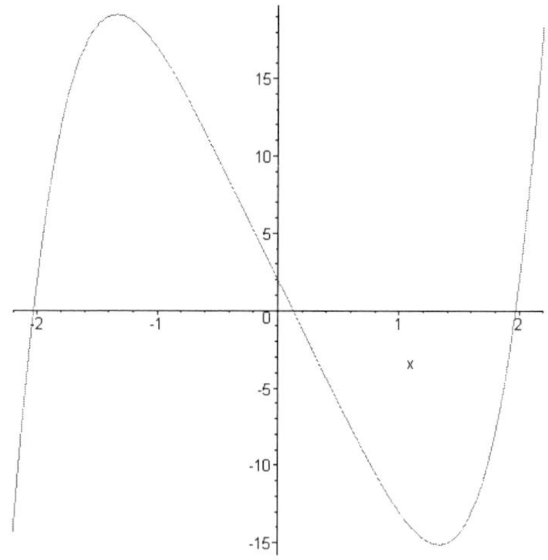

Then, by Corollary 4.8, we know that its roots can not be expressed by radicals over $\mathbb{Q}$.

EXAMPLE 4.16. *The integral polynomial*

$$f(x) = 2x^5 - 10x + 5$$

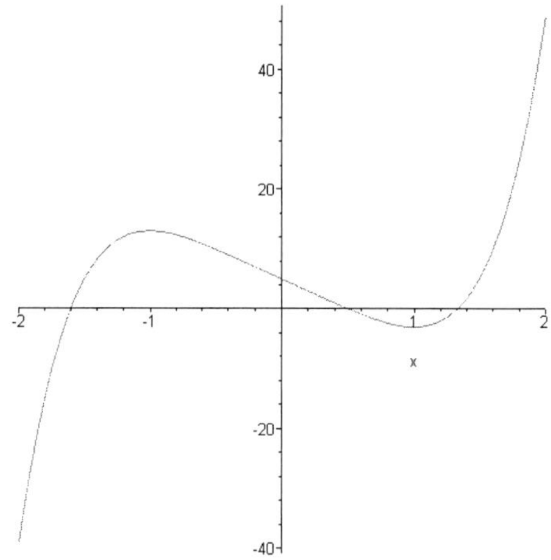

is irreducible over $\mathbb{Q}$ *by Eisenstein Criterion, and we can easily check that it has three distinct* real *roots by the first derivative test. Thus, its roots can not be expressed by radicals over* $\mathbb{Q}$.

3.4.1. *Computation of Discriminants.* In the case of irreducible polynomial equations $f(x) = 0$ over $\mathbb{Q}$ of degree ≤ 4, we have so far seen that the discriminant D of $f(x)$ was very useful in determining the Galois group of $f(x)$. Thus in what follows, we first derive the formula for discriminant D of an irreducible quintic polynomial equation $f(x) = 0$ over $\mathbb{Q}$, and try to use it in determining its Galois group. But, we already needed the resolvent cubic for $f(x)$ when $f(x)$ is an irreducible quartic polynomial over $\mathbb{Q}$, so we may expect that the discriminant alone will not determine the Galois group when $n \geq 4$, and this may require us to figure out all the transitive subgroups of S_n in advance.

Using Computer Algebra System, for instance, REDUCE, we can compute the groebner basis for the ideal generated by

$$\sigma_1 - y_1, \sigma_2 - y_2, \sigma_3 - y_3, \sigma_4 - y_4, \sigma_5 - y_5$$

according to Proposition 4.2 or Algorithm 10:

```
result := {x1 + x2 + x3 + x4 + x5 - y1,
 x2**2 + x2*x3 + x2*x4 + x2*x5 - x2*y1 + x3**2 + x3*x4 + x3*x5 -
x3*y1 + x4**2 + x4*x5 - x4*y1 +x5**2 - x5*y1 + y2,
 x3**3 + x3**2*x4 + x3**2*x5 - x3**2*y1 + x3*x4**2 + x3*x4*x5 -
x3*x4*y1 + x3*x5**2 - x3*x5*y1 + x3*y2 + x4**3 + x4**2*x5 - x4**2*y1
+ x4*x5**2 - x4*x5*y1 + x4*y2 + x5**3 - x5**2*y1 + x5*y2 -y3,
 x4**4 + x4**3*x5 - x4**3*y1 + x4**2*x5**2 - x4**2*x5*y1
 + x4**2*y2 + x4*x5**3 - x4*x5**2*y1 + x4*x5*y2 - x4*y3 + x5**4
 - x5**3*y1 + x5**2*y2 - x5*y3 + y4,
 x5**5 - x5**4*y1 + x5**3*y2 - x5**2*y3 + x5*y4 - y5}$
```

REMARK 4.14. *In Section 2, we have discussed how we can compute the Groebner basis for the ideal efficiently. The direct method introduced there is the most efficient one among the three methods discussed.*

Here,
$\sigma_1 =$ *x1+x2+x3+x4+x5,*
$\sigma_2 =$ *x1*x2+x1*x3+x1*x4+x1*x5+x2*x3+x2*x4+x2*x5 +x3*x4+x3*x5+x4*x5,*
$\sigma_3 =$ *x1*x2*x3+x1*x2*x4+x1*x2*x5+x1*x3*x4+x1*x3*x5 +x1*x4*x5+x2*x3*x4+x2*x3*x5+x3*x4*x5,*
$\sigma_4 =$ *x1*x2*x3*x4+x1*x2*x4*x5+x1*x3*x4*x5 +x2*x3*x4*x5,*
$\sigma_5 =$*x1*x2*x3*x4*x5,*

are the five elementary symmetric functions in the five roots x1, x2, x3, x4,x5 of

$$f(x) = 0.$$

Then we compute:

```
result3:=preduce((x1-x2)^2*(x1-x3)^2*(x1-x4)^2*(x1-x5)^2
*(x2-x3)^2*(x2-x4)^2*(x2-x5)^2*(x3-x4)^2*(x3-x5)^2
*(x4-x5)^2,result,{x1,x2,x3,x4,x5,y1,y2,y3,y4,y5});


result3 := 256*y1**5*y5**3 - 192*y1**4*y2*y4*y5**2
- 128*y1**4*y3**2*y5**2 + 144*y1**4*y3*y4**2*y5
- 27*y1**4*y4**4 + 144*y1**3*y2**2*y3*y5**2
- 6*y1**3*y2**2*y4**2*y5 - 80*y1**3*y2*y3**2*y4*y5
+ 18*y1**3*y2*y3*y4**3 - 1600*y1**3*y2*y5**3
+ 16*y1**3*y3**4*y5 - 4*y1**3*y3**3*y4**2
+ 160*y1**3*y3*y4*y5**2 - 36*y1**3*y4**3*y5
- 27*y1**2*y2**4*y5**2 + 18*y1**2*y2**3*y3*y4*y5
- 4*y1**2*y2**3*y4**3 - 4*y1**2*y2**2*y3**3*y5
+ y1**2*y2**2*y3**2*y4**2 + 1020*y1**2*y2**2*y4*y5**2
```

```
+ 560*y1**2*y2*y3**2*y5**2 - 746*y1**2*y2*y3*y4**2*y5
+ 144*y1**2*y2*y4**4 + 24*y1**2*y3**3*y4*y5
- 6*y1**2*y3**2*y4**3 + 2000*y1**2*y3*y5**3
- 50*y1**2*y4**2*y5**2 - 630*y1*y2**3*y3*y5**2
+ 24*y1*y2**3*y4**2*y5 + 356*y1*y2**2*y3**2*y4*y5
- 80*y1*y2**2*y3*y4**3 + 2250*y1*y2**2*y5**3
- 72*y1*y2*y3**4*y5 + 18*y1*y2*y3**3*y4**2
- 2050*y1*y2*y3*y4*y5**2 + 160*y1*y2*y4**3*y5
- 900*y1*y3**3*y5**2 + 1020*y1*y3**2*y4**2*y5
- 192*y1*y3*y4**4 - 2500*y1*y4*y5**3 + 108*y2**5*y5**2
- 72*y2**4*y3*y4*y5 + 16*y2**4*y4**3 + 16*y2**3*y3**3*y5
- 4*y2**3*y3**2*y4**2 - 900*y2**3*y4*y5**2
+ 825*y2**2*y3**2*y5**2 + 560*y2**2*y3*y4**2*y5
- 128*y2**2*y4**4 - 630*y2*y3**3*y4*y5
+ 144*y2*y3**2*y4**3 - 3750*y2*y3*y5**3
+ 2000*y2*y4**2*y5**2 + 108*y3**5*y5
- 27*y3**4*y4**2 + 2250*y3**2*y4*y5**2
- 1600*y3*y4**3*y5 + 256*y4**5 + 3125*y5**4$
```

```
result4:=sub({y1=0,y2=0,y3=0},result3);
result4:=256*y4**5 + 3125*y5**4$
```

This shows that the discriminant

$$
\begin{aligned}
D &= (x1-x2)^2(x1-x3)^2(x1-x4)^2(x1-x5)^2 \\
&\quad \cdot(x2-x3)^2(x2-x4)^2(x2-x5)^2 \\
&\quad \cdot(x3-x4)^2(x3-x5)^2 \\
&\quad \cdot(x4-x5)^2
\end{aligned}
$$

is indeed a symmetric polynomial by the first part of Theorem 4.1. Moreover, by the second part of Theorem 4.1, we see that

$$
\begin{aligned}
D \ = \ & + 256\sigma_1^5\sigma_5^3 - 192\sigma_1^4\sigma_2\sigma_4\sigma_5^2 - 128\sigma_1^4\sigma_3^2\sigma_5^2 + 144\sigma_1^4\sigma_3\sigma_4^2\sigma_5 \\
& -27\sigma_1^4\sigma_4^4 + 144\sigma_1^3\sigma_2^2\sigma_3\sigma_5^2 - 6\sigma_1^3\sigma_2^2\sigma_4^2\sigma_5 - 80\sigma_1^3\sigma_2\sigma_3^2\sigma_4\sigma_5 \\
& +18\sigma_1^3\sigma_2\sigma_3\sigma_4^3 - 1600\sigma_1^3\sigma_2\sigma_5^3 + 16\sigma_1^3\sigma_3^4\sigma_5 - 4\sigma_1^3\sigma_3^3\sigma_4^2 \\
& +160\sigma_1^3\sigma_3\sigma_4\sigma_5^2 - 36\sigma_1^3\sigma_4^3\sigma_5 - 27\sigma_1^2\sigma_2^4\sigma_5^2 + 18\sigma_1^2\sigma_2^3\sigma_3\sigma_4\sigma_5 \\
& -4\sigma_1^2\sigma_2^3\sigma_4^3 - 4\sigma_1^2\sigma_2^2\sigma_3^3\sigma_5 + \sigma_1^2\sigma_2^2\sigma_3^2\sigma_4^2 + 1020\sigma_1^2\sigma_2^2\sigma_4\sigma_5^2 \\
& +560\sigma_1^2\sigma_2\sigma_3^2\sigma_5^2 - 746\sigma_1^2\sigma_2\sigma_3\sigma_4^2\sigma_5 + 144\sigma_1^2\sigma_2\sigma_4^4 + 24\sigma_1^2\sigma_3^3\sigma_4\sigma_5 \\
& -6\sigma_1^2\sigma_3^2\sigma_4^3 + 2000\sigma_1^2\sigma_3\sigma_5^3 - 50\sigma_1^2\sigma_4^2\sigma_5^2 - 630\sigma_1\sigma_2^3\sigma_3\sigma_5^2 \\
& +24\sigma_1\sigma_2^3\sigma_4^2\sigma_5 + 356\sigma_1\sigma_2^2\sigma_3^2\sigma_4\sigma_5 - 80\sigma_1\sigma_2^2\sigma_3\sigma_4^3 \\
& +2250\sigma_1\sigma_2^2\sigma_5^3 - 72\sigma_1\sigma_2\sigma_3^4\sigma_5 + 18\sigma_1\sigma_2\sigma_3^3\sigma_4^2 \\
& -2050\sigma_1\sigma_2\sigma_3\sigma_4\sigma_5^2 + 160\sigma_1\sigma_2\sigma_4^3\sigma_5 - 900\sigma_1\sigma_3^3\sigma_5^2 \\
& +1020\sigma_1\sigma_3^2\sigma_4^2\sigma_5 - 192\sigma_1\sigma_3\sigma_4^4 - 2500\sigma_1\sigma_4\sigma_5^3 + 108\sigma_2^5\sigma_5^2 \\
& -72\sigma_2^4\sigma_3\sigma_4\sigma_5 + 16\sigma_2^4\sigma_4^3 + 16\sigma_2^3\sigma_3^3\sigma_5 - 4\sigma_2^3\sigma_3^2\sigma_4^2 \\
& -900\sigma_2^3\sigma_4\sigma_5^2 + 825\sigma_2^2\sigma_3^2\sigma_5^2 + 560\sigma_2^2\sigma_3\sigma_4^2\sigma_5 - 128\sigma_2^2\sigma_4^4 \\
& -630\sigma_2\sigma_3^3\sigma_4\sigma_5 + 144\sigma_2\sigma_3^2\sigma_4^3 - 3750\sigma_2\sigma_3\sigma_5^3 \\
& +2000\sigma_2\sigma_4^2\sigma_5^2 + 108\sigma_3^5\sigma_5 - 27\sigma_3^4\sigma_4^2 + 2250\sigma_3^2\sigma_4\sigma_5^2 \\
& -1600\sigma_3\sigma_4^3\sigma_5 + 256\sigma_4^5 + 3125\sigma_5^4 \ .
\end{aligned}
$$

is the discriminant D of an *irreducible* polynomial $f(x)$ over $\mathbb{Q}$:

$$f(x) = x^5 - \sigma_1 x^4 + \sigma_2 x^3 - \sigma_3 x^2 + \sigma_4 x - \sigma_5.$$

EXAMPLE 4.17. *Thus, if* $f(x) = x^5 - 16x + 2$, *then* $\sigma_1 = 0, \sigma_2 = 0, \sigma_3 = 0$, *and hence the discriminant of* $f(x)$ *is*

$$
\begin{aligned}
D &= 256\sigma_4^5 + 3125\sigma_5^4 \\
&= 256 \cdot (-16)^5 + 3125 \cdot (-2)^4 \\
&= -268385456,
\end{aligned}
$$

is not a square in $\mathbb{Q}$, as we expected. In fact, by part a of Proposition 4.3 the Galois group $G = G(K/\mathbb{Q})$ is not a subgroup of the alternating group A_5. But, it could be a proper subgroup of S_5.

On the other hand, since it has exactly two complex roots, it contains a transposition and hence it must be S_5. Thus, its roots can not be expressed by radicals.

EXAMPLE 4.18. *If $f(x) = x^5 - 5x + \frac{5}{2}$ over $\mathbb{Q}$, then its discriminant*

$$\begin{aligned} D &= 256\sigma_4^5 + 3125\sigma_5^4 \\ &= 256 \cdot (-5)^5 + 3125 \cdot (\frac{-5}{2})^4 \\ &= (-10846875)/16, \end{aligned}$$

is not a square in $\mathbb{Q}$, as we expected.

Thus, from these two examples we see that if the equation has three distinct real roots, we do not even need to compute the discriminant D of $f(x)$. Otherwise, it will still provide a clue for finding the Galois group of $f(x)$ in almost all cases:

EXAMPLE 4.19. *If $f(x) = x^5 - 5x + 10$ is the irreducible polynomial over $\mathbb{Q}$, then its discriminant*

$$\begin{aligned} D &= 256\sigma_4^5 + 3125\sigma_5^4 \\ &= 256 \cdot (-5)^5 + 3125(-10)^4 \\ &= 97656250256/3125 \\ &= (2^4 \cdot 3049 \cdot 2001809)/5^5 \end{aligned}$$

is not a square in $\mathbb{Q}$. Thus by part a of Proposition 4.3 the Galois group $G = G(K/\mathbb{Q})$ is not a subgroup of the alternating group A_5. But, it could be still a proper subgroup of S_5, not containing a transposition but containing an odd permutation of order > 2, since it is not a subgroup of

A_5.

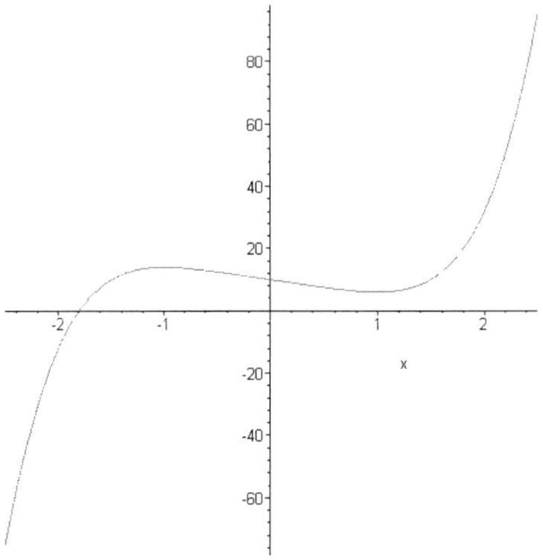

We note that this polynomial has only one real root. In fact, the following lines for REDUCE say that its 5 *numerical solutions and only one of them is a real number:*

———————

```
on fullroots;
on rounded;
solve(x**5 - 5*x + 10);
{x=1.22140106699 + 0.650175291569*i,
x=1.22140106699 - 0.650175291569*i,
x= - 0.320305635197 + 1.67201504509*i,
x= - 0.320305635197 - 1.67201504509*i,
x= - 1.80219086358}$
```

———————

4. The Correspondence Between Geometry and Algebra

In Section 4 of Chapter 1 of this book, we discussed when a polynomial is called a polynomial function in general. In this section, we will further develop such a notion until we explicitly see how Geometry and Algebra can be put together in a contravariant functorial manner.

Let A be an $n \times n$ matrix with entries in a field k. Then we know that left multiplication by A on the column vectors of k^n defines a linear map

$$A\colon k^n \longrightarrow k^n,$$

sending a column vector $X = (x_1, ..., x_n)^t \in k^n$

$$X \mapsto AX.$$

We note that this map is a *polynomial mapping* sending

$$(x_1, ..., x_n) \longmapsto (a_{11}x_1 + \cdots + a_{1n}x_n, ..., a_{n1}x_1 + \cdots + a_{nn}x_n).$$

If moreover A is an invertible matrix, we obviously have the inverse linear map of left multiplication by A.

Now, consider the diagram

$$\begin{array}{ccc} k^n & \xrightarrow{A} & k^n \\ & & \downarrow f \\ & & k \end{array},$$

where the horizontal arrow is left multiplication by the matrix A and the vertical arrow is a *polynomial function* $f = f(x_1, ..., x_n) \in k[x_1, ..., x_n]$, i.e., a k-valued evaluation function on the affine n-space k^n. Then we obviously see that their composition $A \circ f$ as maps is another k-valued evaluation map on k^n, i.e., a polynomial function on k^n, and this *defines* a map

$$A^*\colon k[x_1, ..., x_n] \longrightarrow k[x_1, ..., x_n]$$

on the polynomial ring in n variables over k. The map A^* is thus defined by sending a polynomial f to

$$f \longmapsto f \circ A,$$

thus, if X is a column vector of the domain k^n, then

$$(f \circ A)(X) = f(AX),$$

holds. That is, for each column vector X of the domain k^n,

$$(A^*(f))(X) = (f \circ A)(X) = f(A(X)),$$

is a k-valued evaluation of the domain k^n by the same polynomial but after a change of coordinates.

LEMMA 4.3. *Under this setting, we have:*

(1) *The map A^*: $k[x_1, ..., x_n] \longrightarrow k[x_1, ..., x_n]$ defined above is a k-algebra homomorphism, i.e., both k-linear map and a homomorphism of rings.*
(2) *If the matrix A is invertible, then A^* is an automorphism of k-algebras on $k[x_1, ..., x_n]$.*

PROOF. Part (2) is a consequence of Part (1), so we prove Part (1).

It is very easy to see that $A^*(f+g) = (f+g)\circ A = f\circ A + g\circ A = A^*(f) + A^*(g)$ and $A^*(\alpha f) = (\alpha f)\circ A = \alpha(f\circ A) = \alpha A^*(f)$; thus A^* is k-linear. In addition to this, we have $A^*(fg) = (fg)\circ A$. Then since, for any $X \in k^n$,

$$\begin{aligned} ((fg)\circ A)(X) &= (fg)(AX) = f(AX)g(XA) \\ &= ((f\circ A)(X))((g\circ A)(X)) \\ &= ((f\circ A)(g\circ A))(X), \end{aligned}$$

$A^*(fg) = (fg)\circ A = (f\circ A)(g\circ A) = A^*(f)\circ A^*(g)$ holds for any polynomials $f, g \in k[x_1, ..., x_n]$. Moreover, for the constant polynomial 1 in the domain $k[x_1, ..., x_n]$, we have $A^*(1)(X) = (1\circ A)(X) = 1(A(X)) = 1$ for each $X \in k^n$, i.e., $A^*(1) = 1$. Thus, it is a homomorphism of rings. $\square$

REMARK 4.15. *If the bottom field k is moreover algebraically closed, we can define the inverse map $\hat{}$ of * by first establishing (through the Hilbert Nullstellensatz, Theorem 1.6),*

$$k^n \simeq Spec(k[x_1, ..., x_n]),$$

where $Spec(k[x_1, ..., x_n])$ is, by definition, the set of all k-algebra homomorphisms:

$$\begin{array}{c} k[x_1, ..., x_n] \\ \downarrow \\ k \end{array}$$

and each point X of the domain k^n is, by definition, sent to the evaluation homomorphism ε_X:

$$\begin{array}{c} k[x_1, ..., x_n] \\ \downarrow \varepsilon_X \\ k \end{array}$$

which is also, by definition, simply written as

$$\begin{array}{c} k[x_1, ..., x_n] \\ \downarrow X \\ k \end{array}$$

so that

$$X(g) \equiv \varepsilon_X(g) = g(X) \in k. \tag{4.8}$$

And this direction $X \longmapsto X \equiv \varepsilon_X$ is always defined, without even k being an algebraically closed field. Moreover, one can easily verify, by applying the universally repelling property of the polynomial ring $k[x_1, ..., x_n]$, that any k-algebra homomorphism

$$\begin{array}{c} k[x_1, ..., x_n] \\ \downarrow \varphi \\ k \end{array}$$

is an evaluation homomorphism ε_X for some $X \in k^n$.

[More about this process on evaluation duality is described in detail in Remark 1.10 in Chapter 1].

But, the inverse of this map is not generally well defined, unless the bottom field k is algebraically closed.

If k is algebraically closed, then Hilbert's Nullstellensatz establishes the inverse map as follows: For any k-algebra homomorphism:

$$\begin{array}{c} k[x_1, ..., x_n] \\ \downarrow \varphi \\ k \end{array}$$

its kernel $\ker(\varphi)$ is a maximal ideal that is generated by the linear polynomials $x - \alpha_1, ..., x - \alpha_n$, i.e., $\ker(\varphi) = (x_1 - \alpha_1, ..., x_n - \alpha_n)$ for some

$X = (\alpha_1, ..., \alpha_n) \in k^n$ *provided that the bottom field* k *is algebraically closed. Then these maximal ideals bijectively correspond to the points* $X = (\alpha_1, ..., \alpha_n)$ *of* k^n. *[For more about this, see Proposition 1.14 and Theorem 1.6]. In other words, its inverse map* $\varphi \longmapsto \ker(\varphi) = (x_1 - \alpha_1, ..., x_n - \alpha_n) \longleftrightarrow X = (\alpha_1, ..., \alpha_n)$ *is defined only through Hilbert's Nullstellensatz, by the proof of the* $\Longrightarrow$ *direction of (1) of Theorem 1.6, provided that the bottom field* k *is algebraically closed.*

Once we accept this fundamental notion on ***evaluation duality****, the map* $\hat{}$ *is defined as follows: Given a* k*-algebra homomorphism*

$$\varphi : k[x_1, ..., x_n] \longrightarrow k[x_1, ..., x_n],$$

and a given $X \in k^n$, *we define*

$$\varphi\hat{}\,(X) = X \circ \varphi,$$

where it is moreover defined that, for any f *in the domain* $k[x_1, ..., x_n]$,

$$\varphi\hat{}\,(X)(f) = (X \circ \varphi)(f) = X(\varphi(f)) \equiv \varphi(f)(X). \tag{4.9}$$

Under these two maps, we first have

$$\begin{aligned} ((A^*)\hat{}\,(X))(f) &= (X \circ A^*)(f) = X(A^*(f)) \equiv A^*(f)(X) \\ &= (f \circ A)(X) = f(A(X)) \equiv A(X)(f), \end{aligned}$$

for all $f \in k[x_1, ..., x_n]$, *that is, we have*

$$(A^*)\hat{}\,(X) = A(X)$$

for all $X \in k^n$, *that is, we have*

$$(A^*)\hat{} = A.$$

Conversely, if φ *is a* k*-algebra homomorphism*

$$k[x_1, ..., x_n] \longrightarrow k[x_1, ..., x_n],$$

then we have

$$\begin{aligned}((\varphi\hat{\ })^*(f))(X) &= (f \circ \varphi\hat{\ })(X) = f(\varphi\hat{\ }(X)) \equiv \varphi\hat{\ }(X)(f) \\ &= (X \circ \varphi)(f) = X(\varphi(f)) \equiv \varphi(f)(X),\end{aligned}$$

for all $X \in k^n$, *that is, we have*

$$(\varphi\hat{\ })^*(f) = \varphi(f),$$

for all $f \in k[x_1, ..., x_n]$, *that is, we have*

$$(\varphi\hat{\ })^* = \varphi.$$

This shows that the two maps * *and* $\hat{\ }$ *are indeed inverses to each other, provided that the bottom field* k, *i.e., the place where our actual computations take place, is an algebraically closed field.*

Moreover, we have for all g,

$$\begin{aligned}(\varphi\hat{\ } \circ \phi\hat{\ })(X)(g) &= \varphi\hat{\ }(\phi\hat{\ }(X)(g)) = \varphi\hat{\ }(X \circ \phi)(g) \\ &= ((X \circ \phi) \circ \varphi)(g) = (X \circ (\phi \circ \varphi))(g) \\ &= (\phi \circ \varphi)\hat{\ }(X)(g),\end{aligned}$$

and hence for all X,

$$(\varphi\hat{\ } \circ \phi\hat{\ })(X) = (\phi \circ \varphi)\hat{\ }(X),$$

and hence

$$(\phi \circ \varphi)\hat{\ } = \varphi\hat{\ } \circ \phi\hat{\ }$$

holds. Similarly, we can easily check that

$$(g \circ f)^* = f^* \circ g^*.$$

Furthermore, we may also easily check that id^* *is the identity* k*-algebra homomorphism and* $id\hat{\ }$ *is the identity morphism.*

Therefore, we have: φ *is an isomorphism if and only if* $\varphi\hat{\ }$ *is an isomorphism, or equivalently,* f *is an isomorphism if and only if* f^* *is an isomorphism.*

In Remark 5.6 of Chapter 5, the next material to this subject is presented.

4.1. Symmetric Polynomials Revisited. Among the invertible $n \times n$ matrices, i.e., the elements of the general linear group, $GL_n(k)$, we have a very special subgroup of $GL_n(k)$, which is the subgroup

$$S_n$$

of all permutation matrices. It is a finite subgroup of order $n!$ because there exists a one-to-one correspondence between the set S_n and the set of all permutations on the set of n objects, say, $\{1, 2, ..., n\}$, or the collection of the coordinates, $\{x_1, ..., x_n\}$ of a column vector $X = (x_1, ..., x_n)^t$ of k^n.

EXAMPLE 4.20. *Let E be the elementary row matrix which permutes the pivots of the identity matrix:*

$$E = \begin{bmatrix} 0 & 1 & 0 \\ 0 & 0 & 1 \\ 1 & 0 & 0 \end{bmatrix}.$$

The corresponding permutation $p = (132)$. Thus, $p^{-1} = (123)$. Then for the column vector $X = (x, y, z)^t$, we have

$$EX = \begin{bmatrix} 0 & 1 & 0 \\ 0 & 0 & 1 \\ 1 & 0 & 0 \end{bmatrix} \begin{bmatrix} x \\ y \\ z \end{bmatrix} = \begin{bmatrix} y \\ z \\ x \end{bmatrix}.$$

Or equivalently, if we write the column vector $X = (x_1, x_2, x_3)^t$, we have

$$EX = \begin{bmatrix} 0 & 1 & 0 \\ 0 & 0 & 1 \\ 1 & 0 & 0 \end{bmatrix} \begin{bmatrix} x_1 \\ x_2 \\ x_3 \end{bmatrix} = \begin{bmatrix} x_2 \\ x_3 \\ x_1 \end{bmatrix} = \begin{bmatrix} x_{p^{-1}(1)} \\ x_{p^{-1}(2)} \\ x_{p^{-1}(3)} \end{bmatrix}.$$

Thus, a permutation matrix permutes the variables according to the corresponding permutation p, or equivalently, it permutes the indices of the coordinates of X according to the inverse permutation p^{-1} of p. This means that, although the results are exactly the same, we have in reality *two different ways of interpreting the effect of a permutation on the coordinates of a coordinate vector X, and when we have more than 3 variables, the second expression is a lot better to follow up.*

With this terminology, we see that the elementary symmetric functions σ_i are polynomials fixed by the action of S_n to polynomials in $k[x_1, ..., x_n]$. That is, for any $A \in S_n$,

$$\begin{aligned}(A^*(\sigma_i))(X) &= (\sigma_i \circ A)(X) \\ &= \sigma_i(AX) \\ &= \sigma_i(X),\end{aligned}$$

that is, $A^*(\sigma_i) = \sigma_i$.

DEFINITION 4.6. *We denote by*

$$k[x_1, ..., x_n]^{S_n}$$

the set of all polynomials which are fixed under the action of S_n on $k[x_1, ..., x_n]$.

LEMMA 4.4. *$k[x_1, ..., x_n]^{S_n}$ is a subring of $k[x_1, ..., x_n]$.*

PROOF. The proof is so routine that we leave it as an exercise. □

More generally, we introduce a definition:

DEFINITION 4.7. *Let G be a* finite *subgroup of $GL_n(k)$. Then polynomials in the subring,*

$$k[x_1, ..., x_n]^G,$$

that is, those polynomials which are fixed under the action of G on $k[x_1, ..., x_n]$ are called ***invariant polynomials under (the action of) G****, and this subring is called* ***the ring of invariants*** *under G.*

In particular, if $G = S_n$, the subgroup of permutations, then elements of $k[x_1, ..., x_n]^{S_n}$ are simply called ***symmetric polynomials****.*

Then obviously, all the elementary symmetric functions σ_i are elements of $k[x_1, ..., x_n]^{S_n}$ and moreover, Theorem 4.1 tells us that these elementary symmetric functions σ_i's in fact generate the subring $k[x_1, ..., x_n]^{S_n}$.

That is, $k[x_1, ..., x_n]^{S_n}$ is a finitely generated subring of $k[x_1, ..., x_n]$ with generators $\{\sigma_1, ..., \sigma_n\}$:

$$k[x_1, ..., x_n]^{S_n} = k[\sigma_1, ..., \sigma_n].$$

EXAMPLE 4.21. $k[x_1 x_2, x_3]^{S_3} = k[\sigma_1, \sigma_2, \sigma_3]$, *where*

$$\begin{aligned} \sigma_1 &= x_1 + x_2 + x_3, \\ \sigma_2 &= x_1x_2 + x_1x_3 + x_2x_3, \\ \sigma_3 &= x_1x_2x_3. \end{aligned}$$

Moreover, Proposition 4.1 tells us exactly when a given polynomial f is a symmetric polynomial, and if so, it also tells us how to express it as a polynomial in the generators σ_i's. This was all about what we have done in Section 1 of this chapter.

We also note that each monomial term of the elementary symmetric function σ_i is homogeneous of total degree i.

DEFINITION 4.8. *A polynomial $f \in k[x_1, ..., x_n]$ is called **homogeneous of total degree** k if every term appearing in f has total degree k.*

PROPOSITION 4.4. *A polynomial $f \in k[x_1, ..., x_n]$ is symmetric if and only if each of its homogeneous components are symmetric.*

PROOF. Suppose that f is symmetric. Since any permutation takes a term of f of a total degree l to another term of the same total degree l, the k-th homogeneous component must also be symmetric, and the converse is also trivial.

□

Thus by this proposition, whenever we work with a symmetric polynomial we may assume that it is homogeneous.

Among symmetric polynomials, there is another type which is not elementary symmetric functions, but is quite useful in expressing a given symmetric polynomial in terms of these:

DEFINITION 4.9. *A polynomial of the form*

$$s_k = s_k(n) = x_1^k + x_2^k + \cdots + x_n^k \in k[x_1, ..., x_n],$$

is called a ***power sum*** *of total degree k, for $k = 1, ..., n$.*

Then obviously, each $s_k(n)$ is symmetric and homogeneous, and the following Theorem 4.3 states that an arbitrary symmetric polynomials can be written in terms of these power sums, $s_1, ..., s_n$. We need a well-known lemma:

LEMMA 4.5. *(Newton's Identities) For any $k \geq 1$, we have the identity:*

$$s_k(n) - \sigma_1(n)s_{k-1}(n) + \cdots + (-1)^{k-1}\sigma_{k-1}(n)s_1(n) + (-1)^k k\sigma_k(n) = 0,$$

where we put, by definition, $\sigma_0(n) = 1$ for any $n \geq 1$, $\sigma_i(n) = 0$ if either $i > n$ or $i < 0$, where $\sigma_i(n) = \sigma_i(x_1, ..., x_n)$.

PROOF. We note that for all $n > 1$ and all i and all k,

$$\begin{aligned} s_k(n) &= x_n^k + s_k(n-1), \\ \sigma_i(n) &= \sigma_i(n-1) + x_n \cdot \sigma_{i-1}(n-1), \end{aligned}$$

where $\sigma_i(0) = 0$, $s_k(0) = 0$ by definition. Using these two recursive identities, we leave the proof of the identity by induction on n as an exercise. □

REMARK 4.16. *In mathematics, Newton's identities relate two different ways of describing the roots of a polynomial. They were found by Isaac Newton (1643 - 1727) in about 1666, apparently in ignorance of earlier work (1629) by Albert Girard. These useful identities have immediate applications in many areas of mathematics, including Galois theory, invariant theory, group theory, combinatorics, as well as further applications outside mathematics, including general relativity. They can be considered as applications of ideas in computational algebraic geometry, particularly Groebner bases.*

(Reference: http://en.wikipedia.org/wiki/Newton's_identities).

THEOREM 4.3. *Let k be a field of characteristic 0. Then every symmetric polynomial in $k[x_1, ..., x_n]$ can be written as a polynomial in the power sums $s_1, ..., s_n$.*

PROOF. Since every symmetric polynomial is a polynomial in the elementary symmetric functions σ_i by Theorem 4.1, it is sufficient to prove that each σ_i is a polynomial in $s_1, ..., s_n$. For this we use the Newton's identities in the above lemma.

For $1 \leq k \leq n$, we apply induction on k that σ_k is a polynomial in $s_1, ..., s_n$. If $k = 1$, it is true since $\sigma_1 = s_1$. Assume as induction hypothesis that the claim is true for $1, 2, ..., k-1$. Then by Newton's identities says that

$$\sigma_k(n) = (-1)^{k-1}\frac{1}{k}\left\{s_k(n) - \sigma_1(n)s_{k-1}(n) + \cdots + (-1)^{k-1}\sigma_{k-1}(n)s_1(n)\right\},$$

and the right side is a polynomial in various $s_i(n)$,by induction hypothesis. □

PROPOSITION 4.5. *Let $G \subset GL_n(k)$ be a finite matrix group. Then a polynomial $f \in k[x_1, ..., x_n]$ is invariant under G if and only if so are its homogeneous components.*

PROOF. We first note that since $k[x_1, ..., x_n]$ is a k-vector space with a basis consisting of monomials, $f(X)$ is a unique k-linear combination of the monomials $x_1^{i_1}x_2^{i_2}\cdots x_n^{i_n}$ of various total degree, and hence these monomials of $f(X)$ can be grouped according their total degree, and for any $A \in G$ we can do the same thing to the monomials of $f(AX)$. We note that if $A = (a_{ij})$ then under the coordinate change by A, each monomial $x_1^{i_1}x_2^{i_2}\cdots x_n^{i_n}$ of $f(X)$ of total degree j is mapped to

$$(a_{11}x_1 + \cdots + a_{1n}x_n)^{i_1}\cdots(a_{n1}x_1 + \cdots + a_{nn}x_n)^{i_n},$$

and when we expand this product we easily see that it is a sum of monomials of the same total degree j. Thus for each $A \in G$, $f(X) = f(AX)$ hods if and only if those components of $f(X)$ of total degree j grouped together are the same as those components of $f(AX)$ of the same total degree j grouped together. □

We have a useful lemma:

LEMMA 4.6. *Let $G \subset GL_n(k)$ be a finite group. Suppose that we are given a set of generators $\{A_1, ..., A_m\}$ of G. Then $f \in k[x_1, ..., x_n]$ is invariant under G if and only if*

$$f(X) = f(A_1X) = \cdots = f(A_mX).$$

PROOF. The $\Longrightarrow$ direction is obvious. Conversely, if f is invariant under $A_1, ..., A_m$, then since each element A of G is written as a product

$$A = B_1 \cdots B_s$$

of elements of $\{A_1, ..., A_m\}$, by the definition of the group action, we have:

$$f(X) = f(B_sX) = f(B_{s-1}B_sX) = \cdots = f(B_1 \cdots B_sX).$$

Since this holds for each $A \in G$, f is invariant under G. $\square$

Exercises for Section 4.3

EXERCISE 4.4. *Complete the proof of Lemma 4.5.*

5. Generators for the Ring of Invariants

Let G be a *finite* subgroup of $GL_n(k)$. Our basic question is whether the ring of invariants $k[x_1, ..., x_n]^G$ under the group action of G is finitely generated, that is,

$$k[x_1, ..., x_n]^G = k[f_1, ..., f_m],$$

and if so, we would like to know whether we can uniquely express any invariant polynomial in term of generators of the ring $k[x_1, ..., x_n]^G$ of invariants. We still assume in this section that the characteristic of k is 0.

To answer the first question, we introduce a k-linear operator:

DEFINITION 4.10. *Given a finite subgroup G of $GL_n(k)$, the map R_G: $k[x_1, ..., x_n] \longrightarrow k[x_1, ..., x_n]$ defined by*

$$R_G(f)(X) = \frac{1}{|G|} \sum_{A \in G} f(AX),$$

is called the ***Reynolds operator*** *of G.*

Another name of a subgroup G of $GL_n(k)$ is a **finite matrix group**. The immediate properties the Reynolds operator satisfies are:

PROPOSITION 4.6. *Let R_G: $k[x_1, ..., x_n] \longrightarrow k[x_1, ..., x_n]$ be the Reynolds operator of a finite subgroup G of $GL_n(k)$. Then*

(1) *R_G is k-linear in f.*
(2) *$R_G(f) \in k[x_1, ..., x_n]^G$, for any $f \in k[x_1, ..., x_n]$.*
(3) *$R_G(f) = f$, for any $f \in k[x_1, ..., x_n]^G$.*

PROOF. The proof of Part (3) is trivial, and the proof of Part (1) is also very easy that we leave it as an exercise.

Part (2). We need to verify that $R_G(f)(MX) = R_G(f)(X)$, for each $M \in G$:

$$R_G(f)(MX) = \frac{1}{|G|} \sum_{A \in G} f(AMX) = \frac{1}{|G|} \sum_{A \in G} f(AX),$$

where we have the last equality because, as A runs through the finite set G, so does AM. □

Thus by part (2) of Proposition 4.6 Reynolds operator allows us to generate plenty of invariant polynomials. In particular, we clearly have:

COROLLARY 4.9. *For those monomials x^β with $|\beta| > 0$, $R_G(x^\beta)$ is a homogeneous invariant polynomial of total degree $|\beta|$.* □

THEOREM 4.4. *(Emmy Noether) Let k be a field of characteristic 0. Given a finite matrix group $G \subset GL_n(k)$, the ring $k[x_1, ..., x_n]^G$ of invariants is finitely generated by those homogeneous invariants:*

$$R_G(x^\beta) \qquad \text{such that } |\beta| \leq |G|.$$

PROOF. We first note that if $f = \sum_\alpha c_\alpha x^\alpha \in k[x_1, ..., x_n]^G$, then

$$f = R_G(f) = R_G\left(\sum_\alpha c_\alpha x^\alpha\right) = \sum_\alpha c_\alpha R_G(x^\alpha),$$

and hence each invariant polynomial is a k-linear combination of $R_G(x^\alpha)$. Thus, if we can prove that any $R_G(x^\alpha)$ is a polynomial in those finite $R_G(x^\beta)$'s with $|\beta| \leq |G|$, then we are done.

Given $A = (a_{ij}) \in G$, let A_i denote the i-th row of A, so for each $i = 1, ..., n$,

$$A_i X = a_{i1}x_1 + \cdots + a_{in}x_n.$$

Thus, for $\alpha = (\alpha_1, ..., \alpha_n)$ is an exponent of $x^\alpha = x_1^{\alpha_1} x_2^{\alpha_2} \cdots x_n^{\alpha_n}$, we define

$$(AX)^\alpha := (A_1X)^{\alpha_1} \cdots (A_nX)^{\alpha_n}$$

Using this notation, we have

$$R_G(x^\alpha) = \frac{1}{|G|}\sum_{A \in G}(AX)^\alpha$$

On the other hand, we note that for a fixed k, we have

$$(x_1 + \cdots + x_n)^k = \sum_{|\alpha|=k} a_\alpha x^\alpha,$$

so that when the coordinates $x_1, ..., x_n$ are changed by A to $A_1X, ..., A_nX$, we are going to have

$$(A_1X + \cdots + A_nX)^k = \sum_{|\alpha|=k} a_\alpha (AX)^\alpha.$$

Then we introduce n new variables u_i in order to prevent any possible cancellation among the terns during this summation:

$$(u_1A_1X + \cdots + u_nA_nX)^k = \sum_{|\alpha|=k} a_\alpha (AX)^\alpha u^\alpha.$$

Thus, if we add up these sums for all $A \in G$, we have

$$\begin{aligned} S_k &= \sum_{A \in G} (u_1 A_1 X + \cdots + u_n A_n X)^k \\ &= \sum_{A \in G} \sum_{|\alpha|=k} a_\alpha (AX)^\alpha u^\alpha \\ &= \sum_{|\alpha|=k} a_\alpha \left(\sum_{A \in G} (AX)^\alpha \right) u^\alpha \\ &= \sum_{|\alpha|=k} b_\alpha R_G(x^\alpha) u^\alpha, \end{aligned}$$

where the first equality is by definition, and $b_\alpha = |G|\, a_\alpha$ is non-zero in the bottom field k because the characteristic of k is 0 by hypothesis. Then since

$$S_k = \sum_{A \in G} (u_1 A_1 X + \cdots + u_n A_n X)^k$$

is the k-th power sum of $|G|$ quantities

$$u_1 A_1 X + \cdots + u_n A_n X,$$

so that each S_k is symmetric in these $|G|$ quantities $u_1 A_1 X, \cdots, u_n A_n X$.

Therefore, by Theorem 4.3, we see that each S_k is a polynomial in $S_1, ..., S_{|G|}$:

$$S_k = P(S_1, ..., S_{|G|}),$$

for some polynomial P with coefficients in the bottom field k. Therefore, we have a polynomial identity:

$$\sum_{|\alpha|=k} b_\alpha R_G(x^\alpha) u^\alpha = P\Big(\sum_{|\beta|=1} b_\beta R_G(x^\beta) u^\beta, ..., \sum_{|\beta|=|G|} b_\beta R_G(x^\beta) u^\beta\Big)$$

After expanding the right side and equating the coefficients of u^α of both sides, we see that each $b_\alpha R_G(x^\alpha)$ is indeed a polynomial in the $R_G(x^\beta)$ with $|\beta| \leq |G|$. □

Thus, Noether's theorem, Theorem 4.4 explicitly tells us how to compute the finite number of generators for the ring of invariants. But, this approach requires us to compute too many $R_G(x^\alpha)$'s as n becomes large. The following two examples will tell us this.

EXAMPLE 4.22. *Let G be a cyclic subgroup of $GL_2(k)$ generated by*

$$\begin{bmatrix} 0 & -1 \\ 1 & 0 \end{bmatrix}.$$

Thus other elements of G are:

$$\begin{bmatrix} -1 & 0 \\ 0 & -1 \end{bmatrix}, \begin{bmatrix} 0 & 1 \\ -1 & 0 \end{bmatrix}, \begin{bmatrix} 1 & 0 \\ 0 & 1 \end{bmatrix}.$$

Thus, the column vector $(x,y)^t$ is mapped to $(-y,x)^t, (-x,-y)^t, (y,-x)^t$, and $(x,y)^t$, respectively, by these 4 matrices. Then the following program computes all $R_G(x^iy^j)$ with $i+j \leq 4$:[2]

——————————

```
off echo$
off nat$
load_package assist$
load_package groebner$
torder({x,y,z,u1,u2,u3},lex)$
init:={x^4, x^3*y, x^2*y^2, x^3, x^2*y, x^2, x*y, x}$
input1:=init;
input2:=sub({x=y,y=x},init);
num:=2$
input_all:={}$
j:=1$
while j leq num do
begin
name:=mkid(input,j)$
input_all:=append(input_all,name)$
j:=j+1$
end$
```

[2]A working program for REDUCE for this example is also available from: http://ch.catholic.or.kr/pundang/4/soh/2006_aa_i/Example_4_22.htm

```
input_all:=mkset(input_all);
j:=1$
reynolds:={}$
while j leq length(input_all) do
begin
f:=part(input_all,j)+sub({x=-y,y=x},part(input_all,j))
+sub({x=-x,y=-y},part(input_all,j))
+sub({x=y,y=-x},part(input_all,j))$
reynolds:=f.reynolds$
j:=j+1$
end$
reynolds:=mkset(reynolds);
reynolds := {4*x**2*y**2, - 2*x**3*y + 2*x*y**3,
2*x**2 + 2*y**2,
0,
2*x**3*y - 2*x*y**3, 2*x**4 + 2*y**4}$
```

Since $x^4 + y^4 = (x^2 + y^2)^2 - 2x^2y^2$, *we see that the three polynomials*

$$x^2y^2, x^2 + y^2, x^3y - xy^3$$

generate the ring $k[x, y]^G$ *of invariants by* G.

EXAMPLE 4.23. *For* $S_3 \in GL_3(k)$, *we already know that*

$$k[x, y, z]^{S_3} = k[\sigma_1, \sigma_2, \sigma_3],$$

where

$$\begin{aligned} \sigma_1 &= x + y + z, \\ \sigma_2 &= xy + xz + yz, \\ \sigma_3 &= xyz, \end{aligned}$$

are the three elementary symmetric functions. But, if we just apply the proof of Theorem 4.4 to figure out the generators of the ring of invariants, then since the number of basic monomials $x^iy^jz^k$ *of total degree* ≤ 6 *is* 22, *it will turn out that we in fact have to compute* 22 *number of* $R_G(x^iy^jz^k)$*'s. But, this is obviously too many, because we already know that we need only three generators,* $\sigma_1, \sigma_2,$ *and* σ_3.

At any rate, here is the actual computed results by applying the method in the proof of Noether's theorem, Theorem 4.4. At the end of the following program, we also apply Proposition 4.7 to see that all the computed generators are indeed polynomials in $\sigma_1, \sigma_2, \sigma_3$.[3]

———————

```
off echo$
off nat$
load_package assist$
load_package groebner$
torder({x,y,z,u1,u2,u3},lex)$
init:={x^6, x^5*y, x^4*y^2, x^4*y*z, x^3*y^3, x*3*y^2*z,
x^2*y^2*z^2, x^5, x^4*y, x^3*y^2, x^3*y*z, x^2*y^2*z,
x^4, x^3*y, x^2*y^2, x^2*y*z, x^3, x^2*y, x*y*z, x^2, x*y, x}$
input1:=init;
input3:=sub({x=y,y=x},init);
input4:=sub({x=y,y=z,z=x},init);
input2:=sub({y=z,z=y},init);
input5:=sub({x=z,z=x},init);
input6:=sub({x=z,y=x,z=y},init);
num:=6$
input_all:={}$
j:=1$
while j leq num do
begin
name:=mkid(input,j)$
input_all:=append(input_all,name)$
j:=j+1$
end$
```

[3]A working program for REDUCE for this example is also available from: http://ch.catholic.or.kr/pundang/4/soh/2006_aa_i/Example_4_23.htm

% The following line will produce a set consisting of all distinct 80 *monomials of total degree* ≤ 6.

```
input_all:=mkset(input_all);
length(input_all);
j:=1$
reynolds:={}$
while j leq length(input_all) do
begin
f:=part(input_all,j)+sub({x=y,y=x},part(input_all,j))
+sub({x=z,z=x},part(input_all,j))
+sub({y=z,z=y},part(input_all,j))
+sub({x=z,y=x,z=y},part(input_all,j))
+sub({x=y,y=z,z=x},part(input_all,j))$
reynolds:=f.reynolds$
j:=j+1$
end$
reynolds:=mkset(reynolds);
```

% Here are those 22 *computed distinct* $R_G(x^i y^j z^k)$ *with* $i + j + k \leq 6$, *and here are the results:*

```
reynolds := {6*x*y*z, 6*x**2*y**2*z**2, 6*x**2*y*z
+ 6*x*y**2*z + 6*x*y*z**2,2*x*y + 2*x*z + 2*y*z,
2*x**2*y**2 + 2*x**2*z**2 + 2*y**2*z**2,
2*x**2*y**2*z + 2*x**2*y*z**2 + 2*x*y**2*z**2,
2*x**3*y**3 + 2*x**3*z**3 + 2*y**3*z**3,
2*x + 2*y + 2*z, 2*x**2 + 2*y**2 + 2*z**2,
2*x**3 + 2*y**3 + 2*z**3, 2*x**2*y*z + 2*x*y**2*z
+ 2*x*y*z**2,
2*x**4 + 2*y**4 + 2*z**4, 2*x**3*y*z + 2*x*y**3*z
+ 2*x*y*z**3,
2*x**5 + 2*y**5 + 2*z**5, 2*x**4*y*z + 2*x*y**4*z
+ 2*x*y*z**4,
2*x**6 + 2*y**6 + 2*z**6,
```

```
x**2*y + x**2*z + x*y**2 + x*z**2 + y**2*z + y*z**2,
x**3*y + x**3*z + x*y**3 + x*z**3 + y**3*z + y*z**3,
x**3*y**2 + x**3*z**2 + x**2*y**3 + x**2*z**3
+ y**3*z**2 + y**2*z**3,
x**4*y + x**4*z + x*y**4 + x*z**4 + y**4*z + y*z**4,
x**4*y**2 + x**4*z**2 + x**2*y**4 + x**2*z**4
+ y**4*z**2 + y**2*z**4,
x**5*y + x**5*z + x*y**5 + x*z**5 + y**5*z + y*z**5}$
% Applying Proposition 4.7 below, we now want to check whether these
polynomials are polynomials in the three elementary symmetric functions
σ1, σ2, and σ3, and here are the results:
s1:=x+y+z$
s2:=x*y+x*z+y*z$
s3:=x*y*z$
gb:=groebner({s1-u1,s2-u2,s3-u3},{x,y,z,u1,u2,u3});
gb := {x + y + z - u1,
y**2 + y*z - y*u1 + z**2 - z*u1 + u2,
z**3 - z**2*u1 + z*u2 - u3}$
remainder:={}$
j:=1$
while j leq length(reynolds) do
begin
element:=part(reynolds,j)$
g:=preduce(element,gb,{x,y,z,u1,u2,u3})$
remainder:=g.remainder$
j:=j+1$
end$
remainder:=mkset(remainder);
remainder := {u1**4*u2 - u1**3*u3 - 4*u1**2*u2**2
+ 7*u1*u2*u3 + 2*u2**3 - 3*u3**2,
 - 2*u1**3*u3 + u1**2*u2**2 + 4*u1*u2*u3 - 2*u2**3
- 3*u3**2,
```

```
u1**3*u2 - u1**2*u3 - 3*u1*u2**2 + 5*u2*u3,
 - 2*u1**2*u3 + u1*u2**2 - u2*u3,
u1**2*u2 - u1*u3 - 2*u2**2, u1*u2 - 3*u3,
2*u1**6 - 12*u1**4*u2 + 12*u1**3*u3 + 18*u1**2*u2**2
- 24*u1*u2*u3 - 4*u2**3 + 6*u3**2,
2*u1**3*u3 - 6*u1*u2*u3 + 6*u3**2,
2*u1**5 - 10*u1**3*u2 + 10*u1**2*u3 + 10*u1*u2**2
- 10*u2*u3,
2*u1**2*u3 - 4*u2*u3,
2*u1**4 - 8*u1**2*u2 + 8*u1*u3 + 4*u2**2, 2*u1*u3,
2*u1**3 - 6*u1*u2 + 6*u3, 2*u1**2 - 4*u2,
2*u1, - 6*u1*u2*u3 + 2*u2**3 + 6*u3**2,
2*u2*u3, - 4*u1*u3 + 2*u2**2, 2*u2,
6*u1*u3, 6*u3**2, 6*u3}$
```

% This verifies that each of the 22 generators of $k[x,y,z]^{S_3}$ *computed by the method of the proof of Theorem 4.4 is a polynomial in* σ_1, σ_2*, and* σ_3.

REMARK 4.17. *On the other hand, it is well-known that there are more efficient methods for finding a generating set of the ring of invariants, based on Molien's Theorem in Chapter 2 of B. Sturmfels book,* Algorithms in Invariant Theory.

At any rate, we now know how to compute the generators $\{f_1, ..., f_m\}$ of $k[x_1, ..., x_n]^G$ for a subgroup $G \subset GL_n(k)$ so that

$$k[x_1, ..., x_n]^G = k[f_1, ..., f_m].$$

The next question is whether we can decide a given $f \in k[x_1, ..., x_n]$ is an element of $k[x_1, ..., x_n]^G = k[f_1, ..., f_m]$ or not, and if it is the case, we want to express it as a polynomial in the generators $f_1, ..., f_m$.

For this purpose, we have the following proposition which says more than what Proposition 4.1 says, although its statement is almost the same as Proposition 4.1.

PROPOSITION 4.7. *Let $f_1, ..., f_m \in k[x_1, ..., x_n]$ be given. We choose a monomial ordering on $k[x_1, ..., x_n, y_1, ..., y_m]$ satisfying that any monomial involving one of the x_i's is bigger than all monomials in $k[y_1, ..., y_m]$. [Such a monomial ordering is called an* ***elimination ordering*** *(Definition 5.2 in Chapter 5]. Let G be a Groebner basis for the ideal $(f_1 - y_1, ..., f_m - y_m)$ of the polynomial $k[x_1, ..., x_n, y_1, ..., y_m]$. Given $f \in k[x_1, ..., x_n]$, let $g' = \bar{f}^G$ be the remainder of f on division by G. That is,*

$$f \xrightarrow{G}_+ g'.$$

Then

(1) *$f \in k[f_1, ..., f_m]$ if and only if $g' \in k[y_1, ..., y_m]$.*
(2) *If $f \in k[f_1, ..., f_m]$, then $f = g'(f_1, ..., f_m)$ is a unique polynomial expression of f as a polynomial in $f_1, ..., f_m$.*

REMARK 4.18. *One of such monomial ordering is the dictionary ordering, i.e., the lex order on $k[x_1, ..., x_n, y_1, ..., y_n]$ with*

$$x_1 > x_2 > ... > x_n > y_1 > ... > y_n.$$

PROOF. Part (2) follows immediately from part (1).

Part (1). By Division Algorithm (Algorithm 5), we have

$$f = a_1 g_1 + \cdots + a_t g_t + g$$

for some non-zero polynomials $a_i \in k[x_1, ..., x_n, y_1, ..., y_m]$. We know that the ideal $I = (f_1 - y_1, ..., f_m - y_m)$ is also generated by G, so that each

$$g_i = h_1(f_1 - y_1) + \cdots + h_m(f_m - y_m),$$

for some $h_i \in k[x_1, ..., x_n, y_1, ..., y_m]$. Now, we suppose that the remainder $g \in k[y_1, ..., y_m]$. If we substitute each y_i with f_i in both sides of this expression, each g_i becomes zero, but the $f \in k[x_1, ..., x_n]$ in the left side is the same, because there is no place for substitution. Thus,

$$f = g(f_1, ..., f_m).$$

Conversely, suppose that $f \in k[f_1, ..., f_m]$. Thus there exists

$$g \in k[y_1, ..., y_m]$$

such that

$$f = g(f_1, ..., f_m).$$

Then since f is a polynomial in the f_i's, it is a k-linear combination

$$f = \sum_{\alpha} k_\alpha f_1^{\alpha_1} f_2^{\alpha_2} \cdots f_m^{\alpha_m}$$

of the various monomials

$$f_1^{\alpha_1} f_2^{\alpha_2} \cdots f_m^{\alpha_m}.$$

Now we need to rewrite each of these monomials

$$\begin{aligned} f_1^{\alpha_1} f_2^{\alpha_2} \cdots f_m^{\alpha_m} &= (y_1 + (f_1 - y_1))^{\alpha_1} \cdots (f_m + (f_m - y_m))^{\alpha_m} \\ &= y_1^{\alpha_1} \cdots y_m^{\alpha_m} + \text{ an element } h_\alpha \text{ in } (f_1 - y_1, ..., f_m - y_m). \end{aligned}$$

Thus,

$$\begin{aligned} f &= \sum_{\alpha} k_\alpha y_1^{\alpha_1} \cdots y_m^{\alpha_m} + \sum_{\alpha} k_\alpha h_\alpha \\ &= g(y_1, ..., y_m) + \sum_{\alpha} k_\alpha h_\alpha. \end{aligned}$$

That is, we have

$$f = A_1(f_1 - y_1) + \cdots + A_m(f_m - y_m) + g(y_1, ..., y_m), \tag{4.10}$$

where $A_i \in k[x_1, ..., x_n, y_1, ..., y_m]$.

REMARK 4.19. *We note that the proof of this proposition until this point is exactly the same as that of Proposition 4.1. But, from now on, we can not copy the rest of the proof of Proposition 4.1, because in the case of the proof of Proposition 4.1 we in fact relied on the uniqueness claim in Theorem 4.1, while in the case of a general subgroup $G \subset GL_n(k)$ we do not have such a strong uniqueness claim as in Theorem 4.1.*

We try to divide $g = g(y_1, ..., y_m)$ by those elements g_i of $G = \{g_1, ..., g_t\}$ which are polynomials in $y_1, ..., y_m$. That is, renumbering if necessary, we let

$$G' := G \cap k[y_1, ..., y_m] = \{g_1, ..., g_s\}$$

with $s \leq t$. Then, applying Algorithm 5 (Division Algorithm in $k[y_1, ..., y_m]$), we divide out g by these g_s according to we have

$$g = B_1 g_1 + \cdots + B_s g_s + g',$$

were $B_i, g' \in k[y_1, ..., y_m]$, thus, no term of g' is divisible by any $LT(g_1)$ for $i = 1, ..., s$. Combining this with the above expression for f we have

$$f = A_1'(f_1 - y_1) + \cdots + A_m'(f_m - y_m) + g'.$$

Then since we can express each $f_i - y_i$ in terms of the elements of $G = \{g_1, ..., g_t\}$, we have

$$f = C_1 g_1 + \cdots + C_t g_t + g',$$

where $C_i \in k[x_1, ..., x_n, y_1, ..., y_m]$.

If we can show that this g' is the remainder of f by the Groebner basis G, then we are done, because then it is uniquely determined (Theorem 2.4).

Suppose that there exists $g_i \in G$ such that $LT(g_i)$ divides some term of g'. Then $LT(g_i)$ involves only $y_1, ..., y_m$ and hence $g_i \in G' = G \cap k[y_1, ..., y_m]$. This means that there exists $g_i \in G'$ such that $LT(g_i)$ divides some term of g', but this is impossible because g' was already the remainder of g by those $g_i \in G'$. Therefore, g' is indeed the unique remainder of f by the Groebner basis G, as is claimed in the statement of the proposition, and this finishes the proof. □

EXAMPLE 4.24. *An example of the application of this proposition is given in Example 4.23.*

Exercises for Section 4.4

EXERCISE 4.5. *Applying Theorem 4.4, compute the generators for*

$$k[x_1, x_2, x_3, x_4]^{V_4},$$

where

$$V_4 = \left\{ \pm \begin{pmatrix} 1 & 0 \\ 0 & 1 \end{pmatrix}, \pm \begin{pmatrix} 0 & 1 \\ 1 & 0 \end{pmatrix} \right\} \in GL_2(k).$$

EXERCISE 4.6. *Applying Theorem 4.4, compute the generators for*

$$k[x_1, x_2, x_3, x_4]^{S_4}.$$

Express each generator in terms of the 4 *elementary symmetric functions, as we have done in Example 4.23 in the case of* S_3.

CHAPTER 5

Elimination Theory and Polynomial Mappings

In the previous chapters we have seen the usefulness of the lexicographical ordering, i.e., lex order, or equivalently, the dictionary ordering, on the set of monomials in $k[x_1, ..., x_n]$. We have also discussed very briefly in Section 1 of Chapter 3 that the dictionary ordering has a computational drawback and the idea on how we may overcome such a difficulty. We also discussed in Section 5 of Chapter 3 the importance of eliminating variables in order to express the given parametrically represented object in terms of the coordinates only. Among other things, this point of view on elimination of variables enables us to have a very significant result, namely, we can compute the generators of the intersection ideal for a given two polynomial ideals in $k[x_1, ..., x_n]$, and its consequence is basically all about those results in this chapter.

1. Ideal Theory Without Groebner Basis

We start our discussion by reviewing what we have done in Section 5 in Chapter 1. In particular, Proposition 1.8 tells us that if we know the generating sets of two ideals I and J, respectively, then we also know the generating sets of the ideals, the sum $I + J$ and the product $I \cdot J$. As is explicitly described in Problem 1 there, we do not in general know how to compute the generators for the ideal, the intersection of I and J, even when the ideals I and J are finitely generated polynomial ideals in $k[x_1, ..., x_n]$.

It is usually the case that the canonical projection homomorphism

$$k[x_1, ..., x_n] \longrightarrow k[x_1, ..., x_n]/I$$

behaves very well, but the insertion homomorphism to a bigger ring does not. First of all, insertion homomorphisms are not usually canonically determined; and moreover the inserted one does not satisfy certain property of its preimage:

EXAMPLE 5.1. *In the ring* $\mathbb{Z}[\sqrt{-5}]$ *of Gauss integers of integers, which is an integral domain consisting of the numbers of the form* $a+b\sqrt{-5}$ *with* $a, b \in \mathbb{Z}$, *we have two different factorizations:*

$$6 = 2 \cdot 3 = (1 + \sqrt{-5}) \cdot (1 - \sqrt{-5}).$$

This tells us that both prime integers 2 *and* 3 *are irreducibles but they are not prime elements of* $\mathbb{Z}[\sqrt{-5}]$ *anymore.*

This is really a disturbing evidence related to the insertion homomorphism to a bigger ring.

On the other hand, if we move one level up, i.e., if we consider ideals, then we have a very nice result, which is proved long before 1965 by Commutative Ring theorists, perhaps originally by David Hilbert: Let I and J be two polynomial ideals in $k[x_1, ..., x_n]$. Then I is again a subset of $k[x_1, ..., x_n, y]$ under the insertion homomorphism

$$i : k[x_1, ..., x_n] \longrightarrow k[x_1, ..., x_n, y],$$

sending

$$f(x_1, ..., x_n) \longmapsto f(x_1, ..., x_n),$$

where y is an arbitrary new unknown. But, for an obvious reason, I is not an ideal of $k[x_1, ..., x_n, y]$ because it is not closed under multiplication by the new variable y. On the other hand, if we define, for any polynomial $f(y) \in k[y]$:

DEFINITION 5.1. $f(y)I$ *to be the ideal generated by the set of polynomials* $\{f(y)h(x_1, ..., x_n) : h(x_1, ..., x_n) \in I\}$, *that is,*

$$f(y)I = \{\sum_{i=1}^{finite} q_i(x, y)f(y)h_i(x) : h_i(x) \in I, q_i \in k[x_1, ..., x_n, y]\},$$

where we put $x = (x_1, ..., x_n)$ *for simplicity reason.*

then $f(y)I$ is by definition an ideal of $k[x_1, ..., x_n, y]$. But, its preimage $i^{-1}(f(y)I) = f(y)I \cap k[x_1, ..., x_n] = (0)$, so at a first glance it seems that the ideal $f(y)I$ is nothing to do with the ideal I we started with.

In what follows, we denote $(x_1, ..., x_n)$ by x only for simplicity reason. Thus, $p(x) = p(x_1, ..., x_n)$ for instance.

But, if we go one level down, we have:

LEMMA 5.1. *Under the above setting, we have*

(1) *If* $\{p_1(x), ..., p_r(x)\}$ *is a generator of the ideal* I *of* $k[x_1, ..., x_n]$, *then*

$$\{f(y)p_1(x), ..., f(y)p_r(y)\}$$

is a generator of the ideal $f(y)I$ *of* $k[x_1, ..., x_n, y]$.

(2) *If* $g(x, y) \in g(y)I$, *then* $g(x, \alpha) \in I$ *for any* $\alpha \in k$.

PROOF. Part (2) is a simple result of part (1), and the proof of part (1) is very routine so we leave it as an exercise. □

On the other hand, the sum $yI + (1-y)I$ is the ideal of $k[x_1, ..., x_n, y]$ obviously containing the subset I of $k[x_1, ..., x_n, y]$; hence

$$I = i^{-1}(yI + (1-y)I) = (yI + (1-y)I) \cap k[x_1, ..., x_n]$$

is the ideal I of $k[x_1, ..., x_n]$ we started with. Applying this idea one more time, we have

$$\begin{aligned} I \cap J &= i^{-1}(yI + (1-y)I) \cap i^{-1}(yJ + (1-y)J) \\ &= i^{-1}((yI + (1-y)I) \cap (yJ + (1-y)J)). \end{aligned}$$

Thus, we begin to wonder what the ideal $(yI+(1-y)I) \cap (yJ+(1-y)J)$ of $k[x_1, ..., x_n, y]$ is.

But the surprising result is:

PROPOSITION 5.1. $I \cap J = i^{-1}(yI + (1-y)J)$.

PROOF. Let $f \in I \cap J$. Then since we have a trivial identity,

$$f = yf + (1-y)f,$$

we obviously have $f \in (yI + (1-y)J) \cap k[x_1, ..., x_n]$.

Conversely, suppose that $f \in (yI + (1-y)J) \cap k[x_1, ..., x_n]$. Then we have

$$f(x_1, ..., x_n) = p(x_1, ..., x_n, y) + q(x_1, ..., x_n, y),$$

for some $p(x, y) \in yI$ and $q(x, y) \in (1-y)I$, where $p(x, y)$ is a multiple of y and $q(x, y)$ is a multiple of $1-y$. Thus, putting $y = 1$ and $y = 0$, respectively, the last equality gives, by Lemma 5.1 (b):

$$\begin{aligned} f(x_1, ..., x_n) &= p(x_1, ..., x_n, 1) + 0 \in I, \\ f(x_1, ..., x_n) &= 0 + q(x_1, ..., x_n, 0) \in J. \end{aligned}$$

Thus, $f \in I \cap J$. □

REMARK 5.1. *Thus, we begin to wonder whether there might exist a relationship between* $(yI + (1-y)J)$ *and* $(yI + (1-y)I) \cap (yJ + (1-y)J)$. *Are they the same or not?*

We then recall by Hilbert Basis Theorem, Theorem 1.4 or Theorem 2.2, that polynomial rings over a field k are all Noetherian, that is, polynomial ideals over k are always finitely generated over the bottom field k. Thus, given polynomial ideals I and J, they are finitely generated and moreover the new ideal of $k[x_1, ..., x_n, y]$,

$$yI + (1-y)J$$

is also finitely generated over k, and this is all we can say without the notion of Groebner basis.

As we shall see very soon in what follows immediately, The Extension Theorem, Theorem 5.1 or its corollary, Corollary 5.2, surprisingly enables us to compute the generators of $I \cap J$ from those of the ideals I and J.

Exercises for Section 5.1

EXERCISE 5.1. *Prove Lemma 5.1.*

2. Polynomial Ideal Theory With Groebner Basis

2.1. Computation of Generators for the Intersection of Two Ideals. When elimination of certain variables is our main concern, we need to distinguish those variables to be eliminated and others not to be eliminated.

DEFINITION 5.2. *Let* $X_1 = x_1^{\alpha_1} \cdots x_n^{\alpha_n}$, $X_2 = x_1^{\beta_1} \cdots x_n^{\beta_n}$, $Y_1 = y_1^{\gamma_1} \cdots y_m^{\gamma_m}$, *and* $Y_2 = y_1^{\delta_1} \cdots y_m^{\delta_m}$ *be monomials in the variables* $x_1, ..., x_n, y_1, ..., y_m$. *We define an* ***elimination order*** *with the variables* $x_1, ..., x_n$ *bigger than the variables* $y_1, ..., y_m$ *on the set of monomials in the variables* $x_1, ..., x_n$ *and* $y_1, ..., y_m$ *as follows:*

$$X_1Y_1 > X_2Y_2 \text{ if and only if ``} X_1 >_x X_2\text{'' or ``}X_1 = X_2 \text{ and } Y_1 >_y Y_2\text{'',}$$

where $>_x$*is a monomial orders on the set of monomials in the variables* $x_1, ..., x_n$, *and* $>_y$*is a monomial orders on the set of monomials in the variables* $y_1, ..., y_m$, *respectively.*

COROLLARY 5.1. *Under the setting in Definition 5.2, we have*

1. *An elimination order is a monomial order.*
2. *If* Y *is a monomial in the variables* $y_1, ..., y_m$ *and* Z *is a monomial in the variables* $x_1, ..., x_n, y_1, ..., y_m$ *such that one* x_i *of the variables* $x_1, ..., x_n$ *appears in* Z, *then* $Z > Y$ *under the elimination order with the variables* $x_1, ..., x_n$ *bigger than the variables* $y_1, ..., y_m$.

PROOF. We leave the proofs as an exercise. □

EXAMPLE 5.2. *If* $>_x$ *and* $>_y$ *are the dictionary orders, respectively, then the dictionary order on both variables together is just the elimination order we defined above.*

THEOREM 5.1. ***(The Elimination Theorem)*** *Let* I *be a nonzero ideal of* $k[x_1, ..., x_n, y_1, ..., y_m]$ *under an elimination order with the variables* $x_1, ..., x_n$ *bigger than* $y_1, ..., y_m$. *If* $G = \{g_1, ..., g_t\}$ *is a Groebner basis for* I, *then* $G \cap k[y_1, ..., y_m]$ *is a Groebner basis for the ideal, the* *inverse image* $i^{-1}(I) = I \cap k[y_1, ..., y_m]$, *of the ideal* I *under the insertion*

$$i : k[y_1, ..., y_m] \hookrightarrow k[x_1, ..., x_n, y_1, ..., x_m].$$

PROOF. We first note that the inverse image of an ideal is again an ideal. Then since G is a generating set of the ideal I, we clearly have that $G \cap k[y_1, ..., y_m] \subset I \cap k[y_1, ..., y_m]$. If we can prove that $LM(f)$ is divisible by $LP(g_i)$ for some $g_i \in G \cap k[y_1, ..., y_m]$ for any non-zero polynomial $f \in I \cap k[y_1, ..., y_m]$, then we are done by part (1) of Theorem 2.3.

Let f be a non-zero polynomial in $I \cap k[y_1, ..., y_m]$. Then since it is a polynomial in I with a Groebner basis G, by Theorem 2.3 (1), we know there exists $g_i \in G$ such that $LM(g_i)$ divides $LM(f)$. Then since f is a polynomial in $y_1, ..., y_m$ only, $LM(f)$ is a monomial in the variables $y_1, ..., y_m$; hence so is $LM(g_i)$; hence $g_i \in G \cap k[y_1, ..., y_m]$. Thus, again by Theorem 2.3, we are done. □

REMARK 5.2. *In the above proof, we are in fact using the same trick used at the very last part of the proof of both Proposition 4.1 and Proposition 4.7. The ideal* $i^{-1}(I) = I \cap k[y_1, ..., y_m]$ *is called an* ***elimination ideal*** *of* I*, since the variables* $x_1, ..., x_n$ *are all eliminated.*

COROLLARY 5.2. *Let* I *be a non-zero ideal of* $k[x_1, ..., x_n]$ *under the lex order with* $x_1 > x_2 > \cdots > x_n$. *If* $G = \{g_1, ..., g_t\}$ *is a Groebner basis for* I*, then for each* $j = 0, ..., n$, $G \cap k[x_{j+1}, ..., x_n]$ *is a Groebner basis for the ideal, the inverse image* $I_j = I \cap k[x_{j+1}, ..., x_n]$, *of the ideal* I *under the insertion*

$$i : k[x_{j+1}, ..., y_n] \hookrightarrow k[x_1, ..., x_n].$$

Thus, Lemma 5.1 plus Proposition 5.1 plus Theorem 5.1 (or Corollary 5.2) provides us a very important algorithm for computing the generators of the intersection of two polynomial ideals I and J.

ALGORITHM 11. *(Algorithm for Computing Generators for the ideal* $I \cap J$*) Let* $I = (f_1, ..., f_r)$ *and* $J = (h_1, .., h_s)$ *be two polynomial ideals of*

$k[x_1, ..., x_n]$. *We know by Proposition 5.1 that*

$$I \cap J = (yI + (1-y)J) \cap k[x_1, ..., x_n].$$

To compute the generators of their intersection ideal $I \cap J$, *we do the followings:*

(1) *Compute a Groebner basis* $G = \{g_1, ..., g_t\}$ *of the ideal*

$$(yf_1, ..., yf_r, (1-y)h_1, ..., (1-y)h_s)$$

of $k[y, x_1, ..., x_n]$ *under an elimination order with the variable* y *is bigger than the variables* $x_1, ..., x_n$.

(2) *Then by Theorem 5.1,* $G \cap k[x_1, ..., x_n]$ *is a Groebner basis of the ideal* $I \cap J$.

REMARK 5.3. *The readers should recognize the above algorithm's full significance.*

EXAMPLE 5.3. *Let* $I = (x^2 + y^3 - 1, x - xy + 3), J = (x^2y - 1)$ *be two ideals of* $\mathbb{Q}[x, y]$. *To compute the generators of* $I \cap J$, *we compute a Groebner basis of the ideal*

$$(z(x^2 + y^3 - 1), z(x - xy + 3), (1-z)(x^2y - 1)) \subset k[z, x, y]$$

under an elimination order with the variable z *bigger than the variables* x *and* y *where* $>_z$*the lex order, and* $>_{x,y}$ *is the graded lex order. We use the REDUCE:*[1]

———————————

```
off echo$
off nat$
load_package groebner$
torder({z,x,y},lexgradlex,1)$
f1:=z*(x^2+y^3-1);
f2:=z*(x-x*y+3);
f3:=(1-z)*(x^2*y-1);
```

[1]A working program for REDUCE for this example is also available from: http://ch.catholic.or.kr/pundang/4/soh/2006_aa_i/Example_5_03.htm

```
result:=groebner({f1,f2,f3});
{12853*z + 118*x**4*y + 9*x**2*y**3 - 357*x**3*y
- 972*x**2*y**2 + 2152*x**2*y -
118*x**2 - 9*y**2 + 357*x + 972*y - 2152,
x**5*y + 3*x**2*y**3 + 3*x**2*y**2 - x**3
+ 3*x**2*y - 3*y**2 - 3*y - 3,
x**2*y**4 + x**4*y - x**2*y - y**3 - x**2 + 1,
x**3*y**2 - x**3*y - 3*x**2*y - x*y + x + 3}$
```

———————

Thus, the ideal $I \cap J$ is generated by the following three polynomials:

$$
\begin{aligned}
g_1 &= x^5y + 3x^2y^3 + 3x^2y^2 - x^3 + 3x^2y - 3y^2 - 3y - 3, \\
g_2 &= x^2y^4 + x^4y - x^2y - y^3 - x^2 + 1, \\
g_3 &= x^3y^2 - x^3y - 3x^2y - xy + x + 3.
\end{aligned}
$$

If we choose different elimination order, then the Groebner basis for $I \cap J$ is going to be different:

EXAMPLE 5.4. *The same as Example 5.3, but this time we use a different elimination order. We use the REDUCE:*[2]

———————

```
torder({z,x,y},lex);
result2:=groebner({f1,f2,f3});
result2:= {12853*z + 119*x**2*y**5 - 237*x**2*y**4
+ 9*x**2*y**3 - 1091*x**2*y**2 + 2389*x**2*y
- 119*y**4 + 237*y**3 - 9*y**2 + 1091*y - 2389,
3*x**3*y + x**2*y**5 - x**2*y**4 - x**2*y**2
+ x**2*y - 3*x - y**4 + y**3 + y - 1,
x**2*y**6 - 2*x**2*y**5 + x**2*y**4 - x**2*y**3
+ 2*x**2*y**2 + 8*x**2*y - y**5 + 2*y**4 - y**3
+ y**2 - 2*y - 8}$
```

———————

[2]A working program for REDUCE for this example is also available from: http://ch.catholic.or.kr/pundang/4/soh/2006_aa_i/Example_5_04.htm

Thus, the ideal $I \cap J$ is also generated by the following two polynomials

$$\begin{aligned} h_1 &= 3x^3y + x^2y^5 - x^2y^4 - x^2y^2 + x^2y - 3x - y^4 + y^3 + y - 1, \\ h_2 &= x^2y^6 - 2x^2y^5 + x^2y^4 - x^2y^3 + 2x^2y^2 + 8x^2y - y^5 \\ &\quad +2y^4 - y^3 + y^2 - 2y - 8. \end{aligned}$$

EXAMPLE 5.5. *We now verify that $(g_1, g_2, g_3) = (h_1, h_2)$ using Program 2 of Chapter 3, and here are the computed results:*

$$\begin{aligned} h_1 &= (y-1)g_2 - xg_3, \\ h_2 &= (y^2 - 2y + 1)g_2 + (-xy + 3x - 3)g_3, \\ g_1 &= \frac{1}{9}(3x^2 - xy^4 + xy^3 + xy - x - 3y^3 + 3)h_1 \\ &\quad + \frac{1}{9}(xy^3 - x + 3y^2 + 3y + 3)h_2, \\ g_2 &= \frac{1}{9}(3x - y^4 + y^3 + y - 1)h_1 + \frac{1}{9}(y^3 - 1)h_2, \\ g_3 &= \frac{1}{3}(y-1)h_1 - \frac{1}{3}h_2. \end{aligned}$$

EXERCISE 5.2. *Verify those equalities in the above example.*

2.2. Computation of gcd and lcm of Two Polynomials. We recall that $k[x_1, ..., x_n]$ is not a *PID* but a *UFD*, and hence the greatest common divisor of two polynomials $f, g \in k[x_1, ..., x_n]$ and the least common multiple of $f, g \in k[x_1, ..., x_n]$

$$\begin{aligned} &\gcd(f, g), \\ &\operatorname{lcm}(f, g), \end{aligned}$$

do exist in $k[x_1, ..., x_n]$, and their product is equal to fg,

$$\gcd(f, g) \cdot \operatorname{lcm}(f, g) = fg.$$

We have a small lemma:

LEMMA 5.2. *Let $f, g \in k[x_1, ..., x_n]$. Then*

$$(f) \cap (g) = (\operatorname{lcm}(f, g)).$$

PROOF. Write $l = \text{lcm}(f,g)$. Then by definition, f divides l and g divides l, and hence l belongs to the intersection of the two principal ideals $(f) \cap (g)$. Conversely, if $h \in (f) \cap (g)$, then f divides h and g divides h; hence by definition of $l = \text{lcm}(f,g)$, l divides h. □

ALGORITHM 12. *(Algorithm for* $\text{lcm}(f,g)$ *and* $\gcd(f,g)$*) Let* f, g *be polynomials in* $k[x_1, ..., x_n]$.

(1) *To compute* $\text{lcm}(f,g)$*, apply Algorithm 11.*
(2) *To compute* $\gcd(f,g)$*, apply the division algorithm, Algorithm 5 to the right side of the formula:*

$$\gcd(f,g) = \frac{fg}{\text{lcm}(f,g)}.$$

Moreover, in order to compute the lcm and gcd of more than 2 polynomials, we use this algorithm repeatedly to the relations:

$$\begin{aligned} \text{lcm}(f,g,h) &= \text{lcm}(f, \text{lcm}(g,h)), \\ \gcd(f,g,h) &= \gcd(f, \gcd(g,h)). \end{aligned}$$

REMARK 5.4. *The main point of this algorithm is that it enables us to compute* $\text{lcm}(f,g)$ *and* $\gcd(f,g)$ *of* f, g *in* $k[x_1, ..., x_n]$ *without factorization of* f *and* g.

EXAMPLE 5.6. *Let*

$$\begin{aligned} f &= x^4 + x^3 y + x^3 z^2 - x^2 y^2 + x^2 y z^2 - y^3 z, \\ g &= x^4 + 2x^3 y^2 - y^2 z^4. \end{aligned}$$

We want to compute $\text{lcm}(f,g)$ *and* $\gcd(f,g)$ *using the REDUCE:*[3]

———————–

off echo$
off nat$
load_package groebner$

[3]A working program for REDUCE for this example is also available from: http://ch.catholic.or.kr/pundang/4/soh/2006_aa_i/Example_5_06.htm

torder({w,x,y,z},lexrevgradlex,1)$ % This is an elimination order we choose for the current computation.

```
f1:=w*(x^4+x^3*y+x^3*z^2-x^2*y^2+x^2*y*z^2-y^3*z);
f2:=(1-w)*(x^4+2*x^3*y^2-y^2*z^4);
groebner({f1,f2});
{w*x**7 + 2*w*x**6*y**2 + w*x**6*y + 2*w*x**5*y**3
- w*x**5*y**2 - w*x**4*y**3 + w*x**3*y**4
- w*x**2*y**4*z**2 + w*x*y**5*z - w*y**5*z**3 - x**7
- 2*x**6*y**2 - x**6*y - 2*x**5*y**3 + x**3*y**2*z**4
+ x**2*y**3*z**4,
- w*x**4 - 2*w*x**3*y**2 + w*y**2*z**4 + x**4
+ 2*x**3*y**2 - y**2*z**4,
w*(x**4 + x**3*y + x**3*z**2 - x**2*y**2 + x**2*y*z**2
- y**3*z),
- x**8 - 2*x**7*y**2 - x**7*y - x**7*z**2 - 2*x**6*y**3
- 2*x**6*y**2*z**2 + x**6*y**2 - x**6*y*z**2 + 2*x**5*y**4
- 2*x**5*y**3*z**2 + x**4*y**3*z + x**4*y**2*z**4
+ 2*x**3*y**5*z + x**3*y**3*z**4 + x**3*y**2*z**6
- x**2*y**4*z**4 + x**2*y**3*z**6 - y**5*z**5}$
```

———————

Therefore,

$$\begin{aligned}\operatorname{lcm}(f,g) &= -(-x^8 - 2x^7y^2 - x^7y - x^7z^2 - 2x^6y^3 - 2x^6y^2z^2 + x^6y^2 \\ &\quad -x^6yz^2 + 2x^5y^4 - 2x^5y^3z^2 + x^4y^3z + x^4y^2z^4 + 2x^3y^5z \\ &\quad +x^3y^3z^4 + x^3y^2z^6 - x^2y^4z^4 + x^2y^3z^6 - y^5z^5), \\ \gcd(f,g) &= 1.\end{aligned}$$

Here, $\gcd(f,g) - 1$ *because we easily check that thc product* $fg = \operatorname{lcm}(f,g)$.

If one chooses the lex order instead of the above order as an elimination order, that is, the lex order for the variable w *and the reversed graded lex order for the variables* $x > y > z$, *where* w *is bigger than other variables, then one will find that it takes too much time in computing* $\operatorname{lcm}(f,g)$.

Thus, it is very important to choose a local monomial order other than the lex order, and it is well-known that the reversed graded lex order provides the most efficient computation in many cases.

REMARK 5.5. *The algorithm for* $\gcd(f, g)$ *in Algorithm 12 above is not very efficient and we should point out that more efficient algorithms for* $\gcd(f, g)$ *are implemented in most Computer Algebra Systems.*

For example, if we use the REDUCE command $\operatorname{lcm}(f, g)$ *for the above example, then we will get the same result immediately.*

2.3. Computation of Ideal Quotients. Proposition 5.1 plus Theorem 5.1 (or Corollary 5.2) or Algorithm 11 has another application which is definitely needed in computing a defining equation for the smallest affine variety containing the difference of two affine varieties, $V - W$, in Algebraic Geometry.

DEFINITION 5.3. *Let* I *and* J *be ideals in a ring* R. *The* ***ideal quotient*** $J : I$ *is defined to be*

$$J : I = \{g \in R : gI \subseteq J\}.$$

One easily checks that $J : I$ is an ideal of R. We then have an immediate lemma:

LEMMA 5.3. *Under the above setting, we have:*

(1) $J : R = J$.
(2) $JI \subset K$ *if and only if* $J \subset K : I$.
(3) $I \subset J$ *if and only if* $J : I = R$.
(4) $(K : J) : I = K : JI$.

PROOF. The proofs of these are left as an exercise. □

PROPOSITION 5.2. *Let* $I = (f_1, ..., f_s)$ *and* J *be polynomial ideals in* $k[x_1, ..., x_n]$. *Then*

$$J : I = \bigcap_{i=1}^{s} J : (f_i).$$

PROOF. If $g \in J : I$, then $gI \subset J$ and hence $gf_i \in J$ for each $i = 1, ..., s$; hence $g \in \cap_{i=1}^{s} J : (f_i)$. Conversely, if $g \in \cap_{i=1}^{s} J : (f_i)$, then $g(f_i) \subset J$ for each $i = 1, ..., s$; hence $gI \subset J$; hence $g \in J : I$. □

Moreover, we have

THEOREM 5.2. *Let J be an ideal in a ring R and f be a non-zero element of R. Then*

$$J : (f) = \frac{1}{f}(J \cap (f)).$$

PROOF. If $g \in \frac{1}{f}(J \cap (f))$, then $gf \in J \cap (f) \subset J$; hence $g \in J : (f)$. Conversely, if $g \in J : (f)$, then $gf \in J$; hence $gf \in J \cap (f)$. Therefore, $g \in \frac{1}{f}(J \cap (f))$. □

COROLLARY 5.3. *Let J be an ideal of $k[x_1, ..., x_n]$ and f be a non-zero polynomial in $k[x_1, ..., x_n]$. If $J \cap (f) = (g_1, ..., g_s)$, then*

$$J : (f) = (g_1/f, ..., g_s/f).$$

COROLLARY 5.4. *Let J be an ideal of $k[x_1, ..., x_n]$ and f be a non-zero polynomial in $k[x_1, ..., x_n]$. If $J \cap (f)$ is a monomial ideal, then so is $J : (f)$.*

PROOF. This is true because, $J : (f)$ is generated by those

$$g_i / \gcd(g_i, f), \ i = 1, ..., s.$$

□

Therefore, when $R = k[x_1, ..., x_n]$, Proposition 5.2, Theorem 5.2, and Algorithm 11 enable us to have an algorithm for computing the generators of the quotient ideal $J : I$ in $k[x_1, ..., x_n]$.

ALGORITHM 13. *Let $J = (h_1, ..., h_s)$ and $I = (f_1, ..., f_r)$ be two polynomial ideals in $k[x_1, ..., x_n]$. To compute the generators of $J : I$, we do the followings:*

(1) *Apply Algorithm 11 to compute a Groebner basis for the intersection ideal $(h_1, ..., h_s) \cap (f_i)$ for each $i = 1, ..., r$.*

(2) *Apply the division algorithm, Algorithm 5 to get a generating set for each* $(h_1, ..., h_s) \cap (f_i)$*, using Theorem 5.2.*

(3) *Finally, we apply Algorithm 11* $r - 1$ *times, computing first a generating set for* $J : (f_1, f_2) = (J : (f_1)) \cap (J : (f_2))$*, then* $J : (f_1, f_2, f_3) = (J : (f_1, f_2)) \cap (J : (f_3))$*, etc.*

Exercises for Section 5.2

EXERCISE 5.3. *Prove Corollary 5.1.*

EXERCISE 5.4. *Prove Lemma 5.3.*

3. Polynomial Mappings Versus k-Algebra Homomorphisms?

So far, we have seen several applications of the theory of Groebner basis mostly in Algebra. This means to some of us that our exposition has been too dry or tasteless in a sense, except for those applications to Invariant Theory. In this section, we are going to learn how to compute the generators of the kernel of a k-algebra homomorphism. We note that such a kernel in fact corresponds to the defining equation for an affine variety in Algebraic Geometry.

We recall what we have seen before: In Remark 1.10 in Chapter 1 and Remark 4.15 in Chapter 4, we discussed how Geometry and Algebra are connected. There, we have explicitly seen a map on geometry (which is usually called a morphism), for instance, a linear map, defines the corresponding k-algebra homomorphism, and vice versa. Another thing we are going to see eventually is that there is a correspondence between geometric objects, i.e., affine (algebraic) varieties, and algebraic objects, i.e., finitely generated k-algebras.

DEFINITION 5.4. *Let* k *be a fixed field.*

(1) *A ring* R *with a ring homomorphism from* k *to* R *is called a* k***-algebra****.*

(2) *Let A and B be two k-algebras. A k-linear map: $A \longrightarrow B$ which is also a ring homomorphism is called a k-**algebra homomorphism**.*

(3) *A non-empty subset B of a k-algebra A is called a k-**subalgebra** if it is both a k-subspace of A and a subring of A.*

PROPOSITION 5.3. *Let $k[y_1, ..., y_m]$ and $k[x_1, ..., x_n]$ be polynomial rings over a field k.*

(1) *Any k-algebra homomorphism*

$$\varphi\colon k[y_1, ..., y_m] \longrightarrow k[x_1, ..., x_n]$$

is uniquely determined by the image $f_1, ..., f_m$ of the generators $y_1, ..., y_m$ of $k[y_1, ..., y_m]$.

(2) *Its kernel* $\ker(\varphi)$ *is the ideal of $k[y_1, ..., y_m]$.*

(3) *Its image $im(\varphi)$ is the k-subalgebra $k[f_1, ..., f_m]$.*

PROOF. Part (1). By the universal property of the polynomial ring $k[y_1, ..., y_m]$, we have such a result. That is, let h be any polynomial in $k[y_1, ..., y_m]$. Then,

$$h = \sum_i^{\text{finite}} c_i y_1^{\alpha_1} \cdots y_m^{\alpha_m}$$

is mapped under a given k-algebra homomorphism φ to

$$\begin{aligned}\varphi(h) &= \varphi\left(\sum_i^{\text{finite}} c_i y_1^{\alpha_1} \cdots y_m^{\alpha_m}\right) \\ &= \sum_i c_i \varphi(y_1)^{\alpha_1} \cdots \varphi(y_m)^{\alpha_m}.\end{aligned}$$

Thus, this formula determines the image $f_i = \varphi(y_i)$ of each y_i. Conversely, if we know these $f_i = \varphi(y_i)$ in advance, then for any $h \in k[y_1, ..., y_m]$ its image under a k-algebra homomorphism φ is completely determined as $h(f_1, ..., f_n)$ by this formula.

Proof of part (2) and part (3) are very easy. □

Then we have by the so-called the first isomorphism theorem for k-algebras,

$$k[y_1, ..., y_m]/\ker(\varphi) \approx k[f_1, ..., f_m].$$

Thus we have,

$$h \in \ker(\varphi) \Longleftrightarrow \varphi(h) = h(f_1, ..., f_m) = 0.$$

Because of this reason, $\ker(\varphi)$ is often called the **ideal of relations** among the polynomials $f_1, ..., f_m \in k[x_1, ..., x_n]$. Thus, we are currently talking about "**generators** of a k-subalgebra $k[f_1, ..., f_m]$ and **relations** among them".

3.1. Computation of a Groebner Basis for $\ker(\varphi)$. We need a technical lemma:

LEMMA 5.4. *Let $a_1, ..., a_u, b_1, ...b_u$ be elements of a ring R. Then the elements*

$$a_1a_2\cdots a_u - b_1b_2\cdots b_u \in (a_1 - b_1, ..., a_u - b_u),$$

where $(a_1 - b_1, ..., a_u - b_u)$ is the ideal of R generated by the $a_i - b_i$ over R.

PROOF. We note that

$$\begin{aligned} a_1a_2\cdots a_u - b_1b_2\cdots b_u &= a_1(a_2\cdots a_u - b_2\cdots b_u) \\ &\quad + b_2\cdots b_u(a_1 - b_1). \end{aligned}$$

Thus, we are done by mathematical induction on u. □

The following remark explains how the Theory of Groebner basis finds its place in Algebraic Geometry, as well as in Algebra.

REMARK 5.6. *[Continuation of Remark 4.15] We recall that whenever we are given a map $f\colon A \longrightarrow B$, the graph of f is a subset of the Cartesian product $A \times B$.*

For example, if $A = k^n$, $B = k^m$, the affine spaces, then we have the graph of f in k^{m+n}. After introducing the so-called Zariski topology *on these spaces, we can verify that the graph of* f *is a closed subset of* k^{m+n} *with respect to the Zariski topology; and hence it is defined by a set of polynomial equations; and thus we need to figure out the defining polynomials for the graph of* f *from a given information.*

To talk about these defining equations for an affine variety, graph of f*, from a given information, we sometimes argue in a dual manner. As we shall see it very soon, this approach of entering the subject from the Algebra side will easily enable us to understand that the corresponding morphism on the Geometry side should be the polynomial mapping. Thus, dually, let*

$$\varphi = f^*\colon k[y_1, ..., y_m] \longrightarrow k[x_1, ..., x_n]$$

be the corresponding k-algebra homomorphism.

[In Section 4 of Chapter 4, the reason why this φ must *be a* k*-algebra homomorphism is explained in detail. See Lemma 4.3 there].*

Then by the above Proposition 5.3, or, by the universal property of the polynomial ring $k[y_1, ..., y_n]$*,* φ *is uniquely determined by sending*

$$y_i \longmapsto f_i(x_1, ..., x_n),\ i = 1, ..., m,$$

where f_i *are some polynomials in* $k[x_1, ..., x_n]$.

Suppose that we figure out these polynomials f_i explicitly. *Then since* φ *corresponds to a morphism*

$$\hat{\varphi} : k^n \longrightarrow k^m$$

in geometry side, the point

$$(f_1(x_1, ..., x_n), ..., f_m(x_1, ..., x_n))$$

is in fact the image in k^m *of the point* $(x_1, ..., x_n)$ *of* k^n *under* $\hat{\varphi}$*, that is,* $\hat{\varphi}$ *is naturally* defined *by sending each point*

$$X = (x_1, ..., x_n) \in k^n$$
$$\downarrow \hat{\varphi}$$
$$(f_1(x_1, ..., x_n), ..., f_m(x_1, ..., x_n)) \in k^m.$$

This is because: for any $g \in k[y_1, ..., y_m]$*, we have, for any* $X \in k^n$*,*

$$\begin{aligned} \varepsilon_{X\circ\varphi}(g) &\equiv (X \circ \varphi)(g) = X(\varphi(g)) = X(g(f_1, ..., f_m)) \equiv \varepsilon_X(g(f_1, ..., f_m)) \\ &= g(f_1, ..., f_m)(X) = g(f_1(X), ..., f_m(X)) = \varepsilon_{(f_1(X),...,f_m(X))}(g). \end{aligned}$$

That is, for any $X \in k^n$*, we have:* $\varepsilon_{X\circ\varphi} = \varepsilon_{(f_1(X),...,f_m(X))}$. *Thus we have:*

$$\varphi\hat{}\,(X) = X \circ \varphi \equiv \varepsilon_{X\circ\varphi} = \varepsilon_{(f_1(X),...,f_m(X))} \equiv (f_1(X), ..., f_m(X)).$$

Thus, the corresponding morphism $\varphi\hat{}$ *is nothing but a* polynomial mapping, *And thus, the map* $\varphi\hat{}$ *is usually given to us as a polynomial parametric expression:*

$$\begin{aligned} y_1 &= f_1(x_1, ..., x_n), \\ y_2 &= f_2(x_1, ..., x_n), \\ &\quad .... \\ y_m &= f_m(x_1, ..., x_n), \end{aligned}$$

where the variables $x_1, ..., x_n$ *are parameters and the* y_i *are, this time, not considered as the* unknowns *of* $k[y_1, ..., y_m]$ *but considered as the* coordinate functions *of* k^m*, that is, the canonical projections:* $k^m \longrightarrow k$*, sending*

$$(y_1, ..., y_m) \longmapsto y_i,$$

for each $i = 1, ..., m$*. Thus, we have seen that a polynomial mapping* $\varphi\hat{}$ *corresponds to a given* k*-algebra homomorphism* φ*.*

Conversely, suppose that we are given a polynomial mapping f: $k^n \longrightarrow k^m$ by sending

$$X = (x_1, ..., x_n) \longmapsto (f_1(x_1, ..., x_n), f_2(x_1, ..., x_n), ...f_m(x_1, ..., x_n)).$$

That is, $f(X) = (f_1(X), f_2(X), ...f_m(X))$*.*
Then the corresponding homomorphism f^*: $k[y_1, ..., y_m] \longrightarrow k[x_1, ..., x_n]$

is naturally defined *by sending each*

$$y_i \in k[y_1, ..., y_m]$$
$$\downarrow f^*$$
$$y_i((f_1(x_1, ..., x_n), ...f_m(x_1, ..., x_n)) = f_i(x_1, ..., x_n) \in k[x_1, ..., x_n],$$

for $i = 1, ..., m$. *Thus, the corresponding homomorphism*

$$y_i \longmapsto f_i(x_1, ..., x_n) = f_i,\ i = 1, ..., m,$$

is nothing but a k-algebra homomorphism f^*: $k[y_1, ..., y_m] \longrightarrow k[x_1, ..., x_n]$. *This is because: for any* $X \in k^n$, *we have, for each* $i = 1, ..., m$,

$$f^*(y_i)(X) = (y_i \circ f)(X) = y_i(f(X)) = y_i(f_1(X), f_2(X), ...f_m(X)) = f_i(X).$$

Thus, we have seen that a k*-algebra homomorphism* f^* *corresponds to a given polynomial mapping* f.
Then, one can easily verify that $(\varphi\hat{\ })^* = \varphi$ *and* $(f^*)\hat{\ } = f$, *as we have done in Remark 4.15. We note that this correspondence does not require the bottom field* k *be algebraically closed.*

A more general situation is treated in Subsection 3.4.

Thus, we have proved that

LEMMA 5.5. *There exists a one-to-one correspondence between the set of all polynomial mappings* $k^n \longrightarrow k^m$ *and the set of all* k*-algebra homomorphisms* $k[y_1, ..., y_m] \longrightarrow k[x_1, ..., x_n]$. □

Under a more general setting, those maps, or morphisms, in the Geometry side corresponding to k*-algebra homomorphisms in the Algebra side are usually called* regular maps.

We then note that, under either approach, i.e., either from the Geometry side or from the Algebra side, we will always be given those m *polynomials* $f_i(x_1, ..., x_n)$*'s in* n *unknown* $x_1, ..., x_n$ *explicitly, and hence from the above parametric expression, for instance, we naturally get a system of polynomial equations in the variables* $x_1, ..., x_n$, *and* $y_1, ..., y_m$:

$$(5.1) \qquad \begin{cases} y_1 - f_1(x_1, ..., x_n) & = & 0 \\ y_2 - f_2(x_1, ..., x_n) & = & 0 \\ \quad \cdots & & \\ y_m - f_m(x_1, ..., x_n) & = & 0 \end{cases} .$$

On the Geometry side, we note that the affine variety V *defined by these polynomials,* $y_1 - f_1, ..., y_m - f_m$, *is in fact the* ***graph*** *in* k^{m+n} *of the map* $\varphi\hat{}$. *Thus, we have an affine variety* $V = V(y_1 - f_1, ..., y_m - f_m)$ *in* k^{m+n}.

We then note that the projection $\pi(V)$ *of* V *onto* k^m *under the canonical projection,*

$$\pi \colon k^{m+n} \longrightarrow k^m,$$

is equal to the image $im(\varphi\hat{})$ *of the morphism* $\varphi\hat{}$, *and hence we naturally ask whether the projection* $\pi(V)$ *in* k^m *of this variety* V *onto* k^m, *namely, the image,* $\varphi\hat{}(k^n)$, *is again an affine variety or not.*

Unfortunately, the answer is not true in general.

EXAMPLE 5.7. *For example, the projection of* $V(xy - 1)$ *onto* k^1 *is a line minus one point, so it is not an affine variety.*

We then ask: what is the smallest variety *in* k^m *containing this* V? *That is, what is the* ***Zariski closure of*** $\pi(V) = im(\varphi\hat{})$ *in* k^m? *The answer to this question is: it is equal to*

$$V(\ker(\varphi)),$$

provided that k is an algebraically closed field.

[This should be treated after dealing with the Algebra side first, because once we figure out a generating set of the ideal $\ker(\varphi)$, *then we can explicitly figure out their solution set,* $V(\ker(\varphi))$. *(See Algorithm 21)].*

On the Algebra side, the solution set of equations in (5.1) is a subset of k^{m+n}, *and by "eliminating the variables"* $x_1, .., x_m$, *we can express its solution set in terms of a few polynomial equations* $h_i(y_1, ..., y_m) = 0$,

$$\begin{aligned} 0 &= h_1(y_1, ..., y_m), \\ &\cdots \\ 0 &= h_q(y_1, ..., y_m), \end{aligned}$$

in the variables $y_1, ..., y_m$. *Such a process is usually called "**implicitization**". For this, we use Theorem 5.1 (The Elimination Theorem). Then since each* $y_i = f_i(x_1, ..., x_n)$, *this equivalently says that*

$$\left\{ \begin{aligned} h_1(f_1, ..., f_m) &= 0 \\ h_2(f_1, ..., f_m) &= 0 \\ &\cdots \\ h_q(f_1, ..., f_m) &= 0 \end{aligned} \right. \tag{5.2}$$

That is, each $h_i \in \ker(\varphi)$. *We already know that* $\ker(\varphi)$ *consists of exactly those* h *such that* $h(f_1, ..., f_m) = 0$, *as we have seen it above, and moreover we have:*

$$\ker(\varphi) = (h_1, ..., h_q),$$

as we shall see it in the second part of the proof of Theorem 5.3 below.

In terms of polynomial ideals, our explanation so far given in the above remark, Remark 5.6, corresponds to the followings: Equation in (5.1) corresponds to the ideal

$$J = (y_1 - f_1, ..., y_m - f_m) \subseteq k[x_1, ..., x_n, y_1, ..., y_m]$$

and equations in (5.2) corresponds to

$$\ker(\varphi) = (h_1, ..., h_q) \subseteq k[y_1, ..., y_m].$$

Let I be the inverse image of the ideal,

$$(y_1 - f_1, ..., y_m - f_m),$$

under the insertion homomorphism:

$$i : k[y_1, ..., y_m] \longrightarrow k[x_1, ..., x_n, y_1, ..., y_m],$$

sending anything to itself. Now, our immediate next question is how I and $\ker(\varphi) = (h_1, ..., h_q)$ are related, and this is completely answered by the following theorem. Moreover, thanks to the Theory of Groebner basis, we now can computationally figure out the generating set of $\ker(\varphi)$ completely, as the following algorithm, Algorithm 14 tells us.

THEOREM 5.3. *Let φ: $k[y_1, ..., y_m] \longrightarrow k[x_1, ..., x_n]$ be a k-algebra homomorphism defined by sending*

$$y_i \longmapsto f_i,$$

for each $i = 1, ..., m$, and let

$$J = (y_1 - f_1, ..., y_m - f_m)$$

be the associated polynomial ideal in $k[x_1, ..., x_n, y_1, ..., y_m]$. Then

$$\begin{aligned} \ker(\varphi) &= i^{-1}(J) \\ &= J \cap k[y_1, ..., y_m], \end{aligned}$$

where i: $k[y_1, ..., y_m] \longrightarrow k[x_1, ..., x_n, y_1, ..., y_m]$ is the insertion sending anything to itself.

PROOF. Let $h \in J \cap k[y_1, ..., y_m]$, so that

$$h(y_1, ..., y_m) = \sum_{i=1}^{m} (y_i - f_i) q_i(x, y),$$

for some $q_i(x, y) = q_i(x_1, ..., x_n, y_1, ..., y_m)$. Thus, $h(f_1, ..., f_m) = 0$, i.e., $h \in \ker(\varphi)$.

Conversely, let $h \in \ker(\varphi)$. Since it is a polynomial in the variables $y_1, ..., y_m$, we can write

$$h = \sum_j c_j y_1^{\alpha_1} \cdots y_m^{\alpha_m},$$

where $c_j \in k$. Then since $h(f_1, ..., f_m) = 0$, we have

$$\begin{aligned} h(y_1, ..., y_m) &= h - h(f_1, ..., f_m) \\ &= \sum_j c_j (y_1^{\alpha_1} \cdots y_m^{\alpha_m} - f_1^{\alpha_1} \cdots f_m^{\alpha_m}), \end{aligned}$$

and hence by Lemma 5.4, $h \in J \cap k[y_1, ..., y_m]$. □

In particular, we have a criterion for φ being injective:

COROLLARY 5.5. $J \cap k[y_1, ..., y_m] = (0)$ *if and only if* $\ker(\varphi) = (0)$. □

Therefore, the algebraic process of "elimination of variables $x_1, ..., x_n$", or equivalently, the algebraic process of "implicitization" of these variables, $x_1, ..., x_n$, corresponds to the "inverse image" of the polynomial ideal J in the ambient polynomial ring.

Theorem 5.3 and Theorem 5.1 (The Elimination Theorem) gives

ALGORITHM 14. *(Algorithm for computing a Groebner basis for* $\ker(\varphi)$*)*

(1) *We first compute a Groebner basis G for the ideal*

$$(y_1 - f_1, ..., y_m - f_m)$$

of $k[x_1, .., x_n, y_1, ..., y_m]$ *with respect to an elimination order where the x's are bigger than the y's.*

(2) *Then those polynomials in G without any x variables are a Groebner basis for* $\ker(\varphi)$ *[by Theorem 5.1].*

Thus, once we compute a Groebner basis for the ideal $\ker(\varphi)$, we can apply, for instance, Program 2 to decide a given polynomial $g \in k[y_1, ..., y_n]$ belongs to $\ker(\varphi)$ in an explicit manner.

EXAMPLE 5.8. *Let* φ: $\mathbb{Q}[y_1, y_2, y_3] \longrightarrow \mathbb{Q}[x_1, x_2]$ *be a* $\mathbb{Q}$*-algebra homomorphism defined by sending*

$$\begin{aligned} y_1 &\longmapsto x_1^4, \\ y_2 &\longmapsto x_1^3 x_2, \\ y_3 &\longmapsto x_1 x_2^3. \end{aligned}$$

We first compute a Groebner basis G for the ideal

$$(y_1 - x_1^4, y_2 - x_1^3 x_2, y_3 - x_1 x_2^3)$$

in $k[x_1, x_2, y_1, y_2, y_3]$ *using the REDUCE:*[4]

———————

```
off echo$
off nat$
load_package groebner$
torder({x1,x2,y1,y2,y3},lex)$
f1:=y1-x1^4;
f2:=y2-x1^3*x2;
f3:=y3-x1*x2^3;
gb := {x1**4 - y1, x1**3*x2 - y2, x1**2*x2**2*y1 - y2**2,
x1**2*y3 - x2**2*y2, x1*x2**3 - y3, x1*y1*y3 - x2*y2**2,
x1*y2 - x2*y1, x1*y3**2 - x2**5*y2, x2**8*y2 - y3**3,
x2**4*y1 - y2*y3, x2**4*y2**2 - y1*y3**2,
y1**2*y3 - y2**3}$
```

———————

Thus, a Groebner basis for $\ker(\varphi)$ *consists of the last polynomial of the above gb, so* $\ker(\varphi) = (y_1^2 y_3 - y_2^3)$. *We can also easily verify that*

$$(x_1^4)^2(x_1 x_2^3) - (x_1^3 x_2)^3 = 0.$$

3.2. Computational Criterion for Membership Question to $im(\varphi)$.

THEOREM 5.4. *Let* φ: $k[y_1, ..., y_m] \longrightarrow k[x_1, ..., x_n]$ *be a k-algebra homomorphism defined by sending* $y_i \longmapsto f_i$,*for each* $i = 1, ..., m$, *and let J be a Groebner basis* $G = \{g_1, ..., g_t\}$ *for the associated polynomial ideal*

$$J = (y_1 - f_1, ..., y_m - f_m)$$

in $k[x_1, ..., x_n, y_1, ..., y_m]$ *with respect to an elimination order where the x variables are bigger than the y variables. Then,* $f \in k[x_1, ..., x_n]$ *is in the*

[4]A working program for REDUCE for this example is also available from: http://ch.catholic.or.kr/pundang/4/soh/2006_aa_i/Example_5_08.htm

image $im(\varphi)$ of φ if and only if there exists $h \in k[y_1, ..., y_m]$ such that f is reduced to h under the division by G, i.e., there exists $h \in k[y_1, ..., y_m]$ such that

$$f \xrightarrow{G}_+ h.$$

In this case, $f = \varphi(h) = h(f_1, ..., f_m)$.

PROOF. Let $f \in k[x_1, ..., x_n]$. Suppose that $f \in im(\varphi)$. Then $f = g(f_1, ..., f_m)$ for some polynomial $g \in k[y_1, ..., y_m]$. Then since

$$f(x_1, ..., x_n) - g(y_1, ..., y_m) = g(f_1, ..., f_m) - g(y_1, ..., y_m) \in J,$$

by Lemma 5.4; then Proposition 3.1 says that

$$f(x_1, ..., x_n) - g(y_1, ..., y_m) \in J \Longleftrightarrow N_G(f) = N_G(g).$$

Thus, there exists $h = N_G(f)$, $f \xrightarrow{G}_+ h$, and $g \xrightarrow{G}_+ h$. Moreover, since $g \in k[y_1, ..., y_m]$, g is reduced to h by those $g_i \in G$ such that $LM(g_i)$ is a monomial in the variables $y_1, ..., y_m$ only. Then since the variables x are bigger that the variables y , this means that those g_i used in the reduction are in fact polynomials in the y variables only. Therefore, $h \in k[y_1, ..., y_m]$.

Conversely, if $f \xrightarrow{G}_+ h$ for some $h \in k[y_1, ..., y_m]$, then $f - h \in J$ by Lemma 5.4, and hence

$$f(x_1, ..., x_n) - h(y_1, ..., y_m) = \sum_{i=1}^{m} q_i(x, y)(y_i - f_i).$$

Thus, if we substitute f_i for y_i on both sides, we get

$$f = h(f_1, ..., f_m) = \varphi(h).$$

□

Moreover, we have:

COROLLARY 5.6. *Let $f \in k[x_1, ..., x_n]$. Then $f \in im(\varphi)$ if and only if $N_G(f) \in k[y_1, ..., y_m]$.*

PROOF. The $\Longleftarrow$ direction is true by Theorem 5.4. Conversely, suppose that $f \in im(\varphi)$. Then again by Theorem 5.4, $f \xrightarrow{G}_+ h$ for some polynomial $h \in k[y_1, ..., y_m]$. Thus, by Proposition 3.1 we see that $N_G(f) = N_G(h)$. Since $h \in k[y_1, ..., y_m]$, h is reduced to $N_G(h)$ only by those $g_i \in G$ with $LM(g_i)$ being monomials in the y variables only. This means that those g_i used in the reduction are polynomials in the y variables only. Therefore, $N_G(h) \in k[y_1, ..., y_m]$. □

In particular, we have a criterion for φ being surjective:

COROLLARY 5.7. *For each $i = 1, ..., n$, $N_G(x_i) \in k[y_1, ..., y_m]$ if and only if φ is surjective.*

Corollary 5.6 provides us with a computational criterion for the membership question to $im(\varphi)$.

ALGORITHM 15. *(Algorithm for membership question to $im(\varphi)$) Let f be a polynomial in $k[x_1, ..., x_n]$. To decide whether $f \in im(\varphi)$ or not, we do the followings:*

(1) *Compute a Groebner basis G for the ideal $(y_1 - f_1, ..., y_m - f_m)$.*
(2) *Apply Algorithm 5 to compute the unique remainder $N_G(f)$ of f with respect to G.*
(3) *$N_G(f) \in k[y_1, ..., y_m]$ if and only if $f \in im(\varphi)$.*

EXAMPLE 5.9. *Under the same setting as in Example 5.8, we want to verify whether $x_1^4 + 1 \in im(\varphi)$. [Of course, it is in the subring $im(\varphi)$ because, $1 \in im(\varphi)$ and $x_1^4 \in im(\varphi)$].*[5]

———————

```
off echo$
off nat$
load_package groebner$
torder({x1,x2,y1,y2,y3},lex)$
f1:=y1-x1^4;
f2:=y2-x1^3*x2;
f3:=y3-x1*x2^3;
```

[5]A working program for REDUCE for this example is also available from: http://ch.catholic.or.kr/pundang/4/soh/2006_aa_i/Example_5_09.htm

f:=x1^4+1;

result:=greduce(f,{f1,f2,f3});

y1 + 1$

This means that x_1^4+1 is reduced to y_1+1, and hence $x_1^4+1 \in im(\varphi)$. Moreover, $x_1^4+1=(x_1^4)+1$.

Here is a better criterion for φ to be surjective than Corollary 5.7:

THEOREM 5.5. *Let φ: $k[y_1,...,y_m] \longrightarrow k[x_1,...,x_n]$ be a k-algebra homomorphism defined by sending $y_i \longmapsto f_i$,for each $i=1,...,m$, and let J be the* reduced *Groebner basis $G=\{g_1,...,g_t\}$ for the associated polynomial ideal*

$$J=(y_1-f_1,...,y_m-f_m)$$

in $k[x_1,...,x_n,y_1,...,y_m]$ with respect to an elimination order where the x variables are bigger than the y variables. Then φ is surjective if and only if for each $i=1,...,n$, there exists $g_i \in G$ such that $g_i=x_i-h_i$ with $h_i \in k[y_1,...,y_m]$. If it is the case, then $x_i=h_i(f_1,...,f_m)$.

PROOF. If φ is surjective, then each $x_i \in im(\varphi)$. Renumbering if necessary, we may assume that $x_1<x_2<\cdots<x_n$ under the given order. Then by Theorem 5.4 there exists $h_1' \in k[y_1,...,y_m]$ such that $x_1 \xrightarrow{G}_+ h_1'$. Thus, $x_1-h_1' \in J$; hence $LM(g_1)$ of one $g_1 \in G$ divides $LM(x_1-h_1')=x_1$. Thus, this $g_1=x_1-h_1$ for some $h_1 \in k[y_1,...,y_m]$ since G is the reduced Groebner basis with $x_1<x_2<\cdots<x_n$. Now for x_2, we also see that there exists $h_2' \in k[y_1,...,y_m]$ such that $x_2 \xrightarrow{G}_+ h_2'$. Thus, $x_2-h_2' \in J$; hence $LM(g_2)$ of one $g_2 \in G$ divides $LM(x_2-h_2')=x_2$. Then since G is the reduced Groebner basis with $x_1<x_2$ and $LM(g_1)=x_1$, all the smaller monomial terms of g_2 are polynomials in the y variables only. Thus $g_2=x_2-h_2$ for some $h_2 \in k[y_1,...,y_m]$. Thus, by induction on i for other x_i, we see the right side of the equivalence is true. Conversely, if the right side of the equivalence holds, then each $x_i \xrightarrow{G}_+ h_i$, and $h_i \in k[y_1,...,y_m]$. Thus, each $x_i \in im(\varphi)$ by Theorem 5.4 or Corollary

5.6; hence φ is surjective. Once it is surjective, we have $x_i = h_i(f_1, ..., f_m)$ by Theorem 5.4. □

ALGORITHM 16. *(Algorithm for surjectivity of φ) Theorem 5.5 enables us to verify whether φ is surjective or not.*

EXAMPLE 5.10. *Thus the k-algebra homomorphism in Example 5.8 is not surjective by the criterion in Theorem 5.5.*

Now we consider another example. Let φ: $\mathbb{Q}[y_1, y_2, y_3] \longrightarrow \mathbb{Q}[x]$ be a $\mathbb{Q}$-algebra homomorphism defined by sending

$$\begin{aligned} y_1 &\longmapsto x^5 + x, \\ y_2 &\longmapsto x^4, \\ y_3 &\longmapsto x^3. \end{aligned}$$

We first compute the reduced Groebner basis for the ideal

$$(y_1 - x^5 - x, y_2 - x^4, y_3 - x^3),$$

using the REDUCE:[6]

```
——————————
off echo$
off nat$
load_package groebner$
torder({x,y1,y2,y3},lex)$
f1:=y1-x^5-x;
f2:=y2-x^4;
f3:=y3-x^3;
gb:=groebner({f1,f2,f3});
gb := {x + y1*y2 - y1 - y3**3,
y1**3 - 3*y2**2*y3 - 3*y2*y3 - y3**5 - y3,
y1**2*y2 - y2**2*y3**2 - 2*y2*y3**2 - y3**2,
y1*y2**2 - y2*y3**3 - y3**3,
y1*y3 - y2 **2 - y2, y2**3 - y3**4}$
```

[6]A working program for REDUCE for this example is also available from: http://ch.catholic.or.kr/pundang/4/soh/2006_aa_i/Example_5_10.htm

———————

Since we have only one polynomial

$$x + y_1y_2 - y_1 - y_3^3$$

in G, we conclude by Theorem 5.5 that φ is surjective. Moreover, from this computation, we also see that

$$x = -(x^5 + x) \cdot x^4 + (x^5 + x) + (x^3)^3.$$

3.3. k-Algebra Homomorphisms Between Affine k-Algebras. We start with a definition:

DEFINITION 5.5. *A k-algebra is called an* ***affine k-algebra*** *if it is isomorphic as a k-algebra to $k[x_1, ..., x_n]/I$ for some ideal $I \in k[x_1, ..., x_n]$.*

EXAMPLE 5.11. *The polynomial ring $k[x_1, ..., x_n]$ is such a k-algebra.*

REMARK 5.7. *We recall that, for a chosen set of m polynomials*

$$\{f_1, ..., f_m\}$$

of $k[x_1, ..., x_n]$, we know, by the universal property of the polynomial ring $k[y_1, ..., y_m]$, there exists a unique k-algebra homomorphism

$$\varphi\colon k[y_1, ..., y_m] \longrightarrow k[x_1, ..., x_n],$$

which sends each $y_i \longmapsto f_i$, and hence its image $k[f_1, ..., f_m]$, a k-subalgebra of $k[x_1, ..., x_n]$ is isomorphic to

$$k[y_1, ..., y_m]/\ker(\varphi).$$

Thus, $k[f_1, ..., f_m]$ is an affine k-algebra. Under this setting, if I is an ideal of $k[x_1, ..., x_n]$ then the composition

$$k[y_1, ..., y_m]/\ker(\varphi) \approx k[f_1, ..., f_m] \hookrightarrow k[x_1, ..., x_n] \longrightarrow k[x_1, ..., x_n]/I,$$

is a canonical k-algebra homomorphism:

$$k[y_1, ..., y_m]/\ker(\varphi) \longrightarrow k[x_1, ..., x_n]/I,$$

sending each $y_i + \ker(\varphi) \longmapsto f_i + I$, $i = 1, ..., m$. *Thus, once the* m *polynomials* $f_1, ..., f_m \in k[x_1, ..., x_m]$ *are given to us, we have such a* k-*algebra homomorphism in a canonical manner. But, this process does not tell us how we can computationally verify the* φ *is indeed a* k-*algebra homomorphism.*

Now more generally, let $J = (g_1, ..., g_t)$ be an ideal of $k[y_1, ..., y_m]$ and $I = (h_1, ..., h_s)$ be an ideal of $k[x_1, ..., x_n]$ and we want to define a k-algebra homomorphism

$$\phi\colon k[y_1, ..., y_m]/J \longrightarrow k[x_1, ..., x_n]/I,$$

by sending

$$y_i + J \longmapsto f_i + I, \quad i = 1, 2, ..., m,$$

where the f_i's are some (undetermined) polynomials in $k[x_1, ..., x_n]$.

LEMMA 5.6. *Under this setting, the mapping* φ *is a well-defined* k-*algebra homomorphism if and only if*

$$g_i(f_1, ..., f_m) \in I, \quad i = 1, ..., t,$$

where $J = (g_1, ..., g_t)$.

PROOF. If ϕ is a well-defined k-algebra homomorphism, then for each $i = 1, ..., m$,

$$g_i(y_1, ..., y_m) + J = 0 + J,$$

and hence their image under ϕ must is the same because $\phi(0) = 0$:

$$g_i(f_1, ..., f_m) + I = 0 + I,$$

that is, $g_i(f_1, ..., f_m) \in I$, for each $i = 1, 2, ..., m$.

Conversely, suppose that

$$g_i(f_1, ..., f_m) \in I, \quad i = 1, ..., t,$$

for a given $f_1, ..., f_m$. To prove first that the map ϕ is well-defined, we need to verify that if $\prod_i c_i y_i^{\alpha_i} + J = \prod_i c_i' y_i^{\alpha_i'} + J$, then $\prod_i c_i f_i^{\alpha_i} + I =$

$\prod_i c_i' f_i^{\alpha_i'} + I$. That is, we need to verify that if $\prod_i c_i y_i^{\alpha_i} - \prod_i c_i' y_i^{\alpha_i'} \in J$, then $\prod_i c_i f_i^{\alpha_i} - \prod_i c_i' f_i^{\alpha_i'} \in I$.

But, this is easy: If $\prod_i c_i y_i^{\alpha_i} - \prod_i c_i' y_i^{\alpha_i'} \in J$ holds, then we equivalently have

$$\prod_i c_i y_i^{\alpha_i} - \prod_i c_i' y_i^{\alpha_i'} = \sum_j h_j(y_1, ..., y_m) g_j(y_1, ..., y_m),$$

for some $h_j(y_1, ..., y_m) \in k[y_1, ..., y_m]$. Thus the corresponding term via ϕ,

$$\prod_i c_i f_i^{\alpha_i} - \prod_i c_i' f_i^{\alpha_i'} = \sum_j h_j(f_1, ..., f_m) g_j(f_1, ..., f_m),$$

certainly belongs to the ideal I, *provided that* each $g_j(f_1, ..., f_m)$ belongs to I by hypothesis. Thus we equivalently see that $\prod_i c_i f_i^{\alpha_i} + I = \prod_i c_i' f_i^{\alpha_i'} + I$ holds. That is, the map ϕ is indeed a well-defined map.

It is very routine to verify that ϕ is a ring homomorphism and also a k-linear map, so we leave them as an exercise. □

ALGORITHM 17. *(Algorithm for well definedness of ϕ) Lemma 5.6 provides us with an algorithm for verifying whether the assignments*

$$y_i + J \longmapsto f_i + I, \quad i = 1, 2, ..., m,$$

for arbitrarily chosen $f_i \in k[x_1, ..., x_n]$ is a k-algebra homomorphism or not, because if $I = (h_1, ..., h_s)$, its Groebner basis will tell us whether each $g_i(f_1, ..., f_m)$ belongs to I or not. (See Section 2 of Chapter 3).

Here is a generalization of Theorem 5.3:

Let $J = (g_1, ..., g_t)$ be an ideal of $k[y_1, ..., y_m]$ and $I = (h_1, ..., h_s)$ be an ideal of $k[x_1, ..., x_n]$, and suppose that we are given a k-algebra homomorphism

$$\phi\colon k[y_1, ..., y_m]/J \longrightarrow k[x_1, ..., x_n]/I,$$

which sends $y_i + J \longmapsto f_i + I, \quad i = 1, 2, ..., m$, where the f_i's are some (undetermined) polynomials in $k[x_1, ..., x_n]$. Thus, it satisfies the criterion of Lemma 5.6.

THEOREM 5.6. *Under this setting, let* $K = (I, y_1 - f_1, ..., y_m - f_m)$ *be the ideal of* $k[x_1, ..., x_n, y_1, ..., y_m]$. *Then*

$$\ker(\phi) = (f_1' + J, f_2' + J, ..., f_p' + J),$$

where the ideal $(f_1', ..., f_p')$ *is the inverse image* $K \cap k[y_1, ..., y_m]$ *of* K *under the insertion homomorphism*

$$i\colon k[y_1, ..., y_m] \hookrightarrow k[x_1, ..., x_n, y_1, ..., y_m],$$

which sends anything to itself.

PROOF. Suppose that $f' + J \in (f_1' + J, ..., f_p' + J)$, $f' \in K \cap k[y_1, ..., y_m]$. We need to prove that $\phi(f' + J) = 0 + I$.

Since $f' \in K \cap k[y_1, ..., y_m]$,

$$f'(y_1, ..., y_m) = \sum_{i=1}^{m} q_i(x, y)(y_i - f_i(x)) + w(x, y),$$

for some $q_i(x, y) = q_i(x_1, ..., x_n, y_1, ..., y_m) \in k[x_1, ..., x_n, y_1, ..., y_m]$ and some

$$\begin{aligned} w(x, y) &= w(x_1, ..., x_n, y_1, ..., y_m) \\ &= \sum_{j}^{finite} p_j(x, y) h_j(x_1, ..., x_n) \in I, \end{aligned}$$

for some $p_j(x, y) = p_j(x_1, ..., x_n, y_1, ..., y_m) \in k[x_1, ..., x_n, y_1, ..., y_m]$, where the h_j are generators of the ideal I. Thus,

$$\begin{aligned} \phi(f' + J) &= f'(f_1, ..., f_m) + I \\ &= 0 + w(x, f_1, ..., f_m) + I \\ &= 0 + 0 + I. \end{aligned}$$

Conversely, suppose that

$$\phi(f' + J) = 0 + I, f' = f'(y_1, ..., y_m) \in k[y_1, ..., y_m].$$

Thus $f'(f_1, ..., f_m) \in I$. We need to prove that $f' + J \in (f_1' + J, f_2' + J, ..., f_p' + J)$. For this, we write

$$\begin{aligned} f'(y_1, ..., y_m) &= (f'(y_1, ..., y_m) - f'(f_1, ..., f_m)) + f'(f_1, ..., f_m) \\ &= \sum_{\alpha}^{finite} c_\alpha (y_1^{\alpha_1} \cdots y_m^{\alpha_m} - f_1^{\alpha_1} \cdots f_m^{\alpha_m}) + f'(f_1, ..., f_m). \end{aligned}$$

Then by Lemma 5.4 $\sum_{\alpha}^{finite} c_\alpha (y_1^{\alpha_1} \cdots y_m^{\alpha_m} - f_1^{\alpha_1} \cdots f_m^{\alpha_m}) \in (y_1 - f_1, ..., y_m - f_m)$. Thus,

$$f'(y_1, ..., y_m) \in (I, y_1 - f_1, ..., y_m - f_m) = K,$$

since $f'(f_1, ..., f_m) \in I$. Thus, $f'(y_1, ..., y_m) \in K \cap k[y_1, ..., y_m] = (f_1', ..., f_p')$. Therefore,

$$f'(y_1, ...y_m) + J \in (f_1' + J, f_2' + J, ..., f_p' + J).$$

□

Thus, Theorem 5.6 enables us to explicitly compute the generators of the kernel, $\ker(\phi)$.

ALGORITHM 18. *(Algorithm for computing a Groebner basis for* $\ker(\phi)$*)*

(1) *Compute a Groebner basis G for the ideal $K = (I, y_1 - f_1, ..., y_m - f_m)$ of $k[x_1, .., x_n, y_1, ..., y_m]$ with respect to an elimination order where the x's are bigger than the y's.*
(2) *Then $\ker(\phi) = (g_1 + J, ..., g_u + J)$ where g_i are those polynomials in G with no x variables.*

For this, we apply Theorem 5.1 or the Elimination Theorem, Corollary 5.2.

Here is an application of Theorem 5.4:

THEOREM 5.7. *Under the same setting as in Theorem 5.6, let G be a Groebner basis for the ideal $K = (I, y_1 - f_1, ..., y_m - f_m)$ with respect to an elimination order with the x variables bigger than the y variables. Then $f + I \in k[x_1, ..., x_n]/I$ is in the image $im(\phi)$ if and only if there exists*

$h \in k[y_1, ..., y_m]$ such that the reduction of f with respect to G is h, that is, $f \xrightarrow{G}_+ h$. If it is the case, the $f + I = \phi(h + J) = h(f_1, ..., f_m) + I$.

PROOF. Suppose that $f+I \in im(\phi)$. Thus, there exists $g(y_1, ..., y_m)+J$ such that

$$f + I = \phi(g(y_1, ..., y_m) + J) = g(f_1, ..., f_m) + I.$$

Thus, $f - g(f_1, ..., f_m) \in I \subset k[x_1, ..., x_n, y_1, ..., y_m]$.

On the other hand, the polynomial in $k[x_1, ..., x_n, y_1, ..., y_m]$,

$$\begin{aligned} f(x_1, ...x_n) - g(y_1, ..., y_m) &= (g(f_1, ..., f_m) - g(y_1, ..., y_m)) \\ &\quad +((f(x_1, ..., x_m) - g(f_1, ..., f_m)), \end{aligned}$$

and hence it belongs to the ideal K by Lemma 5.4, because both $g(f_1, ..., f_m)- g(y_1, ..., y_m)$ and $f(x_1, ..., x_m) - g(f_1, ..., f_m)$ belong to K, respectively, by Lemma 5.4. Then Proposition 3.1 says that

$$f(x_1, ..., x_n) - g(y_1, ..., y_m) \in K \Longleftrightarrow N_G(f) = N_G(g).$$

Thus, there exists $h = N_G(f)$, $f \xrightarrow{G}_+ h$, and $g \xrightarrow{G}_+ h$. Moreover, since $g \in k[y_1, ..., y_m]$, g is reduced to h by those $g_i \in G$ such that $LM(g_i)$ is a monomial in the variables $y_1, ..., y_m$ only. Then since the variables x are bigger that the variables y , this means that those g_i used in the reduction are in fact polynomials in the y variables only. Therefore, $h \in k[y_1, ..., y_m]$.

Conversely, suppose there exists $h \in k[y_1, ..., y_m]$ such that $f \xrightarrow{G}_+ h$. Thus, $f - h \in K$; hence

$$f(x_1, ..., x_n) - h(y_1, ..., y_m) = \sum_{i=1}^{m} q_i(x, y)(y_i - f_i(x_1, ..., x_n) + w(x, y),$$

for some $q_i(x, y) = q_i(x_1, ..., x_n, y_1, ..., y_m) \in k[x_1, ..., x_n, y_1, ..., y_m]$ and some

$$\begin{aligned} w(x, y) &= w(x_1, ..., x_n, y_1, ..., y_m) \\ &= \sum_{j}^{finite} p_j(x, y)h_j(x_1, ..., x_n) \in I, \end{aligned}$$

for some $p_j(x,y) = p_j(x_1,...,x_n,y_1,...,y_m) \in k[x_1,...,x_n,y_1,...,y_m]$, where the h_j are generators of the ideal I.

Thus, substituting f_i for y_i, we see that

$$f(x_1,...,x_n) - h(f_1,...,f_m) \in I,$$

that is, $f + I = h(f_1,...,f_m) + I = \phi(h(y_1,...,y_m) + J)$. □

Thus, Theorem 5.7 provides us with a membership algorithm for a coset $f + I \in k[x_1,...,x_n]/I$ to be in the image, $im(\phi)$. Here is the algorithm:

ALGORITHM 19. *(Algorithm for membership question to $im(\phi)$)*

(1) Compute a Groebner basis G for the ideal $K = (I, y_1 - f_1, ..., y_m - f_m)$.
(2) Apply Algorithm 5 to compute the unique remainder $N_G(f) = h$ of f with respect to G.
(3) $N_G(f) \in k[y_1,...,y_m]$ if and only if

$$f + I = h(f_1,...,f_m) + I = \phi(h + I).$$

We also note that this algorithm is virtually the same as Algorithm 15.

COROLLARY 5.8. *$f + I \in k[x_1,...,x_n]/I$ is in the image $im(\phi)$ if and only if $N_G(f) \in k[y_1,...,y_m]$.* □

COROLLARY 5.9. *ϕ is surjective if and only if for each $x_i + I$, $i = 1,...,n$, $N_G(x_i) \in k[y_1,...,y_m]$.* □

REMARK 5.8. *The following theorem provides us with a better computational criterion for ϕ being surjective than this corollary.*

THEOREM 5.8. *Under the same setting as in Theorem 5.6, let G be the* reduced *Groebner basis for the ideal $K = (I, y_1 - f_1, ..., y_m - f_m)$ with respect to an elimination order with the x variables bigger than the y variables. Then ϕ is surjective if and only if for each x_i, $i = 1,...,n$, there exists a polynomial $g_i = x_i - h_i \in G$ with $h_i \in k[y_1,...,y_m]$.*

PROOF. The proof is very similar to that of Theorem 5.5 and leave it as an exercise. □

ALGORITHM 20. *(Algorithm for surjectivity of ϕ) Theorem 5.8 enables to verify whether ϕ is surjective or not.*

EXAMPLE 5.12. *This is a continuation of Example 4.22. Given a cyclic subgroup G of $GL_2(k)$ generated by*

$$\begin{bmatrix} 0 & -1 \\ 1 & 0 \end{bmatrix},$$

we have computed in Example 4.22 that the ring $k[x,y]^G$ of invariants of G is generated as a subring of $k[x,y]$ by three polynomials

$$x^2y^2, x^2+y^2, x^3y-xy^3.$$

Thus we have a k-algebra homomorphism

$$\phi\colon k[u,v,w] \longrightarrow k[x,y],$$

defined by sending

$$\begin{aligned} u &\longmapsto x^2y^2, \\ v &\longmapsto x^2+y^2, \\ v &\longmapsto x^3y-xy^3. \end{aligned}$$

Then its image, $im(\phi) = k[x^2y^2, x^2+y^2, x^3y-xy^3]$, the ring of invariants under the action of G.

In this example, we are going to verify that ϕ is not surjective and explicitly compute the generating set of its kernel, $\ker(\phi)$. *As usual, we use a Computer Algebra System, REDUCE:*[7]

————————

off echo$
off nat$
load_package groebner$

[7]A working program for REDUCE for this example is also available from: http://ch.catholic.or.kr/pundang/4/soh/2006_aa_i/Example_5_12.htm

```
torder({x,y,u,v,w},lex);
f1:=u-x^2*y^2;
f2:=v-(x^2+y^2);
f3:=w-(x^3*y-x*y^3);
gb:=groebner({f1,f2,f3});
gb := {x**2 + y**2 - v,
2*x*y**3 - x*y*v + w,
x*y**2*v - 2*x*u + y*w,
4*x*y*u - x*y*v**2 - 2*y**2*w + v*w,
x*w + y**3*v + 2*y*u - y*v**2,
y**4 - y**2*v + u,
4*u**2 - u*v**2 + w**2}$
```

———————————

Since no leading terms of elements of the computed Groebner basis gb is x or y, we see that ϕ is not surjective. Moreover, $\ker(\phi) = (4u^2-uv^2+w^2)$, *a principal ideal. Thus, we have an isomorphism of k-algebras:*

$$k[u,v,w]/(4u^2 - uv^2 + w^2) \approx k[x^2y^2, x^2+y^2, x^3y - xy^3] \subsetneqq k[x,y].$$

3.4. The One-to-One Correspondence Between Geometry and Algebra. *[Continuation of Remark 5.6]* Before we move into the next section, we are going to finalize our in-depth analysis of the natural correspondence between Geometry and Algebra. After that, we are going to have a few more computed examples.

As we shall see very soon, moving from Geometry to Algebra or moving from Algebra to Geometry, respectively, does not require the bottom field k to be algebraically closed, but moving back to Algebra again in the process of moving from Algebra to Geometry, and then moving back to Algebra do need the bottom field k to be algebraically closed, because we definitely need Hilbert's (Strong) Nullstellensatz, Theorem 1.9, *in order to make the so-called "double-dual of the object" to be equal to the object itself.* For the proof of Hilbert's (Strong) Nullstellensatz for a general algebraically closed field k, the reader may refer to Hungerford's book, Algebra.

In what follows we assume, whenever it is appropriate, that the bottom field k is algebraically closed. We can do this because every field has its algebraic closure as an extension field. We also recall that an algebraically closed field is automatically an infinite field (Lemma 1.3). This means that the bottom field k is always assumed to be an infinite field. We thus note that, if k is an infinite fled, by Corollary 1.4, two different polynomials can not be the same polynomial function, and two different polynomial functions cannot be the same polynomial.

(i) Let V be an affine variety in k^n and W be an affine variety in k^m. That is, $V = V(I), W = V(J)$ for some polynomial ideals I in $k[x_1, ..., x_n]$ and J in $k[y_1, ..., y_m]$. *If we are given a polynomial mapping*

$$f\colon V \longrightarrow W$$

which sends

$$(a_1, ..., a_n) \longmapsto f(a_1, ..., a_n) = (f_1(a_1, ..., a_n), ..., f_m(a_1, ..., a_n)),$$

for some m polynomials $f_i(x_1, ..., x_n) \in k[x_1, ..., x_n]$, then it is in fact from the polynomial mapping $F\colon k^n \longrightarrow k^m$,

$$(x_1, ..., x_n) \longmapsto (f_1(x_1, ..., x_n), ..., f_m(x_1, ..., x_n)).$$

Thus, for any polynomial function $g \in k[y_1, ..., y_m]$, its pull-back

$$F^*(g) \equiv g \circ f = g(f_1, ..., f_m)$$

is a polynomial function in $k[x_1, ..., x_n]$, because for any $(x_1, ..., x_n) \in k^n$, we have

$$\begin{aligned} (g \circ f)(x_1, ..., x_n) &= g(f_1(x_1, ..., x_n), ..., f_m(x_1, ..., x_n)) \\ &= g(f_1, ..., f_m)(x_1, ..., x_n). \end{aligned}$$

In particular, each coordinate function y_i is pulled back to f_i because we have

$$\begin{aligned} (y_i \circ f)(x_1, ..., x_n) &= y_i(f_1(x_1, ..., x_n), ..., f_n(x_1, ..., x_n)) \\ &= f_i(x_1, ..., x_n), \end{aligned}$$

or simply,

$$F^*(y_i) \equiv y_i \circ f = y_i(f_1, ..., f_m) = f_i.$$

Then since the y_i are generators of the polynomial ring $k[y_1, ..., y_n]$, the pull-back process thus naturally defines a k-algebra homomorphism:

$$F^*\colon k[y_1, ..., y_m] \longrightarrow k[x_1, ..., x_n]$$

which sends

$$y_i \longmapsto f_i, \qquad i = 1, 2..., m.$$

Furthermore, we note that among polynomial functions in $k[y_1, ..., y_m]$, those polynomial functions g outside $I(W)$ are non-zero polynomial functions on W, so that their pull-backs

$$f^*(g) \equiv g \circ f = g(f_1, ..., f_m)$$

are again polynomial functions in $k[x_1, ..., x_n]$ and hence they are polynomial functions on V, because for any $(a_1, ..., a_n) \in V$, we have

$$\begin{aligned}(g \circ f)(a_1, ..., a_n) &= g(f_1(a_1, ..., a_n), ..., f_m(a_1, ..., a_n)) \\ &= g(f_1, ..., f_m)(a_1, ..., a_n).\end{aligned}$$

Moreover, for any $g \in I(W)$, its pull-pack vanishes on V, that is, for any $(a_1, ..., a_n) \in V$

$$(g \circ f)(a_1, ..., a_n) = g(f(a_1, ..., a_n)) = 0,$$

because $f(a_1, ..., a_n) = (f_1(a_1, ..., a_n), ..., f_m(a_1, ..., a_n)) \in W$. Thus, we have shown that

$$g \in I(W) \Longrightarrow f^*(g) \equiv g \circ f \in I(V).$$

In addition to this, if $g = y_i$ is the ith coordinate function on k^m, then $y_i \circ f = y_i(f_1, ..., f_m) = f_i$, that is, the pull back of each coordinate function y_i is f_i. If we express all these observations in an algebraically manageable manner, they naturally define a mapping:

$$f^*\colon k[y_1, ..., y_m]/I(W) \longrightarrow k[x_1, ..., x_n]/I(V),$$

which sends the generators

$$y_i + I(W) \longmapsto f_i + I(V).$$

We note that this is a well-defined map because we have already shown that $g \in I(W) \Longrightarrow g \circ f \in I(V)$, that is, the pull-back of a polynomial function vanishing on W also vanishes on V.

Then we can easily check that this map f^* is a k-algebra homomorphism.

In particular, since k is an infinite field, the identity polynomial mapping id: $V \longrightarrow V$ induces the identity k-algebra homomorphism

$$id^* : k[x_1, ..., x_n]/I(V) \longrightarrow k[x_1, ..., x_n]/I(V).$$

(ii) Conversely, suppose that we are given two ideals from Geometry, i.e., $I = I(V)$ for some affine variety $V \subset k^n$ and $J = I(W)$ for some affine variety $W \subset k^m$. *If we are given a (well-defined) k-algebra homomorphism*

$$\varphi\colon k[y_1, ...y_m]/I(W) \longrightarrow k[x_1, ..., x_n]/I(V)$$

which sends the generators

$$y_i + I(W) \longmapsto f_i + I(V), \qquad i = 1, ..., m,$$

where $f_i(x_1, ..., x_n)$ are some polynomials in $k[x_1, ..., x_n]$, it is in fact from the k-algebra homomorphism

$$\Phi : k[y_1, ..., y_m] \longrightarrow k[x_1, ..., x_n]$$

which sends generators

$$y_i \longmapsto f_i(x_1, ..., x_n)\,, \qquad i = 1, 2, ..., m.$$

Thus, it naturally defines a polynomial mapping $\Phi\hat{}\colon\ k^n \longrightarrow k^m$ which sends

$$(x_1, ..., x_n) \longmapsto \Phi\hat{}((x_1, ..., x_n)) = (f_1(x_1, ..., x_n), ..., f_m(x_1, ..., x_n)), \tag{5.3}$$

that is, $\Phi\hat{}(X) := X \circ \Phi \equiv \varepsilon_{X\circ\Phi}$, thus, for any $g \in k[y_1, ..., y_m]$,

$$\begin{aligned}\Phi\hat{}(X)(g) &= (X \circ \Phi)(g) = X(\Phi(g)) = X(g(f_1, ..., f_m)) \\ &= \varepsilon_X(g(f_1, ..., f_m)) = g(f_1(X), ..., f_m(X)) \\ &= (f_1(X), ..., f_m(X))(g),\end{aligned}$$

and hence $\Phi\hat{}(X) = (f_1(X), ..., f_m(X))$, and thus its restriction to

$$V = V(I(V))$$

naturally *defines* a polynomial mapping

$$\varphi\hat{} : V \longrightarrow k^m.$$

Then we note that its image, $im(\varphi\hat{})$ is automatically contained in

$$W = V(I(W)),$$

because, for any point $(a_1, ..., a_n) \in V$ we have, for any $g \in I(W)$

$$\begin{aligned}g(\varphi\hat{}(a_1, ..., a_n)) &= g(f_1(a_1, ..., a_n), ..., f_m(a_1, ..., a_n)) \\ &= g(f_1, ..., f_m)(a_1, ..., a_n),\end{aligned}$$

and then since φ is a well-defined k-algebra homomorphism and since

$$g \in I(W),$$

we see that

$$g(f_1, ..., f_m) = \varphi(g) \in I(V),$$

and hence, for any $(a_1, ..., a_n) \in V$, we have for any $g \in I(W)$,

$$g(\varphi\hat{}(a_1, ..., a_n)) = g(f_1, ..., f_m)(a_1, ..., a_n) = 0.$$

That is,

$$im(\varphi\hat{}) \subseteq W.$$

Thus, the associated polynomial map $\varphi\hat{} : V \longrightarrow k^m$ is in fact

$$\varphi\hat{} : V \longrightarrow W$$

In particular, the identity map id on $k[x_1, ..., x_n]/I(V)$ induces the identity polynomial mapping $id\hat{}$ on V.

(iii) In summary, for a given polynomial mapping $f: V \longrightarrow W$,

$$(a_1, ..., a_n) \longmapsto f(a_1, ..., a_n) = (f_1(a_1, ..., a_n), ..., f_m(a_1, ..., a_n)),$$

where $V = V(I) \subset k^n$ and $W = V(J) \subset k^m$ for some ideal I of $k[x_1, ..., x_n]$ and for some ideal J of $k[y_1, ..., y_m]$, we then easily verify by the two associations described in (i) and (ii) above that

$$(f^*)\hat{} = f,$$

based on the fact that

$$V(I(W)) = W, \qquad V(I(V)) = V,$$

since W and V are affine varieties, and whose proof will be given very soon below. We note that we need these equalities when we move from (ii) $\longrightarrow$ (i) in the (i) $\longrightarrow$ (ii) $\longrightarrow$ (i).

Conversely, for a given (well-defined) k-algebra homomorphism

$$\varphi: \ k[y_1, ..., y_m]/I(W) \longrightarrow k[x_1, ..., x_n]/I(V),$$

$$y_i + I(W) \longmapsto f_i + I(V), \qquad i = 1, ..., m,$$

we then easily verify by the two associations described in (ii) and (i) above that

$$(\varphi\hat{})^* = \varphi,$$

based on the fundamental result,

$$I(V(I(W))) = I(W), \qquad I(V(I(V))) = I(V),$$

by Hilbert's Strong Nullstellensatz, Theorem 1.9. We note that we need these equalities when we move from (i) $\longrightarrow$ (ii) in the (ii) $\longrightarrow$ (i) $\longrightarrow$ (ii).

(iv) In particular, If $V = k^n$ and $W = k^m$, then $I(V) = (0)$ and $I(W) = (0)$ because k is an infinite field, and hence (iii) is reduced to the explanation given in Remark 5.6.

PROPOSITION 5.4. *Let k be an algebraically closed field. Let S be a subset of k^n. Then $V(I(S))$ is the smallest affine variety containing S.*

PROOF. Let $W = V(J)$ for some ideal $J = (f_1, ..., f_s)$ of $k[x_1, ..., x_n]$ contain S, $S \subseteq W$. Thus we have

$$I(S) \supseteq I(W) = I(V(J)) = \sqrt{J},$$

by Hilbert's Strong Nullstellensatz, Theorem 1.9, and hence

$$V(I(S)) \subseteq V(I(W)) = V(\sqrt{J}) = V(J),$$

where the last equality holds, because $f_1, ..., f_s \in I(W)$ so that $V(I(W)) \subseteq V(J)$, and because every $f \in I(W)$ vanishes on W so that $W \subseteq V(I(W))$. That is, $V(\sqrt{J}) = V(J)$, provided that k is an infinite field. □

DEFINITION 5.6. *$V(I(S))$ is called the **Zariski closure** of S.*

THEOREM 5.9. ***(The Closure Theorem)*** *Let k be an algebraically closed field. Let $V(I)$ be an affine variety defined by an ideal I of the polynomial ring $k[x_1, ..., x_n, y_1, ..., y_m]$ and let π: $k^{m+n} \longrightarrow k^m$ be the projection map. Then the Zariski closure of $\pi(V(I))$ is $V(I \cap k[y_1, ..., y_m])$.*

PROOF. For a simplicity reason, we write $J = I \cap k[y_1, ..., y_m]$. We first need to prove that $\pi(V(I)) \subseteq V(J)$. Let $g \in J$ and let $(b_1, ..., b_m)$ be a point of $\pi(V(I)) \subset k^m$. Thus, there exists a point

$$(a_1, ..., a_n, b_1, ..., b_m)$$

in $V(I)$ such that its image under π is $(b_1, ..., b_m)$. As a set J is obviously contained in I, we have $0 = g(a_1, ..., a_n, b_1, ..., b_m) = g(b_1, ..., b_m)$. Thus, $\pi(V(I)) \subseteq V(J)$ holds.

Now, by Proposition 5.4 we already know that the smallest affine variety containing $\pi(V(I))$ is $V(I(\pi(V(I))))$, thus we only need to prove that

$$V(J) \subseteq V(I(\pi(V(I)))).$$

For this, let $g \in I(\pi(V(I)))$. Since $g \in I(\pi(V(I)))$, $g(b_1, ..., b_m) = 0$ for any $(b_1, ..., b_m) \in \pi(V(I))$ and hence $0 = g(b_1, ..., b_m) = g(a_1, ..., a_n, b_1, ..., b_m)$

for any point $(a_1, ..., a_n, b_1, ..., b_m) \in V(I)$, because g has only y variables. That is, $g \in I(V(I))$ and hence $g \in \sqrt{I}$ by Hilbert's Nullstellensatz, Theorem 1.9. Thus, there exists $N \geq 1$ such that $g^N \in I \cap k[y_1, ..., y_m] = J$. That is, $g \in \sqrt{J}$ holds; hence we have proved that $I(\pi(V(I))) \subseteq \sqrt{J}$. Therefore,

$$V(J) = V(\sqrt{J}) \subseteq V(I(\pi(V(I)))),$$

which is what we wanted to prove. □

More generally, we have:

THEOREM 5.10. ***(Polynomial Implicitization)*** *Suppose that k is an infinite field. Let $f : k^n \longrightarrow k^m$ be the polynomial mapping defined parametrically by polynomial functions in (6.2). Let*

$$I = (y_1 - f_1, ..., y_m - f_m) \subset k[x_1, ..., x_n, y_1, ..., y_m]$$

be the defining ideal for the graph $\Gamma(f)$ of f and let $I_n = I \cap k[y_1, ..., y_m]$ be the nth elimination ideal of I. Then $V(I_n)$ is the smallest affine variety in k^m containing $f(k^n)$.

This result is proved in Section 4 of Chapter 6 (Theorem 6.3).

PROPOSITION 5.5. *Let $f \colon V \longrightarrow k^m$ be a polynomial mapping of an affine variety V in k^n to an ambient m-space k^m.*

(1) *Then a polynomial $g \in k[y_1, ..., y_m]$ belongs to $I(im(f))$ if and only if $g \in \ker(f^*)$. That is,*

$$I(im(f)) = \ker(f^*).$$

(2) *The Zariski closure $V(I(im(f))$ of the image $im(f)$ is*

$$V(\ker(f^*)).$$

PROOF. Part (1). $g \in I(im(f))$ if and only if

$$0 = g(f(a_1, ..., a_n)) = (g \circ f)(a_1, ..., a_n) = f^*(g)(a_1, ..., a_n)$$

for any $(a_1, ..., a_n)$ point in V if and only if $f^*(g) \in I(V)$ if and only if $f^*(g) = 0 + I(V)$ if and only if $g \in \ker(f^*)$.

Part (2) is an immediate consequence of part (1). □

Thus, we have

ALGORITHM 21. *(Algorithm for the Zariski closure of the image)*

(1) *Theorem 5.9 plus Theorem 5.6 enables us to compute the defining equations for the Zariski closure of* $\pi(V(I))$.
(2) *Proposition 5.5 plus Theorem 5.6 enables us to compute the defining equations for the Zariski closure of* $im(f)$.

THEOREM 5.11. *Under the above setting (i) - (iv), the two polynomial mappings* $f\colon V \longrightarrow W$, $g\colon W \longrightarrow V$ *are such that* $f \circ g = id$ *and* $g \circ f = id$ *if and only if* $g^* \circ f^* = id^*$ *and* $f^* \circ g^* = id^*$.

That is, V *and* W *are isomorphic as a polynomial mapping if and only if* $k[y_1, ..., y_m]/I(W)$ *and* $k[x_1, ..., x_n]/I(V)$ *are isomorphic as a* k*-algebra homomorphism.*

PROOF. We only need to check, for instance, that

$$(f \circ g)^* = g^* \circ f^* \qquad \text{and} \qquad (\phi \cdot \varphi)\hat{} = \varphi\hat{} \cdot \phi\hat{},$$

because we have already verified that id^* is the identity k-algebra homomorphism and $id\hat{}$ is the identity polynomial mapping in (i) and (ii) above. Now, if $f(a_1, ..., a_n) = (f_1(a), ..., f_m(a)))$ and $g(b_1, ..., b_m) = (g_1(b), ..., g_n(b))$, then, on the one hand, by definition, for any $X \in k^m$, we have

$$\begin{aligned}(f \circ g)^*(y_i)(X) &= y_i \circ (f \circ g)(X) = y_i(f(g(X))\\ &= y_i(f_1(g_1(X), ..., g_n(X)), ..., f_m(g_1(X), ..., g_n(X)))\\ &= f_i(g_1(X), ..., g_n(X)) = f_i(g_1, ..., g_n)(X),\end{aligned}$$

and hence $(f \circ g)^*(y_i) = f_i(g_1, ..., g_n)$, that is, $(f \circ g)^*$ sends

$$y_i + I(W) \mapsto f_i(g_1, ..., g_n) + I(W),$$

and, on the other hand, by definition, for any $X \in k^m$, we have

$$\begin{aligned}(g^* \circ f^*)(y_i)(X) &= g^*(y_i(f(X))) = g^*(f_i(X)) = (f_i \circ g)(X)\\ &= f_i(g_1(X), ..., g_n(X)) = f_i(g_1, ..., g_n)(X),\end{aligned}$$

and hence $(g^* \circ f^*)(y_i) = f_i(g_1, ..., g_n)$, that is, $f^* \circ g^*$ sends

$$y_i + I(W) \mapsto f_i + I(V) \mapsto f_i(g_1, ...g_m),$$

and therefore, they are the same.
On the other hand, if $\varphi(y_i + I(W)) = f_i + I(V)$, $i = 1, ..., m$ and $\phi(x_j + I(V)) = g_j + I(W)$, $j = 1, ..., n$, then by definition, $\varphi\hat{}\,(a_1, ..., a_n) = (f_1(a), ..., f_m(a))$, and $\phi\hat{}\,(b_1, ..., b_m) = (g_1(b), ..., g_n(b))$, and $(\phi \circ \varphi)(y_i + I(W)) = \phi(f_i + I(V)) = f_i(g_1, ..., g_n) + I(W)$, and hence

$$(\phi \circ \varphi)\hat{}\,(b_1, ..., b_m) = (f_1(g_1, ..., g_n)(b), ..., f_m(g_1, ..., g_n)(b))$$

and

$$\begin{aligned} \varphi\hat{}\, \circ \phi\hat{}\,(b_1, ..., b_m) &= \varphi\hat{}\,(g_1(b), ..., g_n(b)) \\ &= (f_1(g_1, ..., g_n)(b), ..., f_m(g_1, ..., g_n)(b)), \end{aligned}$$

and therefore, they are the same. Since $(f \circ g)^* = g^* \circ f^*$ and $(\phi \circ \varphi)\hat{}\, = \varphi\hat{}\, \circ \phi\hat{}\,$ also hold, we have such an equivalence. □

ALGORITHM 22. *(Algorithm for isomorphism) Theorem 5.11 together with Theorem 5.6 and Theorem 5.8 enables us to verify whether a morphism of two affine varieties is an isomorphism or not.*

EXAMPLE 5.13. *This example is rather artificial in the sense that it is prepared for a demonstration purpose. Let V be an affine variety in $\mathbb{C}^2$ defined by two equations: $x^2 - 4xy = 0$ and $y^2 = 0$, that is, the origin of $\mathbb{C}^2$, and let W be an affine variety containing the origin of $\mathbb{C}^3$, say, by two equations: $uv - w = 0, u^2 - v^2 - w^2 - w = 0$. Define a polynomial mapping $\mathbb{C}^2 \longrightarrow \mathbb{C}^3$ as follows:*

$$(x, y) \longmapsto f(x, y) = (x + y, x - y, 4xy),$$

so that the origin of the domain is mapped to the origin of the target. Thus, the image of the origin of the domain is contained in W, and this defines a polynomial map $f\colon V \longrightarrow W$. At this moment, we have two choices. That is, if we explicitly know the generators of both radical ideals

$I(V(I))$ and $I(V(J))$, then the corresponding

$$f^*\colon \mathbb{C}[u,v,w]/I(W) \longrightarrow \mathbb{C}[x,y]/I(V)$$

is going to be automatically well-defined., because the ideal $I(V)$ and $I(W)$ are uniquely determined.

Otherwise, that is, if we do not know them explicitly then we have to stick to the ideals $I = (x^2-4xy, y^2)$ and $J = (uv-w, u^2-v^2-w^2-w)$ but we can still move to the Algebra side, because, for instance, a polynomial ideal I and its radical $\sqrt{I}$ generates the same affine variety $V(I)$.

Returning back to our case, we easily see that

$$I(V((x^2-4xy,y^2)) = \sqrt{(x^2-4xy,y^2)} = (x,y),$$

but we do not easily compute the generating set of the radical ideal

$$I(V(J)) = \sqrt{(uv-w, u^2-v^2-w^2-w)} = ?$$

On the other hand, we do know that for any polynomial ideal, I and $\sqrt{I}$ define the same affine variety $V(I)$ by the last part of the proof of Proposition 5.4, so we instead consider

$$f^* : \mathbb{C}[u,v,w]/J \longrightarrow \mathbb{C}[x,y]/I,$$

$$\begin{aligned} u+J &\longmapsto x+y+I, \\ v+J &\longmapsto x-y+I, \\ w+J &\longmapsto 4xy+I. \end{aligned}$$

Here we note that the number 4 in the above polynomial ideal I and this polynomial mapping is particularly chosen to satisfy the requirement that f^* to be well-defined by Lemma 5.6. Indeed, this k-algebra homomorphism f^* is well-defined, because for generators $uv-w$ and $u^2-v^2-w^2-w$ of J, we have:

$$\begin{aligned} (x+y)(x-y)-4xy &= x^2-y^2-4xy \in I, \\ (x+y)^2-(x-y)^2-(4xy)^2-4xy &= -16x^2y^2 \in I. \end{aligned}$$

For this purpose, we may use the REDUCE command, greduce(f,{f1,f2}).

Now in order to compute the kernel, $\ker(f^*)$*, we compute the reduced Groebner basis for the following ideal:*

$$(x^2 - 4xy, y^2, u - (x+y), v - (x-y), w - 4xy),$$

under the lex order with $x > y > u > v > w$.[8]

```
off echo$
off nat$
load_package groebner$
torder({x,y,u,v,w},lex);
f1:=x^2-4*x*y;
f2:=y^2;
f3:=u-(x+y);
f4:=v-(x-y);
f5:=w-4*x*y;
gb:=groebner({f1,f2,f3,f4,f5});
gb := {2*x - u - v, 2*y - u + v,
2*u**2 - 3*w, u*v - w, u*w, 2*v**2 - w, v*w, w**2}$
```

Thus, by Theorem 5.6, we see that

$$\begin{aligned}\ker(f^*) &= \begin{pmatrix} 2u^2 - 3w + J, uv - w + J, uw + J, \\ 2v^2 - w + J, vw + J, w^2 + J \end{pmatrix} \\ &= (2u^2 - 3w + J, uw + J, 2v^2 - w + J, vw + J, w^2 + J).\end{aligned}$$

Moreover, if we define an ideal L to be

$$L = (2u^2 - 3w, uw, 2v^2 - w, vw, w^2, uv - w, u^2 - v^2 - w^2 - w),$$

then

$$\ker(f^*) = L/J.$$

---

[8]A working program for REDUCE for this example is also available from: http://ch.catholic.or.kr/pundang/4/soh/2006_aa_i/Example_5_13.htm

On the other hand, by Theorem 5.8, we see that f^* is surjective, because of the first two elements of the computed Groebner basis, gb, above. Therefore, we have an isomorphism of k-algebras:

$$\mathbb{C}[u,v,w]/L \approx (\mathbb{C}[u,v,w]/J)/(L/J) \approx \mathbb{C}[x,y]/I.$$

Thus by this isomorphism, we must have

$$V(L) = \{(0,0,0)\},$$

because $V(L)$ is the image $f(V)$. Moreover, L cannot be a maximal ideal because I is contained in the maximal ideal (x,y), so that $\mathbb{C}[x,y]/I$ is not a field.

EXAMPLE 5.14. *This example is a continuation of the last example with different coefficients for polynomials and polynomial mapping. Let V be an affine variety in $\mathbb{C}^2$ defined by two equations: $x^2 - xy = 0$ and $y^2 = 0$, that is, the origin of $\mathbb{C}^2$, and let W be an affine variety containing the origin of $\mathbb{C}^3$, say, by two equations: $uv - w = 0, u^2 - v^2 - w^2 - w = 0$. Thus, $I = (x^2 - xy, y^2)$ and $J = (uv - w, u^2 - v^2 - w^2 - w)$ in what follows. Define a polynomial mapping as follows:*

$$(x,y) \longmapsto f(x,y) = (x+y, x-y, 2xy),$$

so that the origin of the domain is mapped to the origin of the target. Since

$$I(V(I)) = \sqrt{(x^2 - xy, y^2} = (x,y),$$

this time we consider

$$f^* : \mathbb{C}[u,v,w]/J \longrightarrow \mathbb{C}[x,y]/I(V)$$

defined by

$$\begin{aligned} u + J &\longmapsto x + y + (x,y), \\ v + J &\longmapsto x - y + (x,y), \\ w + J &\longmapsto 2xy + (x,y). \end{aligned}$$

Then f^* is a well-defined map by Lemma 5.6, because

$$\begin{aligned}(x+y)(x-y)-2xy &= x^2-y^2-2xy \in (x,y),\\ (x+y)^2-(x-y)^2-(2xy)^2-2xy &= -4x^2y^2+2xy \in (x,y).\end{aligned}$$

If we compute the kernel, $\ker(f^*)$, *we first compute the reduced Groebner basis for the ideal*

$$(x^2-xy, y^2, u-(x+y), v-(x-y), w-2xy),$$

and we obtain:

gb := {2*x - u - v, 2*y - u + v,
2*u**2 - 3*w, 2*u*v - w, u*w, 2*v**2 + w, v*w, w**2}$

Thus,

$$\ker(f^*) = (2u^2-3w+J, 2uv-w+J, uw+J, 2v^2+w+J, vw+J, w^2+J),$$

and f^* is surjective by Theorem 5.8.

If we define

$$L = (2u^2-3w, 2uv-w, uw, 2v^2+w, vw, w^2, uv-w, u^2-v^2-w^2-w),$$

then

$$\ker(f^*) = L/J,$$

and hence

$$\mathbb{C}[u,v,w]/L \approx (\mathbb{C}[u,v,w]/J)/(L/J) \approx \mathbb{C}[x,y]/(x,y) \approx \mathbb{C}.$$

This shows that L is a maximal ideal, and hence it must be a radical ideal, i.e., $\sqrt{L} = L$. Of course, $V(L)$ must be $\{(0,0,0)\}$, since it is the image $f(V)$ by this isomorphism.

REMARK 5.9. *In particular, this example suggests us how we may produce lots of maximal ideals.*

REMARK 5.10. *When we actually do a computation, we are always working over $\mathbb{Q}$, so that it will be OK even if we replace $\mathbb{C}$ with $\mathbb{Q}$ in our arguments by the last part of the proof of Proposition 5.4.*

Exercises for Section 5.3

EXERCISE 5.5. *Let V be an affine variety in $\mathbb{C}^2$ defined by an equation: $x^2-xy+1=0$, and let W be an affine variety in $\mathbb{C}^2$ defined by an equation: $u^4+u^3+2u^2v+v^2+uv+1=0$. Prove that V is isomorphic to W. [Hint: For instance, consider the following polynomial mapping:*

$$(x,y) \longmapsto f(x,y) = (y, -y^2 - x),$$

to see if its restriction to V gives an isomorphism].

4. Minimal Polynomials of Elements in Field Extensions

As an application of Section 3, we are going to study how to compute the minimal polynomial of an algebraic element β in an extension field K of finite degree of a given field F. For this purpose, we review most elementary facts about field extensions.

4.1. A Review on Field Extensions.

DEFINITION 5.7. *An* ***extension field*** *or an* ***extended field*** *of F is a field which contains F as a subfield.*

An extension K of a field F is always a vector space over F. Addition is just the addition law in K, and scalar multiplication of an element α of K by an element c of F is the product $c\alpha$ in K.

DEFINITION 5.8. *Let K be an extension of a field F, and an element α of K is called* ***algebraic*** *over F if it is a root of some non-zero polynomial with coefficients in F. Otherwise, it is called* ***transcendental*** *over F.*

Thus every element α of F is automatically algebraic over itself, because it is a root of $x - \alpha$, which has coefficients in F. Moreover, the unknown x or the variable x in the field $F(x)$ of rational functions = the field of fractions of the *UFD* $F[x]$ is automatically transcendental over F.

On the other hand, we note that the two properties, algebraic and transcendental, are generally dependent on the given bottom field F.

EXAMPLE 5.15. *The complex number $2\pi i$ is algebraic over the field $\mathbb{R}$ of real numbers, but it is transcendental over the field $\mathbb{Q}$ of rational numbers.*

We also note that for a given $\alpha \in K$, the injection $F \hookrightarrow K$ gives rise to the natural ring homomorphism, called the **substitution homomorphism**:

$$\varphi : F[x] \longrightarrow K\,, \tag{5.4}$$

defined by sending $x \mapsto \alpha$, and hence sending $f(x)$ to $f(\alpha)$, and whose restriction on F is the identity map. Thus, in terms of this map, $\alpha \in K$ is transcendental if and only if $\ker(\varphi) = 0$, and is algebraic if and only if $\ker(\varphi) \neq 0$.

If $\alpha \in K$ is algebraic over F, then since $F[x]$ is a *PID*, its kernel $\ker(\varphi)$, which is an ideal in $F[x]$, is generated by a non-zero single element $f(x)$, the monic polynomial of lowest degree having α as a root. Since $f(\alpha) = 0$ in a field K, $f(x)$ must be necessarily an irreducible polynomial over F. Thus, $f(x)$ is the only irreducible monic polynomial in $\ker(\varphi)$, because all other polynomials in $\ker(\varphi)$ are multiples of $f(x)$. We call this polynomial $f(x)$ the **irreducible polynomial for α over F**, or the **minimal polynomial for α over F**.

REMARK 5.11. *It is important to note that this irreducible polynomial $f(x)$ is dependent on F as well as on α, because irreducibility of a polynomial is dependent on the bottom field.*

EXAMPLE 5.16. *Let $F = \mathbb{Q}[i]$, and let α be the complex number $\sqrt{i} = \frac{1}{2}\sqrt{2}(1+i)$. Then the irreducible polynomial for α over $\mathbb{Q}$ is x^4+1, but this polynomial factors in F: $x^4+1 = (x^2+i)(x^2-i)$; hence the irreducible polynomial for α over F is $x^2 - i$.*

Thus, where there are several fields around us, we must make it sure to which field we refer. It is always better to say that $f(x)$ is **irreducible over F**, or that it is an **irreducible element of $F[x]$**.

DEFINITION 5.9. *Let $\alpha \in K$. The field extension of F which is generated by an element $\alpha \in K$ is denoted by*

$$F(\alpha).$$

Thus by definition, $F(\alpha)$ is the smallest field containing F and α. This is true because it is the intersection of all subfields of K containing both K and α.

More generally, if $\alpha_1, ..., \alpha_n$ are elements of an extension field K of F, then the notation

$$F(\alpha_1, ..., \alpha_n)$$

will denote for the smallest subfield of K containing both F and the elements $\alpha_1, ..., \alpha_n$.

We denote the ring generated by α over F by

$$F[\alpha].$$

Thus, it consists of those elements of K which can be written as a polynomial in α with coefficients in F:

$$a_n\alpha^n + a_{n-1}\alpha^{n-1} + \cdots + a_1\alpha + a_0, \tag{5.5}$$

with $a_i \in F$. Then since $F[\alpha]$ is automatically an integral domain, we have its field FR of fractions, and moreover, the subfield $F(\alpha)$ is isomorphic to the field FR of fractions of $F[\alpha]$ under the map

$$FR \ni a/b \longmapsto ab^{-1} \in F(\alpha),$$

where $a, b \in F[\alpha]$. Thus, elements of $F(\alpha)$ are ratios of elements of the form (5.5).

Now if $\alpha \in K$ is transcendental over F, then $\varphi(F[x]) = F[\alpha]$ is isomorphic to $F[x]$, where $\varphi : F[x] \longrightarrow K$ is the map in (5.4).

PROPOSITION 5.6. *If α is transcendental over F, then the map*

$$F[x] \longrightarrow F[\alpha]$$

in (Cf. 5.4) is an isomorphism, and hence $F(\alpha)$ is isomorphic to the field $F(x)$ of rational functions. $\square$

This simple fact has the consequence that the field extensions $F(\alpha)$ are isomorphic for all transcendental element α, because they are all isomorphic to the field of rational functions $F(x)$.

EXAMPLE 5.17. *We recall that π and e are both transcendental over $\mathbb{Q}$. Therefore, $\mathbb{Q}(\pi)$ and $\mathbb{Q}(e)$ are isomorphic fields, sending π to e.*

REMARK 5.12. *We know how these two transcendental numbers are defined. But, both of them are transcendental simply means that there is no way of differentiating them algebraically.*

PROPOSITION 5.7.

a: *Suppose that α is algebraic over F, and let $f(x)$ be its irreducible polynomial over F. Then the map*

$$F[x]/(f(x)) \longrightarrow F[\alpha]$$

is an isomorphism, and $F[\alpha]$ is a field. Thus, $F[\alpha] = F(\alpha)$.

b: *More generally, let $\alpha_1, ..., \alpha_n$ be algebraic element of a field extension K of F. Then $F[\alpha_1, ..., \alpha_n] = F(\alpha_1, ..., \alpha_n)$.*

PROOF. Let φ be the map in (5.4), with $K = F(\alpha)$. Since $f(x)$ generates $\ker(\varphi)$, $F[x]/(f(x))$ is isomorphic to its image, which is $F[\alpha]$. Since $f(x)$ is irreducible over F, it generates a maximal ideal, and hence $F[\alpha]$ is in fact a field; hence in particular, its field of fractions is itself. Then since $F(\alpha)$ is the smallest subfield of K containing both F and α, and $F[\alpha]$ is already a subfield of K, $F(\alpha) = F[\alpha]$. This proves part *a*. For part *b*, we note that $F[\alpha_1, ..., \alpha_n] = F[\alpha_1, ..., \alpha_{n-1}][\alpha_n]$ and hence induction on n will finish its proof. □

PROPOSITION 5.8. *Let α be an algebraic element over F, and let $f(x)$ be its irreducible polynomial over F. Suppose $f(x)$ has degree n. Then $(1, \alpha, \alpha^2, ..., \alpha^{n-1})$ is a basis for $F[\alpha]$ as a vector space over F.*

PROOF. Since we have a projection $F[x] \longrightarrow F[x]/(f(x))$, which sends $g(x)$ to $g(x) \operatorname{mod}(f(x))$, and since $F[x]$ is a vector space over F with a basis $(1, x, x^2, ...)$ the images $\bar{x}^j = x^j \operatorname{mod} f(x) = x^j$ of the x^j under this projection for $j = 0, 1, ..., n-1$ is a basis of $F[x]/(f(x))$. For this, one just checks the images of x^j are (i) linearly independent over F and (ii) they span $F[x]/(f(x))$. Then since the isomorphism $F[x]/(f(x)) \longrightarrow F[\alpha]$ is induced from the ring homomorphism $F[x] \longrightarrow F[\alpha]$, which sends x to α and is identity on F, $(1, \alpha, ..., \alpha^{n-1})$ is a basis of the F-vector space $F[\alpha]$. For this, we have to use the relation $f(\alpha) = 0$ in $F[\alpha]$. □

Now, let α, β be two algebraic elements over F. If $F[\alpha]$ is isomorphic to $F[\beta]$, then the irreducible polynomial for α over F and the irreducible polynomial for β necessarily have the same degree over F, because this degree is the dimension of the field extension as an F-vector space. But, as the following example shows, having the same degree is not sufficient for different reasons.

EXAMPLE 5.18. *Let $f(x) = x^2 + 1$, and $g(x) = x^2 + 5$. We have already seen that in the field $\mathbb{Q}[x]/(f(x))$ is the ring $\mathbb{Z}[i]$ of Gauss integers, which is the collection of all algebraic integers over $\mathbb{Z}$ there, has 4 units: $\pm 1, \pm i$.But, in the field $\mathbb{Q}[x]/(g(x)) = \mathbb{Q}[\sqrt{-5}]$ is the ring $\mathbb{Z}[\sqrt{-5}]$, which is the collection of all algebraic integers over $\mathbb{Z}$ there, has only 2 units: ± 1.Hence, these two fields can not be isomorphic. In fact, $\mathbb{Z}[i]$ is a PID but $\mathbb{Z}[\sqrt{-5}]$ is not a UFD, so it is not a PID.*

On the other hand, one easily checks that if α is a root of $x^3 - x + 1$, then $\beta = \alpha^2$ is a root of $x^3 - 2x^2 + x - 1$. [Hint: $\alpha^3 - \alpha = -1$ and square it.] Since their reduction $\bmod 2$ is the same irreducible polynomial $x^3 + x + 1$ over the field $\mathbb{Z}_p = \mathbb{Z}/p\mathbb{Z}$ with $p = 2$, these two polynomials are irreducible over $\mathbb{Z}$, and hence irreducible over $\mathbb{Q}$. We obviously have $\mathbb{Q}[\beta] \subseteq \mathbb{Q}[\alpha]$.Then since $\mathbb{Q}[\alpha]$ is a $\mathbb{Q}$-vector space of dimension 3 and $\mathbb{Q}[\beta]$ is its subspace with the same dimension 3, two fields $\mathbb{Q}[\alpha]$ and $\mathbb{Q}[\beta]$ are actually the same.

An extension K of a field F is always a vector space over F. Addition is just the addition law in K, and scalar multiplication of an element α of K by an element c of F is the product $c\alpha$ in K.

DEFINITION 5.10. *The dimension of K as an F-vector space over F is called the* **degree** *of the field extension $F \subset K$. It is denoted by*

$$[K : F],$$

and an extension $F \subset K$ is called a ***finite extension*** *if $[K : F]$ is finite.*

The degree is the simplest but important invariant of an extension of a field.

EXAMPLE 5.19. *$\mathbb{C}$ has the $\mathbb{R}$-basis $(1, i)$, and hence $[\mathbb{C} : \mathbb{R}] = 2$.*

Extensions of degree 2 are called *quadratic extensions,* those of degree 3 are called *cubic,* and so on.

We note that the degree of an extension $F \subset K$ is 1 if and only if $F = K$.

In fact, the term degree comes from the case that $K = F(\alpha)$ is generated by one algebraic element over F. In this case, K has a basis $(1, \alpha, \alpha^2, ..., \alpha^{n-1})$, where n is the degree of the irreducible polynomial for α over F. This establishes the first important property of the degree.

PROPOSITION 5.9. *If α is algebraic over F, then $[F(\alpha) : F]$ is the degree of the irreducible polynomial for α over F.* □

This degree is also called the **degree of α over F**.

COROLLARY 5.10. *α has degree 1 if and only if it is an element of the bottom field F, and α has degree ∞ if and only if it is transcendental over F.*

PROOF. The proof of the first part is very easy. For the second part, if α is transcendental over F then the subfield $F[\alpha]$ of $F(\alpha)$ is infinite dimensional and hence so is $F(\alpha)$. If α is algebraic over F, then by Proposition 5.8, α has a finite degree. □

THEOREM 5.12. *Let $F \subset K \subset L$ be fields. Then*

$$[L : F] = [L : K][K : F].$$

PROOF. Let $\mathbf{B} = (y_1, ..., y_n)$ be a basis for L as a K-vector space, and let $\mathbf{C} = (x_1, ..., x_m)$ be a basis for K as an F-vector space. Thus, $[L : K] = n$ and $[K : F] = m$. We will prove that their mn products $(..., x_i y_j, ...)$ is a basis $\mathbf{P}$ of L as an F-vector space. The same reasoning will work if B or $\mathbf{C}$ is infinite.

Let α be an element of L. Since $\mathbf{B}$ is a basis for L over K, we can write

$$\alpha = \beta_1 y_1 + \cdots + \beta_n y_n,$$

with $\beta_i \in K$ in a unique way. Then since $\mathbf{C}$ is a basis for K over F, we can uniquely express each β_i as a linear combination of the x_j's:

$$\beta_j = a_{1j} x_1 + \cdots + a_{mj} x_m,$$

with unique $a_{ij} \in F$. Thus,

$$\alpha = \sum_{i,j}\sum a_{ij} x_i y_j,$$

shows that the products $x_i y_j$ spans L as an F-vector space. Given α, we have seen that each β_i is uniquely determined and each a_{ij} is again uniquely determined from each β_j; hence the a_{ij} are uniquely determined for a given α. Thus, $\mathbf{P} = (..., x_i y_j, ...)$ is a basis for L over F. □

One important case of a tower of field extensions is that K is a given extension of a field F and α is an element of K. Then $F(\alpha)$ generated by α is an *intermediate* field:

$$F \subset F(\alpha) \subset K.$$

Here, α could be an element of the bottom field F. Now, if we add one more finite condition that $[K : F]$ is finite to theses settings, we have an immediate corollary:

COROLLARY 5.11. *Let α be an element of a finite extension K of F of degree n. Then α is algebraic over F, and its degree divides n.*

PROOF. Since the whole degree $[K : F]$ is finite, its divisor $[F(\alpha) : F]$ must be also finite by Theorem 5.12; hence α is algebraic over F of degree $[F(\alpha) : F]$ by the second part of Corollary 5.10. □

DEFINITION 5.11. *An extension field K of a field F is called an* ***algebraic extension*** *if every element of K is algebraic over F, and K is said to be* ***algebraic over*** *F if it is an algebraic extension of F.*

Thus, this corollary also says:

COROLLARY 5.12. *A finite extension of F is an algebraic extension.*

But, the converse does not hold in general: The algebraic closure $\bar{F}$ of a finite field F is not a finite extension, by Lemma 1.3.

THEOREM 5.13. *Let $F \subset K \subset L$ be fields. If L is algebraic over K and K is algebraic over F, then L is algebraic over F.*

PROOF. We have to show that every element $\alpha \in L$ is algebraic over F. Since α is algebraic over K, it satisfies some equation of the form

$$\alpha^n + a_{n-1}\alpha^{n-1} + \cdots a_1\alpha + a_0 = 0,$$

with coefficients $a_0, ..., a_{n-1} \in K$. Thus, α is algebraic over the subfield $F(a_0, ..., a_{n-1})$ generated by $a_0, ..., a_{n-1}$ over F. Then since each coefficients a_i is algebraic over F, we can consider a tower or an increasing chain of subfields

$$F \subset F(a_0) \subset F(a_0, a_1) \subset \cdots \subset F(a_0, ...a_{n-1}) \subset F(a_0, ...a_{n-1}, \alpha),$$

obtained by adjoining these coefficients a_i, and α in succession. Then since each a_{i+1} is algebraic over F, it is algebraic over $F(a_0, ...a_i)$ and also α is algebraic over $F(a_0, ...a_{n-1})$. Thus, each intermediate extension in the chain is a finite extension. Therefore, by Theorem 5.12, the degree of $F(a_0, ...a_{n-1}, \alpha)$ over F is finite. Thus, by Corollary 5.11 α is indeed algebraic over F. □

Now, we would to examine some explicit applications of Corollary 5.11.

COROLLARY 5.13. *Let α be an element of a field extension K of F of prime degree p and $\alpha \notin F$. Then α has degree p over F; hence $K = F(\alpha)$.*

PROOF. This is because the prime integer p divides the degree products $[K : F(\alpha)][F(\alpha) : F]$ and $[F(\alpha) : F] > 1$ since $\alpha \notin F$. Thus, $p = [F(\alpha) : F]$. Then since the subfield $F(\alpha)$ has the same dimension as K, it must be the same as K. □

If we further specialize to the case where $F = \mathbb{R}$ and $p = 2$, we obtain

COROLLARY 5.14. *Every irreducible real polynomial has degree* 1 *or* 2.

PROOF. Let $g(x)$ be an irreducible polynomial over $\mathbb{R}$. Then $g(x)$ has a root α in $\mathbb{C}$ by Gauss. Then since $[\mathbb{C} : \mathbb{R}] = 2$, the degree of the intermediate field $\mathbb{R}(\alpha)$ must be either 1 or 2. □

But as the following example shows, the degree related questions could be quite complicated when the bottom field F is smaller than $\mathbb{R}$, i.e., for instance, $F = \mathbb{Q}$.

EXAMPLE 5.20. *Let $\alpha = \sqrt[3]{2}, \beta = \sqrt[4]{5}$. Consider the field $\mathbb{Q}(\alpha, \beta)$, obtained by adjoining α and β to the field $\mathbb{Q}$ of rational number. We are going to explicitly show that $\mathbb{Q}(\alpha, \beta)$ is a finite extension of degree $12 = 3 \cdot 4$.*

Since the irreducible polynomials for α, β over $\mathbb{Q}$ respectively are $x^3 - 2$ and $x^4 - 5$, $[\mathbb{Q}(\alpha) : \mathbb{Q}] = 3$ and $[\mathbb{Q}(\beta) : \mathbb{Q}] = 4$. Thus, $\mathbb{Q}(\alpha, \beta)$ contains subfields of degree 3 and 4, respectively; hence $[\mathbb{Q}(\alpha, \beta) : \mathbb{Q}]$ must be a common multiple of 3 and 4. Then since $x^4 - 5$ may not be irreducible over $\mathbb{Q}(\alpha)$, $[\mathbb{Q}(\alpha, \beta) : \mathbb{Q}(\alpha)]$ is at most 4. Thus,

$$[\mathbb{Q}(\alpha, \beta) : \mathbb{Q}(\alpha)][\mathbb{Q}(\alpha) : \mathbb{Q}] = [\mathbb{Q}(\alpha, \beta) : \mathbb{Q}] \leq 12.$$

But since no number < 12 is both divisible by 3 and 4, $[\mathbb{Q}(\alpha, \beta) : \mathbb{Q}]$ must be 12.

REMARK 5.13. *We note that* $\gcd(3, 4) = 1$ *in this example; hence,*

$$\operatorname{lcm}(3, 4) = 3 \cdot 4 / \gcd(3, 4) = 12.$$

Thus, the arguments made in this example should work in a general situation of $\gcd(n, m) = 1$.

THEOREM 5.14. *The collection of all algebraic elements over F in an extension field K of a field F is a subfield of K.*

PROOF. Let α, β be two algebraic elements of K. We must show that $\alpha+\beta, \alpha\beta, -\alpha$, and α^{-1} (if $\alpha^{-1} \neq 0$) are all algebraic too. Since α is algebraic over F, $[F(\alpha) : F] < \infty$. Then since β is also algebraic over F, β is also algebraic over the bigger field $F(\alpha)$. Thus, $F(\alpha, \beta)$ is generated by β over $F(\alpha)$ so that it is a finite extension of $F(\alpha)$, i.e., $[F(\alpha, \beta) : F(\alpha)] < \infty$. Thus, by Theorem 5.12 $[F(\alpha, \beta) : F]$ is finite too. Then since every finite extension is an algebraic extension, we see that $F(\alpha, \beta)$ is algebraic over F. Then since $\alpha+\beta, \alpha\beta, -\alpha$, and α^{-1} (if $\alpha \neq 0$) are in $F(\alpha, \beta)$, they are all algebraic. □

REMARK 5.14. *Now suppose that the irreducible polynomials for α and β over F are given to us. Then by this proof, we know that $\alpha+\beta$ is algebraic over F. But, the proof does not tell us how to compute the irreducible polynomial for $\alpha+\beta$ over F. We are going to discuss a computational method, which enables us to compute a polynomial over F whose root is $\alpha+\beta$. This method is historically known as* ***a method of undetermined coefficients****.*

Suppose that the irreducible polynomials for α of degree d_1 and β of degree d_2 over field F are known. Let $n = d_1 d_2$. Then Since $F(\alpha, \beta) = F[\alpha, \beta]$, each element of $F(\alpha, \beta)$ is an F-linear combination of the n monomials $\alpha^i \beta^j$, $0 \leq i \leq d_1 - 1, 0 \leq j \leq d_2 - 1$. Thus, given $\gamma \in F(\alpha, \beta)$, we write the powers $1, \gamma, \gamma^2, ..., \gamma^n$ as F-linear combinations of these monomials. Since we have $n+1$ number of the powers γ^j and only n monomials $\alpha^i \beta^j$, the powers must be linearly dependent; hence a linear dependence relation among the powers γ^j determines a polynomial with coefficients in F of which root is γ. Thus, we only need to figure out a linear dependence relation among the powers γ^j by relating these powers in one way or another.

Here is such an example.

EXAMPLE 5.21. *Suppose that* $\alpha = \sqrt{a}, \beta = \sqrt{b}$, *where* $a, b \in F$. *We want to determine a polynomial whose root is* $\gamma = \alpha + \beta$. *To do this, we square* γ *and square its square, etc., and try to relate these powers together in order to find a linear relations among the powers* γ^j.

$$\begin{aligned} \gamma^2 &= (\alpha+\beta)^2 = (a+b) + 2\alpha\beta, \\ (\gamma^2)^2 &= (a^2+6ab+b^2) + 4(a+b)\alpha\beta. \end{aligned}$$

In this case, these two equations are enough since we can relate these two equations by getting rid of $\alpha\beta$:

$$\gamma^4 - 2(a+b)\gamma^2 + (a-b)^2 = 0.$$

Thus, γ *is a root of the polynomial*

$$g(x) = x^4 - 2(a+b)x^2 + (a-b)^2$$

over F.

But there is one point which complicates matters. Let $g(x)$ be the polynomial having γ as a root which we find in this way. This polynomial may be reducible. For instance, it may happen that γ is actually in the bottom field F, though α, β are not in F. If so, the method of undermined coefficients described above is unlikely to produce its irreducible equation $x - \gamma$. It is harder to determine the *irreducible* polynomial for γ over F.

REMARK 5.15. *In the above it says that it is harder to determine the irreducible polynomial for* $\gamma = \alpha + \beta$ *for given algebraic elements of* K *over* F. *Yes, it is hard if we do not have a strong computational means.*

The thing is that even though we study field extensions in terms of polynomials in one variable over a field F, *we some times have to explicitly figure out certain coefficients by solving a system of polynomial equations, which we have seen in the explanation of the method of unknown coefficients given above.*

We can explicitly compute the irreducible polynomial for γ, *using the theory of Groebner bases, a fairly recently developed computation theory for polynomials in several variables since* 1965. *One of the things that this book aims is to try to combine effectively axiomatic treatments, whose main purpose is to pursuit certain properties qualitatively, and computational treatments, whose main purpose is to compute and compare everything quantitatively, into a one piece of Algebra. This is because no one can actually apply purely abstract theory if it is only fancy for those axiomists, unless it is well-supported by a theory of actual computation.*

DEFINITION 5.12. *An element γ which generates a field extension K/F is called a* **primitive element** *for K over F.*

THEOREM 5.15. *(Existence of a primitive element) Let K be a finite extension field of a field F of characteristic* 0. *Then there exists an element $\gamma \in K$ such that $K = F(\gamma)$.*

REMARK 5.16. *This theorem is not true for fields of characteristic p, where p is a prime number > 1.*

PROOF. By hypothesis K is finite over F, so $K = F(\alpha_1, ..., \alpha_n)$. We use induction on the number n of generators α_i's of K over F. If $n = 1$, there is nothing to prove. For $n > 1$, by induction hypothesis we assume that the theorem is true for the intermediate field $K_1 = F(\alpha_1, ..., \alpha_{n-1})$. Thus, $K_1 = F(\beta)$ for some $\beta \in K_1$; hence $K = K_1(\alpha_n) = F(\beta, \alpha_n)$. Thus, we are reduced to the case when $n = 2$, so that K is generated over F by two elements α, β.

Thus, we need to prove that this

$$K = F(\alpha, \beta)$$

is generated by a single element over F:

First of all, since any finite extension is an algebraic extension, we know that α and β are algebraic over F. Thus we let $f(x), g(x)$ be the irreducible polynomials for α, β over F, respectively. Then we have an extension field K' of $K = F(\alpha, \beta)$ where the product $f(x) \cdot g(x)$ splits completely; hence in which $f(x)$ and $g(x)$ split completely. Let $\alpha = \alpha_1, \alpha_2, ..., \alpha_n$ be roots in K' of $f(x) = 0$ and $\beta = \beta_1, \beta_2, ..., \beta_m$ be roots in

K' of $g(x) = 0$. Then these roots are all distinct because the characteristic of F is 0 by hypothesis.

We are now going to show that for an undetermined coefficient $c \in F$ (which is going to be determined later), the linear combination

$$\gamma = \beta + c\alpha$$

generates $K = F(\alpha, \beta)$:

Obviously we have $F(\gamma) \subseteq K = F(\alpha, \beta)$, and hence it is sufficient to show that $\alpha \in F(\gamma)$, because if so, then $\beta = \gamma - c\alpha \in F(\gamma)$.

In order to show that $\alpha \in F(\gamma)$, we are going to determine its irreducible polynomial over $F(\gamma)$ and it goes as follows. We know that α is a root of $f(x) = 0$. Consider

$$h(x) := g(\gamma - cx).$$

It is clearly a polynomial over $F(\gamma)$, and $h(\alpha) = 0$. Thus, if we can show that the greatest common divisor of $f(x)$ and $h(x)$ in $F(\gamma)[x]$ is $x - \alpha$, then it will follow that $-\alpha$, the constant term of $x - \alpha$, is an element of $F(\gamma)$:

For this, we recall that the monic greatest common divisor of $f(x)$ and $h(x)$ is the same, whether it is computed in $F(\gamma)[x]$ or $K'[x]$. Thus, we may make our computation in $K'[x]$. In that polynomial ring, $f(x)$ is linearly factored:

$$f(x) = (x - \alpha)(x - \alpha_2) \cdots (x - \alpha_n) \in K'[x],$$

and hence it is sufficient to show that none of $x - \alpha_i$, $i = 2, ..., n$, except for $\alpha = \alpha_1$, divides $h(x)$ in $K'[x]$. For this, we need to figure out all the roots in K' of $h(x) = 0$ and it goes as follows: Since the β_j's are roots in K' of $g(x) = 0$, the roots of $h(x) = g(\gamma - cx) = 0$ are determined by solving the equations for x:

$$\gamma - cx = \beta_j.$$

Then since $\gamma = \beta + c\alpha$ holds, the roots are

$$\frac{\gamma - \beta_j}{c} = \frac{\beta - \beta_j}{c} + \alpha.$$

Since we are looking for a reasonable condition on c under which

$$\gcd(f(x), h(x)) = x - \alpha$$

holds, we want these roots to be different from each α_i, for $i = 2, ..., n$, and this will be the case *provided that* c is not one of the finitely many values,

$$-\frac{\beta_j - \beta}{\alpha_i - \alpha}, \tag{5.6}$$

with $i = 2, ..., n$ and $j = 2, ..., m$. □

REMARK 5.17. *Since finitely many inadmissible values for $c \in F$ are explicitly described, this proof is very constructive in nature, and consequently, it also shows us that there should be many primitive elements γ in general.*

EXAMPLE 5.22. *Let K be the field $\mathbb{Q}[i, \sqrt[3]{2}]$. This field has degree 6 over $\mathbb{Q}$, because no smaller numbers than 6 are divisible by both by 2 and 3. For a similar argument, see Example 5.20. Adapting the same notation used in the previous proof, $\beta = \beta_1 = i, \beta_2 = -i$, and $\alpha = \alpha_1 = \sqrt[3]{2}, \alpha_2 = \omega\sqrt[3]{2}$, and $\alpha_3 = \omega^2\sqrt[3]{2}$, where $\omega = e^{2\pi i/3}$. Then condition (5.6) becomes*

$$\sqrt[3]{2}c \neq -\frac{\pm i - 1}{\omega^\nu - 1}, \qquad \nu = 1, 2.$$

We note that this condition holds for all $c \in \mathbb{Q}$ except for $c = 0$. Therefore, $\gamma = i + c\sqrt[3]{2}$ generates K over $\mathbb{Q}$ for all rational numbers $c \neq 0$.

EXAMPLE 5.23. *(This is the continuation of Example 5.22.) Of course, many other combinations of the two elements α, β will generate $K = F[\alpha, \beta]$. In the case of this example, the product $i\sqrt[3]{2}$ also generates K, because its irreducible polynomial over $\mathbb{Q}$ is $x^6 + 4$. This is very easy since $\left(i\sqrt[3]{2}\right)^2 = 2^{2/3}, \left(i\sqrt[3]{2}\right)^3 = -2i, \left(i\sqrt[3]{2}\right)^6 = -4$, and hence,*

$$x^6 + 4$$

is obviously the minimal polynomial over $\mathbb{Q}$ *of the product* $i\sqrt[3]{2}$, *since no other polynomials in* $\mathbb{Q}[x]$ *of less degree than* 6 *can have* $i\sqrt[3]{2}$ *as its root.*

REMARK 5.18. *Therefore, Theorem 5.15 is very important for at least two reasons.*

First, explicit computation in an extension of the form $F(\gamma)$ *is very easy if the irreducible polynomial for* γ *over* F *is known to us or computable.*

Second, since finite extensions have the form $F(\gamma)$, *we can derive their properties from facts about algebraic elements. The second aspect of Theorem 5.15 is most important for us.*

REMARK 5.19. *But, as for the first reason, the next two examples suggest that there is a lot more general and powerful method based on the theory of Groebner Basis.*

4.2. Minimal Polynomials of Elements in $F(\alpha_1, ..., \alpha_n)$**.** We first study the case when $n = 1$. Let β be an element of $F(\alpha)$, a finite extension of F for some algebraic element α over F. Since every finite extension is an algebraic extension, β is automatically algebraic over F. Let $f(x)$ be the minimal polynomial of α of degree d over F Then by Proposition 5.7 we have a F-algebra isomorphism

$$F[x]/(f(x)) \cong F(\alpha).$$

Thus, all the computations among elements in $F(\alpha)$ are going to be done in $F[x]/(f(x))$ through this isomorphism. Since $f(x)$ is irreducible over F the principal ideal $(f(x))$ in the polynomial ring $F[x]$, which is a *PID* (Principal Ideal Domain), is a maximal ideal; hence $F(\alpha)$ is a field containing the bottom field F. Moreover, $F(\alpha)$ is an F-vector space with a basis consisting of $\{1, \alpha, \alpha^2, ..., \alpha^{d-1}\}$, every element of $F(\alpha)$ can be uniquely expressed as a quotient of two polynomial expressions for α.

THEOREM 5.16. *Under this setting, let* $\beta \in F(\alpha)$ *be a non-zero element,*

$$\beta = \frac{p(\alpha)}{q(\alpha)},$$

for some polynomials $p(x) = a_n x^n + \cdots + a_1 x + a_0$, $q(x) = b_m x^m + \cdots + b_1 x + b_0 \in F[x]$. *Then the minimal polynomial of* β *over* F *is the monic polynomial in* y *that generates the inverse image* $J \cap F[y]$ *in* $F[y]$ *of the ideal*

$$J = (f(x), q(x) \cdot y - p(x)) \in F[x, y]$$

under the insertion homomorphism i: $F[y] \hookrightarrow F[x, y]$.

PROOF. Since $q(\alpha) \neq 0$ in a field $F(\alpha)$, it has the inverse $r(\alpha)$, so $q(\alpha)r(\alpha) - 1 = 0$ in $F(\alpha)$. Thus, we have $r(x) \in F[x]$ such that

$$q(x)r(x) - 1 \in (f(x)).$$

Define

$$h(x) = p(x)r(x).$$

Then $\beta = h(\alpha)$, because

$$\begin{aligned} h(\alpha) &= p(\alpha)r(\alpha) = (\frac{p(\alpha)}{q(\alpha)} \cdot q(\alpha)) \cdot r(\alpha) \\ &= \frac{p(\alpha)}{q(\alpha)} = \beta. \end{aligned}$$

This means that the isomorphism of F-algebras,

$$F[x]/(f(x)) \xrightarrow{\cong} F(\alpha),$$

sends $h(x) + (f(x)) \longmapsto \beta$. Moreover, by the universal property of the polynomial ring $F[y]$, we have a unique F-algebra homomorphism:

$$\varphi\colon F[y] \longrightarrow F[x]/(f(x)),$$

which sends the generator $y \longmapsto h(x) + (f(x))$. Thus, if we consider the composition of these two F-algebra homomorphisms,

$$\phi\colon F[y] \longrightarrow F[x]/(f(x)) \xrightarrow{\cong} F(\alpha),$$

we easily see that

$$\ker(\phi) = \ker(\varphi).$$

Thus, we have: $g(y) \in \ker(\phi) = \ker(\varphi)$ if and only if

$$g(h(x)) + (f(x)) = 0 + (f(x))$$

if and only if $0 = g(h(\alpha)) = g(\beta)$.

On the other hand, we know by Theorem 5.6 that

$$\ker(\varphi) = (f(x), y - h(x)) \cap F[y].$$

Therefore, if we can prove that

$$(f(x), y - h(x)) = (f(x), q(x)y - p(x)),$$

then we are done.

Now, since $q(x)r(x) \equiv 1 \pmod{f(x)}$, we have

$$\begin{aligned} y - h(x) &= y - p(x)r(x) \\ &\equiv r(x)(q(x)y - p(x)) \pmod{f(x)}, \end{aligned}$$

and hence $y - h(x) \in (f(x), q(x)y - p(x))$; consequently we have

$$(f(x), y - h(x)) \subseteq (f(x), q(x)y - p(x)).$$

For the reverse inclusion, we note that

$$\begin{aligned} q(x)y - p(x) &\equiv q(x)(y - p(x)r(x)) \\ &= q(x)(y - h(x)) \pmod{f(x)}, \end{aligned}$$

and hence $q(x)y - p(x) \in (f(x), y - h(x))$; consequently we have,

$$(f(x), q(x)y - p(x)) \subseteq (f(x), y - h(x)).$$

□

ALGORITHM 23. *(Algorithm for computing the minimal polynomial of an algebraic element over F) Given $\alpha, \beta = p(\alpha)/q(\alpha)$, $p(x), q(x) \in F[x]$, and the minimal polynomial $f(x)$ of α over F,*

(1) *Compute a Groebner basis G for the ideal $(f(x), q(x)y - p(x))$ under the lex order with $x > y$.*

(2) *Then among the elements of G, the only one polynomial in y is the minimal polynomial of β over F.*

The following is a new result which could be practically very helpful:

COROLLARY 5.15. *Under the same setting as in Theorem 5.16, suppose that we only know that α is a root of $h(x) = 0$ for some polynomial $h(x)$ over F. Then the single Groebner basis element of the computed Groebner basis for $F[y] \cap (h(x), q(x) \cdot y - p(x))$ is a multiple of the minimal polynomial of $\beta = p(\alpha)/q(\alpha)$ over F.*

PROOF. Suppose that $h(x)$ is not irreducible over F. Then since $h(\alpha) = 0$, $h(x)$ is a polynomial multiple of the minimal polynomial $f(x)$ for α. Then since

$$F[y] \cap (h(x), q(x) \cdot y - p(x)) \subseteq F[y] \cap (f(x), q(x) \cdot y - p(x)),$$

and since $F[y]$ is a *PID*, the only single generator for the Groebner basis for the ideal $F[y] \cap (h(x), q(x) \cdot y - p(x))$ is divisible by the only single generator for the Groebner basis for the ideal $F[y] \cap (f(x), q(x) \cdot y - p(x))$. □

REMARK 5.20. *The main point of Corollary 5.15 is that we can still compute the minimal polynomial of*

$$\beta = \frac{p(\alpha)}{q(\alpha)},$$

over k, once we are given a possibly non-irreducible polynomial $h(x)$ such that $h(\alpha) = 0$.

EXAMPLE 5.24. *Let $\beta = \frac{1+\sqrt{2}}{\sqrt{2}} \in \mathbb{Q}(\sqrt{2})$. Thus,*

$$\begin{aligned} f(x) &= x^2 - 2, \\ p(x) &= 1 + x, \\ q(x) &= x. \end{aligned}$$

We use the Computer Algebra System, REDUCE:[9]

[9] A working program for REDUCE for this example is also available from: http://ch.catholic.or.kr/pundang/4/soh/2006_aa_i/Example_5_24.htm

---

```
off echo$
off nat$
load_package groebner$
torder({x,y},lex);
f:=x^2-2;
p:=1+x;
q:=x;
gb:=groebner({f,q*y-p});
gb := {x - 2*y + 2,2*y**2 - 4*y + 1}$
```

Thus, $y^2 - 2y + 1/2$ *is the minimal polynomial of* β *over* $\mathbb{Q}$*. We easily check that*

$$\left(\frac{1+\sqrt{2}}{\sqrt{2}}\right)^2 - 2\frac{1+\sqrt{2}}{\sqrt{2}} + \frac{1}{2} = 0.$$

Thus, $\mathbb{Q}(\beta) = \mathbb{Q}(\sqrt{2})$.

REMARK 5.21. *Applying the method of undetermined coefficients, we can get a polynomial* $4x^4 - 12x^2 + 1$ *of which solution is* β*. But it may not be easy to see manually that*

$$4x^4 - 12x^2 + 1 = (2x^2 - 4x + 1)(2x^2 + 4x + 1).$$

REMARK 5.22. *In the above example, we happen to know that* $x^2 - 2$ *is the minimal polynomial of* $\sqrt{2}$ *over* $\mathbb{Q}$*. What will happen if we only know that* $\sqrt{2}$ *is a root of a polynomial over* $\mathbb{Q}$*? For example,* $\sqrt{2}$ *is certainly a root of* $f = (x^2 - x^3)(x^2 - 2)$*. If we compute a Groebner basis for the ideal* $(f, q * y - p)$ *then do you think we are still able to figure out the minimal polynomial of* $\beta = \frac{1+\sqrt{2}}{\sqrt{2}}$ *over* $\mathbb{Q}$*? The answer is Yes, according to Corollary 5.15*

EXAMPLE 5.25. *This is a continuation of the above example. Here we replace the previous* $f = x^2 - 2$ *with a polynomial multiple*

$$f = (x^2 - x + 3)(x^2 - 2)$$

of f in order to see whether we can still figure out the minimal polynomial of $\beta = \frac{1+\sqrt{2}}{\sqrt{2}}$ over $\mathbb{Q}$. We use the Computer Algebra System, REDUCE:[10]

———————

```
off echo$
off nat$
load_package groebner$
torder({x,y},lex)$
f:=(x^2-x+3)*(x^2-2);
p:=1+x;
q:=x;
gb:=groebner({f,q*y-p});
gb := {x - 6*y**3 + 20*y**2 - 21*y + 6,
6*y**4 - 26*y**3 + 41*y**2 - 27*y + 5}$
result:=factorize(part(gb,length(gb)));
result := {{3*y**2 - 7*y + 5,1},{2*y**2 - 4*y + 1,1}}$
on rounded$
result1:=part(sub({y=(1+sqrt(2))/sqrt(2)},part(result,1)),1);
result1:=1.79289321881$
result2:=part(sub({y=(1+sqrt(2))/sqrt(2)},part(result,2)),1);
result2 := 0$
```

———————

In the above computation, we actually verify that the last polynomial

$$6y^4 - 26y^3 + 41y^2 - 27y + 5$$

of the computed result, gb, is a multiple of the minimal polynomial of $\beta = \frac{1+\sqrt{2}}{\sqrt{2}}$ over $\mathbb{Q}$, by first factorizing it and then numerically evaluating each of its two factors at $\beta = \frac{1+\sqrt{2}}{\sqrt{2}}$. Thus, we conclude by Corollary 5.15 that

$$2y^2 - 4y + 1$$

is indeed the minimal polynomial of $\beta = \frac{1+\sqrt{2}}{\sqrt{2}}$ over $\mathbb{Q}$.

[10]A working program for REDUCE for this example is also available from: http://ch.catholic.or.kr/pundang/4/soh/2006_aa_i/Example_5_25.htm

EXAMPLE 5.26. *Let* $\beta = \frac{1+2\sqrt[3]{3}+3(\sqrt[3]{3})^2}{1+\sqrt[3]{3}}$. *Thus,*

$$\begin{aligned} f(x) &= x^3 - 3, \\ p(x) &= 1 + 2x + 3x^2, \\ q(x) &= 1 + x. \end{aligned}$$

We use the Computer Algebra System, REDUCE:[11]

———————

```
off echo$
off nat$
load_package groebner$
torder({x,y},lex);
f:=x^3-3;
p:=1+2*x+3*x^2;
q:=1+x;
gb:=groebner({f,q*y-p});
gb := {61*x + 2*y**2 - 23*y - 27, 2*y**3 + 3*y**2 - 21*y -
107}$
```

———————

Thus, the minimal polynomial of β *over* $\mathbb{Q}$ *is*

$$y^3 + (3/2)y^2 - (21/2)y - 107/2\,.$$

Thus, $\mathbb{Q}(\beta) = \mathbb{Q}(\sqrt[3]{3})$.

The following theorem, which is a generalization of Theorem 5.16, enables us to compute the minimal polynomial of an element $\beta \in F(\alpha_1, ..., \alpha_n)$, $n \geq 1$, where the α_i are algebraic over F.

THEOREM 5.17. *Let* $F(\alpha_1, ..., \alpha_n)$ *be an algebraic extension of* F, *and for each* $i = 1, ..., n$, *let* $f_i' \in F[x_1, ..., x_i]$ *be a polynomial such that* $f_i = f_i(\alpha_1, ..., \alpha_i, x_i) \in F(\alpha_1, ..., \alpha_{i-1})[x_i]$ *is the minimal polynomial of* α_i *over*

[11]A working program for REDUCE for this example is also available from: http://ch.catholic.or.kr/pundang/4/soh/2006_aa_i/Example_5_26.htm

$F(\alpha_1, ..., \alpha_{i-1})$. Let

$$\beta = \frac{p(\alpha_1, ..., \alpha_n)}{q(\alpha_1, ..., \alpha_n)}$$

be an element of $F(\alpha_1, ..., \alpha_n)$ for some polynomials $p = p(x_1, ..., x_n), q = q(x_1, ..., x_n) \in F[x_1, ..., x_n]$. Then the minimal polynomial of β over F is the monic polynomial in y that generates the inverse image $J \cap F[y]$ in $k[y]$ of the ideal

$$J = (f_1', ..., f_n', q \cdot y - p) \in F[x_1, ..., x_n, y]$$

under the insertion homomorphism $i\colon k[y] \hookrightarrow k[x_1, ..., x_n, y]$.

PROOF. By the universal property of the polynomial ring $F[x_1, ..., x_n]$ we have a unique surjective F-algebra homomorphism

$$\varphi_n\colon F[x_1, ..., x_n] \longrightarrow F(\alpha_1, ..., \alpha_n),$$

which sends each generator $x_i \longmapsto \alpha_i$, $i = 1, 2, ..., n$, and each polynomial f_i' is clearly an element of $\ker(\varphi_n)$; hence

$$(f_1', ..., f_n') \subseteq \ker(\varphi_n).$$

Then by induction on n, we are going to prove the reverse inclusion. Once we prove that they are in fact the same, then we apply exactly the same arguments of the proof of Theorem 5.16 to finish the proof..

When $n = 1$, we have proved it as Theorem 5.16. By induction hypothesis, we assume the the result holds for $n-1$.That is, we assume that the kernel $\ker(\varphi_{n-1})$ of the surjective k-algebra homomorphism,

$$\varphi_{n-1}\colon F[x_1, ..., x_{n-1}] \longrightarrow F(\alpha_1, ..., \alpha_{n-1}),$$

which sends each $x_i \longmapsto \alpha_i$, is $(f_1', ..., f_{n-1}')$. Let $g = g(x_1, ..., x_n) \in \ker(\varphi_n)$. Then $g(\alpha_1, ..., \alpha_n) = 0$, so α_n is a root of the polynomial $h = h(x_n)$ in x_n only:

$$h(x_n) = g(\alpha_1, ..., \alpha_{n-1}, x_n) \in F(\alpha_1, ..., \alpha_{n-1})[x_n].$$

Thus by hypothesis, the minimal polynomial

$$f_n = f_n(x_n) \in F(\alpha_1, ..., \alpha_{n-1})[x_n]$$

of α_n should divide h. Thus,

$$h = f_n q_n$$

for some $q_n = q_n(x_n) \in F(\alpha_1, ..., \alpha_{n-1})[x_n]$.

Now, let $q'_n(x_1, ..., x_n)$ be a polynomial in $F[x_1, ..., x_{n-1}, x_n]$ such that $q'_n(\alpha_1, ..., \alpha_{n-1}, x_n) = q_n$, and consider the polynomial $g - f'_n q'_n \in F[x_1, ..., x_n]$. If we write it as a F-linear combination of the monomials of $F[x_1, ..., x_n]$ in a unique manner and arrange it as a polynomial in x_n, we have

$$g - f'_n q'_n = \sum_j^{finite} h_j(x_1, ..., x_{n-1}) x_n^j,$$

for some polynomials $h_j = h_j(x_1, ..., x_{n-1}) \in k[x_1, ..., x_{n-1}]$. Then since

$$(g - f'_n q'_n)(\alpha_1, ..., \alpha_{n-1}, x_n) = h - f_n q_n = 0,$$

we see that

$$\sum_j^{finite} h_j(\alpha_1, ..., \alpha_{n-1}) x_n^j = 0.$$

But, this happens only when each $h_j(\alpha_1, ..., \alpha_{n-1}) = 0$ for each j. Thus, each $h_j \in \ker(\varphi_{n-1}) = (f'_1, ..., f'_{n-1})$ by induction hypothesis. Thus, $g - f'_n q'_n \in (f'_1, ..., f'_{n-1})$; consequently $g \in (f'_1, ..., f'_{n-1}, f'_n) = \ker(\varphi_n)$. This proves $\ker(\varphi_n) = (f'_1, ..., f'_n)$.

Thus, we have an isomorphism of F-algebras:

$$\varphi\colon F[x_1, ..., x_n]/(f'_1, , , .f'_n) \cong F(\alpha_1, ..., \alpha_n).$$

To finish the proof, we then just apply exactly the same arguments of the entire proof of Theorem 5.16, this time replacing x with $(x_1, ..., x_n)$, α with $(\alpha_1, ..., \alpha_n)$, $\{f(x)\}$ with $\{f'_1(x_1, ..., x_n), ..., f'_n(x_1, ..., x_n)\}$, etc. □

The following corollary is a new result:

COROLLARY 5.16. *Under the same setting as in Theorem 5.17, suppose that we only know that each* α_i *is a root of* $h_i(x) = 0$ *for some polynomial* $h_i(x)$ *over* F. *Then the single Groebner basis element of the computed Groebner basis for* $F[y] \cap (h_1, ..., h_s, q(x) \cdot y - p(x))$ *is a multiple of the minimal polynomial of* $\beta = \frac{p(\alpha_1,...,\alpha_n)}{q(\alpha_1,...,\alpha_n)}$ *over* F.

PROOF. Suppose that the $h_i(x)$ are not irreducible over F. Then since $h_i(\alpha_i) = 0$, $h_i(x)$ is a polynomial multiple of the minimal polynomial $f_i(x)$ for each α_i. Then since

$$F[y]\cap(h_1(x), ..., h_s(x), q(x)\cdot y - p(x)) \subseteq F[y]\cap(f_1(x), ..., f_s(x), q(x)\cdot y - p(x)),$$

and since $F[y]$ is a *PID*, the only single generator for the Groebner basis for the ideal $F[y]\cap(h_1(x), ..., h_s(x), q(x)\cdot y - p(x))$ is divisible by the only single generator for the Groebner basis for the ideal $F[y] \cap (f_1(x), ..., f_s(x), q(x) \cdot y - p(x))$. □

REMARK 5.23. *The main point of Corollary 5.16 is that we can still compute the minimal polynomial of*

$$\beta = \frac{p(\alpha_1, ..., \alpha_n)}{q(\alpha_1, ..., \alpha_n)},$$

over k, *once we are given* n *possibly non-irreducible polynomials* $h_i(x)$ *over the bottom field* k *such that* $h_i(\alpha_i) = 0$ *for* $i = 1, 2, ..., n$.

We should note that Corollary 5.16 could be in fact very useful, because we sometimes do not know whether a given polynomial $h(x)$, *which is irreducible over the bottom field* k, *is irreducible over the intermediate field properly containing the bottom field* k.

EXAMPLE 5.27. *(This is the continuation of Example 5.22 and Example 5.23.) We now know that any* $\gamma = i + c\sqrt[3]{2}$ *with a non-zero* $c \in \mathbb{Q}$ *is a primitive element for* $K = \mathbb{Q}(i, \sqrt[3]{2})$. *Then, for instance, what is the minimal polynomial over* $\mathbb{Q}$ *of the primitive element* $i + \sqrt[3]{2}$*?*

The minimal polynomial of $\sqrt[3]{2}$ *over* $\mathbb{Q}$ *is clearly* $x^3 - 2$, *and since* $\mathbb{Q}(\sqrt[3]{2})$ *is a subfield of* $\mathbb{R}$, *the minimal polynomial* $y^2 + 1$ *of* $i = \sqrt{-1}$ *over* $\mathbb{R}$ *is again the minimal polynomial of* i *over* $\mathbb{Q}(\sqrt[3]{2})$.

Then the following codes for the Computer Algebra System, REDUCE, will provide the right answer very quickly:[12]

```
off echo$
off nat$
load_package groebner$
q1:=x^3-2$
q2:=y^2+1$
q3:=z-(x+y)$
result1:=groebner({q1,q2,q3},{x,y,z});
result1:={22*x - 12*z**5 + 9*z**4 - 40*z**3 + 78*z**2 - 100*z - 91,
22*y + 12*z**5 - 9*z**4 + 40*z**3 - 78*z**2 + 78*z + 91, z**6 +
3*z**4 - 4*z**3 + 3*z**2 + 12*z + 5}$
```

Thus, the last equation in result1 with only variable z,

$$x^6 + 3x^4 - 4x^3 + 3x^2 + 12x + 5,$$

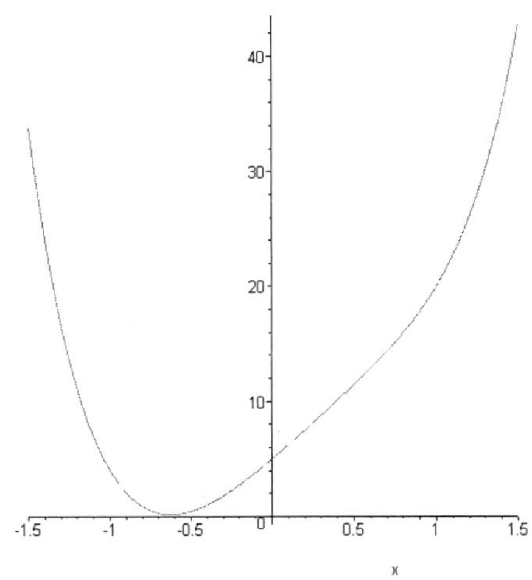

is the minimal polynomial over $\mathbb{Q}$ *of* $i+\sqrt[3]{2}$, *i.e., the irreducible polynomial over* $\mathbb{Q}$ *of least degree of* $i + \sqrt[3]{2}$.

[12]A working program for REDUCE for this example is also available from: http://ch.catholic.or.kr/pundang/4/soh/2006_aa_i/Example_5_27.htm

REMARK 5.24. *As the following example shows, we can still compute the minimal polynomial for*

$$\beta = \frac{p(\alpha_1, ..., \alpha_n)}{q(\alpha_1, ..., \alpha_n)}$$

even if we are only given those n *polynomials* $h_i(x)$ *whose roots are* α_i *for each* $i = 1, ..., n$. *This is due to Corollary 5.16.*

EXAMPLE 5.28. *This is a continuation of Example 5.27, and here in this example we replace the previous q1 and q2 with q1:=(x^2+x+3)*(x^3-2) and q2:=(y^2-5)*(y^2+1) respectively, and try to compute the minimal polynomial of* $i + \sqrt[3]{2}$ *over* $\mathbb{Q}$ *by examining factors of the single computed Groebner basis element for the ideal generated by the minimal polynomial of* $i + \sqrt[3]{2}$ *over* $\mathbb{Q}$. *For this purpose, we use the Computer Algebra System, REDUCE:*[13]

———————

```
off echo$
off nat$
load_package groebner$
q1:=(x^2+x+3)*(x^3-2)$
q2:=(y^2-5)*(y^2+1)$
q3:=z-(x+y)$
gb:=groebner({q1,q2,q3},{x,y,z});
gb := {50489431988836402759687426996290*x
+ 3095258729228333012806617272*z**19
+ 10198221201773034274293916768 3*z**18
- 13311101996726075072801189299 4*z**17
- 11434520255846816517863417796 42*z**16
- 14928267681981471567799910280 12*z**15
- 11949433893333044097081471607 19*z**14
+ 78761209500434649111274534940 02*z**13
- 22429036301678533720167170950 186*z**12
```

[13]A working program for REDUCE for this example is also available from: http://ch.catholic.or.kr/pundang/4/soh/2006_aa_i/Example_5_28.htm

```
- 184573475373052854675305769748372*z**11
- 223666069980306259653843349498034*z**10
+ 323328951421891594176856616528896*z**9
- 389175831927428247706848234280618*z**8
+ 482489260304885965628816085644336*z**7
- 8685548976924879853597244415161183*z**6
- 2433606052742041127312389396996166*z**5
+ 8489372129018375617485573909160772*z**4
- 4494704998405643900942650170046004*z**3
- 23031530552886999073809013354571883*z**2
- 24780211243078099995093245069112650*z
- 8364890289805272717075320973725150,
504894319888364027596874269962900*y
- 30952587292283330128066172724*z**19
- 101982212017730342742939167683*z**18
+ 133111019967260750728011892994*z**17
+ 1143452025584681651786341779642*z**16
+ 1492826768198147156779991028012*z**15
+ 1194943389333304409708147160719*z**14
- 7876120950043464911127453494002*z**13
+ 22429036301678533720167170950186*z**12
+ 184573475373052854675305769748372*z**11
+ 223666069980306259653843349498034*z**10
- 323328951421891594176856616528896*z**9
+ 389175831927428247706848234280618*z**8
- 482489260304885965628816085644336*z**7
+ 8685548976924879853597244415161183*z**6
+ 2433606052742041127312389396996166*z**5
- 8489372129018375617485573909160772*z**4
+ 4494704998405643900942650170046004*z**3
+ 23031530552886999073809013354571883*z**2
+ 24275316923189735967496370799149750*z
```

```
+ 8364890289805272717075320973725150,
z**20 + 4*z**19 - 2*z**18 - 40*z**17 - 74*z**16
- 72*z**15 + 226*z**14 - 548*z**13 - 6481*z**12
- 11392*z**11 + 5482*z**10 - 5324*z**9 + 6065*z**8
- 268504*z**7 - 275738*z**6 + 228660*z**5
+ 39940*z**4 - 859844*z**3 - 1293584*z**2
- 790300*z - 178475}$
result:=factorize(part(gb,length(gb)));
result := {{z**6 + 3*z**4 - 4*z**3 + 3*z**2 + 12*z + 5,1},
{z**6 - 15*z**4 - 4*z**3 + 75*z**2 - 60*z - 121,1},
{z**4 + 2*z**3 + 9*z**2 + 8*z + 5,1},
{z**4 + 2*z**3 - 3*z**2 - 4*z + 59,1}}$
on rounded$
j:=1$
while j leq length(result) do
begin
write "result",j," :=",part(sub({z=i+2^(1/3)},part(result,j)),1);
j:=j+1$
end$
result1 :=0$
result2 :=56.1429467773*i - 66.3392840344$
result3 :=41.1633010103*i + 10.8018873555$
result4 := - 1.07480418714*i + 42.6340221331$
```

Thus, we conclude from the above computation that

$$z^6 + 3z^4 - 4z^3 + 3z^2 + 12z + 5$$

is indeed the minimal polynomial of $i + \sqrt[3]{2}$ *over* $\mathbb{Q}$ *by Corollary 5.16.*

REMARK 5.25. *As in the above example, even if we are going to use the algorithm for factorization of polynomials, it would be a lot better if we start our computation with all irreducible polynomial inputs. That is, if we start our computation with those irreducible polynomials* h_i *over* *the bottom field k each of which has the given α_i as a root, then it will certainly reduce computation time a lot.*

Even in this case, it could happen that some of the h_i's are not irreducible over the intermediate fields properly containing k, so that we will still need Corollary 5.16 and the codes in Example 5.28.

For example, if we take

$$q1 := (x - i) * (x\hat{}3 - 2)$$

instead of the old

$$q1 := (x\hat{}2 + x + 3) * (x\hat{}3 - 2)$$

in the above example, then this new q1 is not irreducible over the intermediate field $\mathbb{Q}(i)$. Of course, the codes in the above example surely compute the minimal polynomial of $i + \sqrt[3]{2}$ over $\mathbb{Q}$ from these new inputs.

EXAMPLE 5.29. *This is the continuation of Example 5.27. More generally,*[14]

———————

```
load_package groebner$
off nat$
q1:=x^3-2$
q2:=y^2+1$
q3:=z-(c*x+y)$
result1:=groebner({q1,q2,q3},{x,y,z});
result1 := {(54*c**7 - 32*c)*x - 12*z**5 + (9*c**3)*z**4 - 40*z**3
+ (78*c**3)*z**2 - (72*c**6 + 28)*z - 91*c**3, (54*c**6 - 32)*y +
12*z**5 - (9*c**3)*z**4 + 40*z**3 - (78*c**3)*z**2 + (18*c**6+ 60)*z
+ 91*c**3,
z**6 + 3*z**4 - (4*c**3)*z**3 + 3*z**2 + (12*c**3)*z +
(4*c**6 + 1)}$
```

———————

[14]A working program for REDUCE for this example is also available from: http://ch.catholic.or.kr/pundang/4/soh/2006_aa_i/Example_5_29.htm

Thus, the last equation in result1 with only variable z*,*

$$x^6 + 3x^4 - 4c^3x^3 + 3x^2 + 12c^3x + 4c^6 + 1$$

is the minimal polynomial over $\mathbb{Q}$ *of* $i + c\sqrt[3]{2}$*, i.e., the irreducible polynomial over* $\mathbb{Q}$ *of least degree of* $i + c\sqrt[3]{2}$*, for a non-zero* $c \in \mathbb{Q}$*.*

In this way, we can generate infinitely many irreducible polynomials over $\mathbb{Z}$ *by choosing any non-zero integers* c*.*

We also note that when $c = 1$*, it agrees well with the result in Example 5.27, which should be the case.*

REMARK 5.26. *If you understand the meaning of the assignment for q3 in the above program, then you can use this program for other situations as well. That is, if* $K = F(\alpha_1, ..., \alpha_n)$ *and if*

$$\gamma = \frac{f(\alpha_1, ..., \alpha_n)}{g(\alpha_1, ...\alpha_n)} \in K,$$

with $f(x_1, ..., x_n), g(x_1, ..., x_n) \in F[x_1, ..., x_n]$*, then*

$$q3 := g(x_1, ..., x_n)x_{n+1} - f(x_1, ..., x_n)$$

is the needed expression.

In the case of the above example, $\alpha_1 = i$ *and* $\alpha_2 = \sqrt[3]{2}$*, and* $i + c\sqrt[3]{2}$ *is a polynomial in* i *and* $\sqrt[3]{2}$ *with coefficients* 1 *and* $c \in \mathbb{Q}$*. Thus,* $g(x, y) \equiv 1$ *and* $f(x, y) = cx + y$*.*

EXAMPLE 5.30. *Let* $\alpha = \sqrt{5 + 1 \cdot \sqrt{21}} \in \mathbb{R}$*. Its irreducible polynomial* $f(x)$ *is*

$$f(x) = x^4 + bx + c,$$

where $b = -2r = -2 \cdot 5 = 10$ *and* $c = r^2 - s^2t = 5^2 - 21 = 4$*. Then since none of the divisors* $\pm 1, \pm 2, \pm 4$ *of the constant term* 4 *is a root of* $f(x) = 0$*,* $f(x)$ *is irreducible over* $\mathbb{Q}$*.*

Now, since $\alpha' = \sqrt{5 - 1 \cdot \sqrt{21}}$*, we see that*

$$\alpha\alpha' = 2,$$

and hence

$$(\alpha+\alpha')^2 = 5+\sqrt{21}+5-\sqrt{21}+2\alpha\alpha' = 14.$$

Then since $\alpha > \alpha' > 0$, their sum is also positive; hence $\alpha+\alpha' = \sqrt{14}$, and the intermediate subfield L is $\mathbb{Q}(\sqrt{14})$.

In the same way, we obtain

$$(\alpha-\alpha')^2 = 5+\sqrt{21}+5-\sqrt{21}-2\alpha\alpha' = 6.$$

Thus, $\alpha-\alpha' = \sqrt{6}$.

Thus, we have obtained a system of two linear equations for α and α':

$$\begin{cases} \alpha+\alpha' &= \sqrt{14} \\ \alpha-\alpha' &= \sqrt{6} \end{cases}.$$

Solving this system, we find that

$$\begin{aligned} \alpha &= \frac{1}{2}(\sqrt{14}+\sqrt{6}), \\ \alpha' &= \frac{1}{2}(\sqrt{14}-\sqrt{6}), \end{aligned}$$

respectively. We note that both α and α' are elements of the newly constructed field $\mathbb{Q}(\sqrt{14},\sqrt{6})$, and hence the splitting field $K = \mathbb{Q}(\alpha,\alpha')$ of $f(x)$ is in fact the same as $\mathbb{Q}(\sqrt{14},\sqrt{6})$. That is,

$$\mathbb{Q}(\alpha,\alpha') = \mathbb{Q}(\sqrt{14},\sqrt{6}).$$

Now, y^2-6 is the minimal polynomial of $\sqrt{6}$ over $\mathbb{Q}$, and since $\sqrt{14} \notin \mathbb{Q}(\sqrt{6})$ and $\sqrt{6} \notin \mathbb{Q}(\sqrt{14})$, $\big[\mathbb{Q}(\sqrt{14},\sqrt{6}) : \mathbb{Q}\big] > 2$, and since the minimal polynomial x^2-14 of $\sqrt{14}$ over $\mathbb{Q}$ is irreducible over $\mathbb{Q}(\sqrt{6})$, we see that x^2-14 is still the minimal polynomial of $\sqrt{14}$ over $\mathbb{Q}(\sqrt{6})$, and consequently $\big[\mathbb{Q}(\sqrt{14},\sqrt{6}) : \mathbb{Q}\big] = 4$.

Then the following lines are another way of verifying that $\mathbb{Q}(\frac{1}{2}(\sqrt{14}+\sqrt{6})) = \mathbb{Q}(\frac{1}{2}(\sqrt{14}-\sqrt{6}))$ and it is the splitting field of the following irreducible polynomial over $\mathbb{Q}$,

$$x^4-10x^2+4,$$

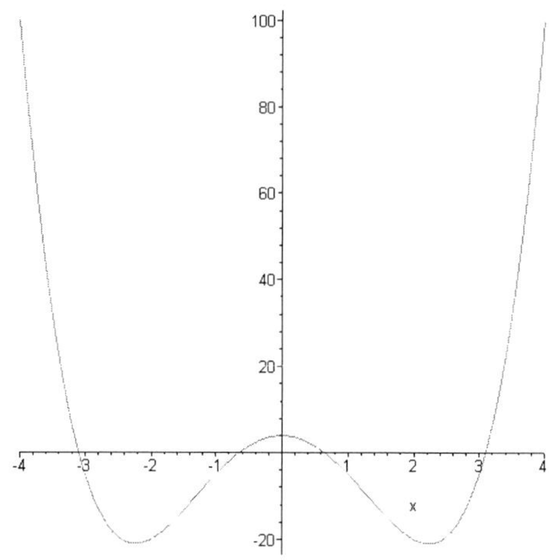

where we use a Computer Algebra System, REDUCE:[15]

————————

```
load_package groebner$
p1:=x^2-14;
p2:=y^2-6;
p3:=z-(x+y)/2;
p4:=z-(x-y)/2;
result1:=groebner({p1,p2,p3},{x,y,z});
result1 := {2*x + z**3 - 12*z, 2*y - z**3 + 8*z, z**4 - 10*z**2 + 4}$
result2:=groebner({p1,p2,p4},{x,y,z});
result2 := {2*x + z**3 - 12*z, 2*y + z**3 - 8*z, z**4 - 10*z**2 + 4}$
```

————————

More precisely, the last polynomials of result1 and resut2 in the above computation are exactly the same, and hence the irreducible polynomial over $\mathbb{Q}$ *for both* $\alpha = \frac{1}{2}(\sqrt{14} + \sqrt{6})$ *and* $\alpha' = \frac{1}{2}(\sqrt{14} - \sqrt{6})$ *are the same polynomial*

$$x^4 - 10x^2 + 4.$$

[15] A working program for REDUCE for this example is also available from: http://ch.catholic.or.kr/pundang/4/soh/2006_aa_i/Example_5_30.htm

In addition to this, we already know that

$$(x-\alpha)\cdot(x+\alpha)(x-\alpha')(x+\alpha') = x^4 - 10x^2 + 4,$$

which is the irreducible polynomial for $\alpha = \sqrt{5+1\cdot\sqrt{21}}$ *over* $\mathbb{Q}$.

Thus, we have double-checked that

$$K = \mathbb{Q}(\alpha,\alpha'),\ \mathbb{Q}(\sqrt{14},\sqrt{6}),\ \mathbb{Q}(\frac{1}{2}(\sqrt{14}+\sqrt{6})), \mathbb{Q}(\frac{1}{2}(\sqrt{14}-\sqrt{6}))$$

are in fact the same subfield of $\mathbb{R}$.

REMARK 5.27. *In general, the splitting field* $K = F(\alpha,\alpha')$ *of the polynomial,*

$$f(x) = (x^2-(r+s\sqrt{t}))(x^2-(r-s\sqrt{t})) = x^4+bx^2+c \in F[x],$$

whose roots are $\alpha = \sqrt{r+s\sqrt{t}}$ *and* $\alpha' = \sqrt{r-s\sqrt{t}}$ *can be constructed by the sequence* $\sqrt{t},\alpha,\alpha'$ *of three square root adjunctions, so that the degree* $[K:F]$ *divides* 8. *Such a sequence of extensions is an example of the so-called* radical extension.

EXAMPLE 5.31. *Let* $p=7$. *Then* 6 *is divisible by* 3 *and* 2, *so that we expect that it has two intermediate subfields of degree* 3 *and* 2, *respectively. In the next example we will figure out the intermediate subfield of degree* 2. *For the case of degree* 3, *we let* $\omega = e^{2\pi i/7}$, *and let*

$$\begin{aligned}\alpha_1 &= \omega+\omega^6 = 2\cos(2\pi/7) \in \mathbb{Q}(\omega),\\ \alpha_2 &= \omega^2+\omega^5 \in \mathbb{Q}(\omega),\\ \alpha_3 &= \omega^3+\omega^4 \in \mathbb{Q}(\omega).\end{aligned}$$

They are all real numbers. In what follows, we are going to show that they are algebraic over $\mathbb{Q}$.

We note that generators of $\mathbb{Q}(\omega)$ *are:*

$$\omega,\omega^2,\omega^3,\omega^4,\omega^5,\omega^6,$$

and from these elements we obtain the three real numbers $\alpha_1,\alpha_2,$*and* α_3.

Here, we use a Computer Algebra System, REDUCE:[16]

```
——————————
% off echo$
off nat$ load_package groebner$
q1:=(x^7-1)/(x-1); q2:=factorize(q1);
p:=part(part(q2,1),1);
——————————
```

This tells us that (the 7-th cyclotomic polynomial)

$$\Phi_7(x) = x^6 + x^5 + x^4 + x^3 + x^2 + x + 1$$

is the irreducible polynomial for ω over $\mathbb{Q}$. *[This can be also checked using Eisenstein Criterion, Proposition 1.1].*

Then the following REDUCE commands compute what we are looking for:

```
p1:=y-(x+x^6);
p2:=y-(x^2+x^5);
p3:=y-(x^3+x^4);
result1:=groebner({p,p1},{x,y});
result1 := {x**2 - x*y + 1,y**3 + y**2 - 2*y - 1}$
result2:=groebner({p,p2},{x,y});
result2 := {x**2 + x*y**2 + x*y - x + 1, y**3 + y**2 - 2*y - 1}$
result3:=groebner({p,p3},{x,y});
result3 := {x**2 - x*y**2 + 2*x + 1, y**3 + y**2 - 2*y - 1}$
——————————
```

We note that the last polynomials of result1, result2, and result3 are the same. Thus, α_1, α_2, and α_3 are distinct roots of the same irreducible polynomial $f(x)$ *over* $\mathbb{Q}$*:*

$$f(x) = x^3 + x^2 - 2x - 1 = (x - \alpha_1)(x - \alpha_2)(x - \alpha_3).$$

[16]A working program for REDUCE for this example is also available from: http://ch.catholic.or.kr/pundang/4/soh/2006_aa_i/Example_5_31.htm

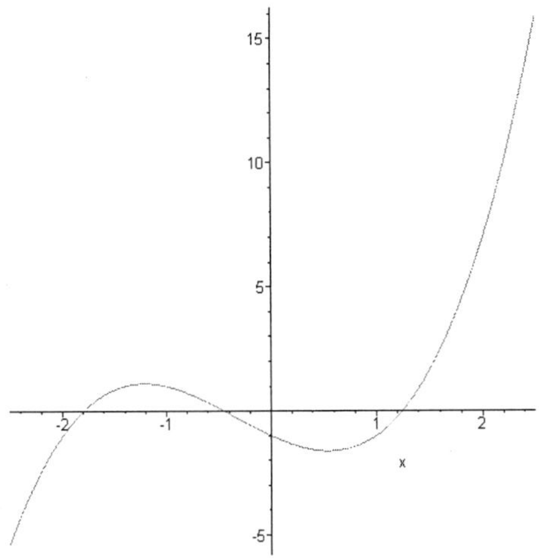

We recall that $x^3 + x^2 - 2x - 1 \pmod 2 = x^3 + x^2 + 1$ *is irreducible over* $\mathbb{F}_2 = \mathbb{Z}/2\mathbb{Z}$*, and hence* $f(x)$ *is irreducible over* $\mathbb{Q}$*. Thus, the splitting field* L *of* $f(x)$ *contains all the roots* $\alpha_1, \alpha_2,$ *and* α_3*; consequently we see that the intermediate subfield* $L = \mathbb{Q}(\alpha_1)$ *has degree* 3.

THEOREM 5.18. *Under the same setting as in Theorem 5.17, Let* G *be the* reduced *Groebner basis for the ideal,*

$$J = (f_1', ..., f_n', q \cdot y - p) \in F[x_1, ..., x_n, y],$$

with respect to an elimination order with the x *variables bigger than the* y *variables. Then* $F(\alpha_1, ..., \alpha_n) = F(\beta)$ *if and only if, for each* $i = 1, ..., n$*, there exists a polynomial* $g_i \in G$ *such that* $g_i = x_i - h_i$ *for some* $h_i \in F[y]$*. If it is the case, then each* $\alpha_i = h_i(\beta)$, $i = 1, 2, ..., n$.

PROOF. We have constructed a composition of F-algebra homomorphism:

$$\phi\colon F[y] \xrightarrow{\varphi} F[x_1, ..., x_n]/(f_1', ..., f_n') \xrightarrow{\cong} F(\alpha_1, ..., \alpha_n),$$

which sends $y \longmapsto h + (f_1', ..., f_n') \longmapsto \beta$. Moreover, we have also verified, in the proof of Theorem 5.17 that

$$J = (f_1', ..., f_n', y - h).$$

Thus, $F(\beta) = F(\alpha_1, ..., \alpha_n)$ if and only if φ is surjective, and thus we finish our proof by Theorem 5.8 with this expression for J. □

REMARK 5.28. *We recall that, if the bottom field F has characteristic 0, then the method introduced in the proof of Theorem 5.15 tells us which elements β of $F(\alpha_1, \alpha_2)$ are going to be a primitive element for $F(\alpha_1, \alpha_2)$ over F, but it does not tell us how to compute the irreducible polynomial for β over F.*

On the other hand, Theorem 5.18 is a lot better than the method explained in the proof of Theorem 5.15, because Theorem 5.18 explicitly states us that we only need to choose a right side of the equivalence relation in the theorem. The only thing we need to do is to choose the right candidate for a primitive element β for $F(\alpha_1, ..., \alpha_n)$ over F, even when $n \geq 2$.

EXAMPLE 5.32. *If we examine the computed reduced Groebner basis of Example 5.27 above, we immediately see that $i+\sqrt[3]{2}$ is a primitive element for $\mathbb{Q}(i, \sqrt[3]{2})$ over $\mathbb{Q}$, and*

$$\begin{aligned}
\sqrt[3]{2} &= \frac{-1}{22}\begin{pmatrix} -12\left(i+\sqrt[3]{2}\right)^5 + 9\left(i+\sqrt[3]{2}\right)^4 - 40\left(i+\sqrt[3]{2}\right)^3 \\ +78\left(i+\sqrt[3]{2}\right)^2 - 100(i+\sqrt[3]{2}) - 91 \end{pmatrix}, \\
i &= \frac{-1}{22}\begin{pmatrix} 12\left(i+\sqrt[3]{2}\right)^5 - 9\left(i+\sqrt[3]{2}\right)^4 + 40\left(i+\sqrt[3]{2}\right)^3 \\ -78\left(i+\sqrt[3]{2}\right)^2 + 78\left(i+\sqrt[3]{2}\right) + 91 \end{pmatrix}.
\end{aligned}$$

Here are the actual codes for such computation for the REDUCE:[17]

———————

```
off echo$
off nat$
f := ((-1)/22)*( - 12*z**5 + 9*z**4 - 40*z**3 + 78*z**2 - 100*z -
91);
g:=((-1)/22)*(12*z**5 - 9*z**4 + 40*z**3 - 78*z**2 + 78*z + 91);
result1:= sub({z=i+2^(1/3)},f);
2**(1/3)$
```

[17]A working program for REDUCE for this example is also available from: http://ch.catholic.or.kr/pundang/4/soh/2006_aa_i/Example_5_32.htm

```
result2:= sub({z=i+2^(1/3)},g);
i$
```

————————

Exercises for Section 5.4

EXERCISE 5.6. *Let* $\alpha = \sqrt[4]{2}$ *and* $\beta = i$. *Apply Corollary 5.16 to compute the minimal polynomial of* $\alpha + \alpha\beta$ *over* $\mathbb{Q}$, *using the fact that the minimal polynomial of* α *over* $\mathbb{Q}$ *is* $x^4 - 2$ *and the minimal polynomial for* β *over* $\mathbb{Q}$ *is* $y^2 + 1$.

SOLUTION 1. *Here is the program for Exercise 5.6:*[18]

```
off echo$
off nat$
load_package groebner$
q1:=x^4-2;
q2:=y^2+1;
q3:=z-(x+x*y);
gb:=groebner({q1,q2,q3},{x,y,z});
subs_set:={z=2^(1/4)+i*2^(1/4)};
% End of Inputs.
result:=factorize(part(gb,length(gb)));
on rounded$
j:=1$
while j leq length(result) do
begin
if length(result)=1 then
  begin
  write "result",1," :=",part(result,1);
      end
   else
      begin
```

[18]A working program for REDUCE for this example is also available from: http://ch.catholic.or.kr/pundang/4/soh/2006_aa_i/Exercise_5_06.htm

```
answer:=part(sub(subs_set,part(result,j)),1)$
      write "result",j," :=",answer;
       if answer=0 then
          begin
write "Thus, the answer is: ",part(part(result,j),1);
          end
          else begin end$
      end$
   j:=j+1$
  end$
```

EXERCISE 5.7. *Let* $\alpha = \sqrt[4]{2}$ *and* $\beta = i\sqrt[4]{2}$. *Apply Corollary 5.16 to compute the minimal polynomial of* $\alpha + \beta$ *over* $\mathbb{Q}$, *using the fact that the minimal polynomial of* α *over* $\mathbb{Q}$ *is* $x^4 - 2$ *and the minimal polynomial for* β *over* $\mathbb{Q}$ *is* $y^4 - 2$.

SOLUTION 2. *Here is the program for Exercise 5.7:*[19]

```
off echo$
off nat$
load_package groebner$
q1:=x^4-2;
q2:=y^4-2;
q3:=z-(x+y);
gb:=groebner({q1,q2,q3},{x,y,z});
subs_set:={z=2^(1/4)+i*2^(1/4)};
% End of Inputs.
result:=factorize(part(gb,length(gb)));
on rounded$
j:=1$
while j leq length(result) do
begin
   if length(result)=1 then
```

[19]A working program for REDUCE for this example is also available from: http://ch.catholic.or.kr/pundang/4/soh/2006_aa_i/Exercise_5_07.htm

```
      begin
      write "result",1," :=",part(result,1);
      end
   else begin
answer:=part(sub(subs_set,part(result,j)),1)$
      write "result",j," :=",answer;
       if answer=0 then
          begin
write "Thus, the answer is: ",part(part(result,j),1);
          end
          else begin end$
      end$
 j:=j+1$
end$
```

CHAPTER 6

Elimination Theory and Resultants

Here is a brief history of a Jewish-English mathematician, J.J. Sylvester, and Resultants: It was J.J. Sylvester (1814 - 1897) who first discovered in 1851 the discriminant of a cubic equation and he was also the first mathematician who used the name "discriminant" for such expressions of quadratic equations and those of higher order. He published important papers in 1852 and 1853, namely "*On the principle of the calculus of forms*" and "*On the theory of syzygetic relations and two rational integer functions*". In particular he used matrix theory to study higher dimensional geometry. In 1877 Sylvester accepted a chair at Johns Hopkins University and he founded in 1878 the American Journal of Mathematics, the first mathematical journal in the United States.

Resultants were introduced and studied by Euler, Bezout, Sylvester, Cayley, Dixon and Macaulay to extend linear algebra techniques for solving linear equations to nonlinear polynomial equation solving. Recently, the use of resultants as a computational tool for elimination of variables has renewed the interest in finding explicit formulas for their computation. One of the central problems in elimination theory is to find "nice" determinantal formulas for resultants. The best one can hope for is a *Sylvester-type formula*, that is, a square matrix whose non-zero entries are the coefficients of the given polynomial equation and whose determinant equals precisely the resultant. Some Computer Algebra System, for example, MAPLE, provides a command for producing Sylvester matrix.

1. Resultants in a *PID*, $k[x]$

When we are given a general polynomial, say,

$$f(x) = a_0x^l + a_1x^{l-1} + \cdots + a_{l-1}x + a_l,$$

we see two different objects, the powers of the unknown x and the known quantities, the coefficients a_i's. But, it is entirely our decision that which symbols are assigned for the unknown and which symbols are assigned for the known. We know that a polynomial expression in terms of the known is again a known quantity to us, but a polynomial expression containing, for instance, one unknown is again unknown to us. Now, suppose that we are given a polynomial identity

$$Af(x) + Bg(x) = \det(M(x)),$$

where A, B are known quantities to us and $f(x), g(x)$ are polynomials in the unknown x with known coefficients, and $\det(M(x))$ is the determinant of a square matrix whose entries are polynomials in x. Since the monomials x^{α_i} form a basis for $k[x]$ where $k =$ the collection of the known, a field, having such a polynomial identity is exactly the same as having a multiple number of equalities between coefficients of both sides, one for each monomial x^{α_i}. In this case, it is quite straight forward. But, if we assume that those known quantities in the field k are in fact polynomials in terms of other quantities, we suddenly begin to have a serious problem especially when we want to substitute some of the symbols with others, mainly because certain symbols are algebraically independent over the bottom field k, but others are not, and sometimes we do not know in advance whether those symbols under consideration are algebraically independent or not over k.

REMARK 6.1. *Another thing we should practically keep in mind is that symbols in a Computer Algebra System are always assumed to be algebraically independent over the bottom field, unless an explicit algebraic relation is provided as an input. Of course, those symbols pre-defined by a Computer Algebra System are exceptional to this remark.*

1.1. GCD and LCM Computations in a *UFD*, $k[x, y]$, and in a *PID*, $k[x]$. We recall that prime elements are by definition irreducible elements but not conversely in a ring R, and the converse also holds if R is a Unique Factorization Domain. Thus, we are able to compute first the least common multiple, LCM, of two given polynomials $f(x) = f(x_1, ..., x_n), g(x) = g(x_1, ..., x_n) \in k[x_1, ..., x_n]$, because

$$(f(x)) \cap (g(x)) = (\text{lcm}(f(x), g(x)).$$

See Algorithm 12 of Chapter 5. We also recall that computation of

$$\text{lcm}(f(x), g(x))$$

is done through the computation of the generating set of the intersection ideal $I \cap J$ of two polynomial ideals I and J, thanks to B. Buchberger's algorithm, Algorithm 11, for computing a Groebner basis for polynomial ideals. Then since

$$\text{gcd}(f(x), g(x)) \cdot \text{lcm}(f(x), g(x)) = f(x) \cdot g(x)$$

holds, so we are able to compute $\text{gcd}(f(x), g(x))$ using this formula:

$$\text{gcd}(f(x), g(x)) = \frac{f(x) \cdot g(x)}{\text{lcm}(f(x), g(x))}.$$

If $n = 1$, we even do not need this approach since Euclid's division algorithm, Algorithm 2, of Chapter 1 will efficiently compute $\text{gcd}(f(x), g(x))$. We note that this is all possible because $k[x]$ is a Principal Ideal Domain (*PID*). See below immediately after the next paragraph.

But if $n > 1$, then $k[x_1, ..., x_n]$ is not a *PID* anymore. It is a Unique Factorization Domain (*UFD*) and since it involves the division process, this approach is now perfectly OK only when we can factorize both polynomials $f(x)$ and $g(x)$ quite easily. Otherwise, it will take too much time or in the worst case, we will not be able to compute $\text{gcd}(f(x), g(x))$ at all.

On the other hand, if $n = 1$, then since $k[x]$ is a *PID* we have

$$(f(x), g(x)) = (\text{gcd}(f(x), g(x))),$$

and hence we can write $\gcd(f(x), g(x))$ as a $k[x]$-linear combination of $f(x)$ and $g(x)$. That is, thanks to Euclid's division algorithm, we can write $\gcd(f(x), g(x))$ as a $k[x]$-linear combination of $f(x), g(x)$:

$$\gcd(f(x), g(x)) = A(x) \cdot f(x) + B(x) \cdot f(x),$$

for some $A(x), B(x) \in k[x]$.

But again very unfortunately, if $n > 1$ then since $k[x_1, ..., x_n]$ not a *PID* but is only a *UFD*, such a $k[x]$-linear combination is not automatically guaranteed, and this is the right moment why the resultants play their role.

Even when $n = 1$, if we are required to compute the $\gcd(f(x), g(x))$ without division in the bottom field k, then we need to find another method for computing $\gcd(f(x), g(x))$, other than Euclid's division algorithm. Very fortunately, it is possible to compute $\gcd(f(x), g(x))$ without division at all in terms of resultants.

Before we move to the new subject, resultants, we need to prove the following small corollary:

COROLLARY 6.1. *Suppose $f, g \in k[x_1, ..., x_n]$ have positive degree in x_1. Then f and g have a common factor in $k[x_1, ..., x_n]$ of positive degree in x_1 if and only if they have a common factor in $k(x_2, ..., x_n)[x_1]$.*

PROOF. The $\Longrightarrow$ direction is obvious. Conversely, suppose that f and g have a common factor $h' \in k(x_2, ..., x_n)[x_1]$, so that we write

$$\begin{aligned} f &= h' f_1', \\ g &= h' g_1', \end{aligned}$$

for some $f_1', g_1' \in k(x_2, ..., x_n)[x_1]$. Here, h', f_1', and g_1' are considered as polynomials in x_1 with coefficients in $k(x_2, ..., x_n)$. Let $d \in k[x_2, ..., x_n]$ be the common denominator of the coefficients in $k(x_2, ..., x_n)$ of h', f_1', and g_1'. Then clearing out these common denominators we have in $k[x_1, ..., x_n]$

$$\begin{aligned} d^2 f &= h f_1, \\ d^2 g &= h g_1. \end{aligned}$$

On the other hand, we note that, since $h' = h/d$ has positive degree in x_1, h is a non-constant polynomial of positive degree in x_1; hence h has an irreducible factor h_1 of positive degree in x_1. Then this h_1 divides $d^2 f$ but it cannot divide d^2, so it must divide f. (For this, we recall that an irreducible polynomial over k is a prime element of $k[x_1, ..., x_n]$). Similarly, the same h_1 divides g, and this finishes the proof. □

1.2. Resultants in $k[x]$.

LEMMA 6.1. *Let $f, g \in k[x]$ be polynomials of degree $p > 0$ and $q > 0$, respectively. Then f and g have a common factor if and only if there exist polynomials $A, B \in k[x]$ such that*

(1) *A and B are not both zero;*
(2) *A has degree at most $q - 1$ and B has degree at most $p - 1$;*
(3) *$Af + Bg = 0$.*

PROOF. If h is a common factor of f and g, write $f = hf_1$ and $g = hg_1$. Then we clearly have:

$$g_1 f + (-f_1)g = 0,$$

with $\deg(g_1) \leq q - 1$ and $\deg(f_1) \leq p - 1$.

Conversely, suppose there exist $A, B \in k[x]$ satisfying the three conditions. We may assume that B is not the zero polynomial. If f and g were not divisible by a common factor, then we would have $\gcd(f, g) = 1$ and then since $k[x]$ is a *PID*, we would have A' and $B' \in k[x]$ such that

$$1 = A'f + B'g.$$

Thus, multiplying both sides by this non-zero polynomial B, we would get

$$\begin{aligned} B &= BA'f + BB'g \\ &= BA'f + B' \cdot (-Af) = (BA' - B'A)f, \end{aligned}$$

in $k[x]$, and consequently we would have $\deg(B) \geq p$, which is impossible. □

Now, the next question is how to explicitly figure out those polynomials A and B, and this is the right place where the theory of results shines. To express explicitly these polynomials A and B, we really have to write down everything involved in detail, utilizing linear algebra as much as we could, and it goes as follows. We first recall that $k[x]$ is a k-vector space with a basis consisting of monomials x^j, $j = 0, 1, 2, \ldots.$ We write

$$\begin{aligned} f &= a_0x^p + a_1x^{p-1} + \cdots + a_p, \qquad a_0 \neq 0, \\ g &= b_0x^q + b_1x^{q-1} + \cdots + b_q, \qquad b_0 \neq 0, \\ A &= c_0x^{q-1} + c_1x^{q-2} + \cdots + c_{q-1}, \\ B &= d_0x^{p-1} + d_1x^{p-2} + \cdots + d_{p-1}. \end{aligned}$$

After that, we want to solve the following equation for c_i and $d_i \in k$:

$$Af + Bg = 0,$$

where the monomials x^j enables us to establish a system of $p+q$ linear equations in $p+q$ unknowns c_i and d_i:

$$\begin{array}{llllll} a_0c_0 & & + & b_0d_0 & & = 0 \text{ coefficient of } x^{p+q-1} \\ a_1c_0 & + \ a_0c_1 & & b_1d_0 & + \ b_0d_1 & = 0 \text{ coefficient of } x^{p+q-2} \\ & \ddots & & \ddots & & \vdots \\ & & a_pc_{q-1} & + & b_qd_{p-1} & = 0 \text{ coefficient of } x^0. \end{array}$$

Thus, in matrix notation, we have a system of homogeneous linear equations:

$$Syl(f,g,x)\begin{bmatrix} c_0 \\ \vdots \\ c_{q-1} \\ d_0 \\ \vdots \\ d_{p-1} \end{bmatrix} = 0,$$

where the matrix of size $(p+q)\times(p+q)$

$$Syl(f,g,x) = \begin{bmatrix} a_0 & & & & b_0 & & & \\ a_1 & a_0 & & & b_1 & b_0 & & \\ a_2 & a_1 & \ddots & & b_2 & b_1 & \ddots & \\ \vdots & & \ddots & a_0 & \vdots & & \ddots & b_0 \\ & \vdots & & a_1 & & \vdots & & b_1 \\ a_p & & & & & & & \\ & a_p & & \vdots & & b_q & & \vdots \\ & & \ddots & & & & \ddots & \\ & & & a_p & & & & b_q \end{bmatrix}$$

is called the **Sylvester matrix** of f and g with respect to x.

DEFINITION 6.1. *The* ***resultant*** *of f and g with respect to x, denoted by $Res(f,g,x)$, is the determinant of the Sylvester matrix:*

$$Res(f,g,x) = \det(Syl(f,g,x)) \in k.$$

From the equation $Af + Bg = 0$ we have:

PROPOSITION 6.1. *Let $f, g \in k[x]$ be two polynomials of positive degree.*

(1) *The resultant $Res(f,g,x)$ in the bottom field k is an integer polynomial in the coefficients of f and g.*
(2) *f and g have a common factor in $k[x]$ if and only if*

$$Res(f,g,x) = 0.$$

PROOF. Part (1). For a square matrix $A = (a_{ij})$, $\det(A)$ is:

$$\det(A) = \sum_{\sigma \in Perm(n)} sgn(\sigma) a_{1\sigma(1)} \cdots a_{n\sigma(n)},$$

so that the determinant is a polynomial in its entries with coefficients either $+1$ or -1. Thus, $Res(f,g,x)$ is an integer polynomial in the coefficients a_i's and b_j's of f and g.

Part (2). We note that the resultant is zero if and only if the Sylvester matrix is not invertible if and only if the system of homogeneous linear equations has a non-zero solution if and only if the equation $Af + Bg = 0$ in Lemma 6.1 has non-zero solutions for A and B. Thus, by Lemma 6.1 we finish our proof. □

EXAMPLE 6.1. *Let* $f = x^2 + 3x + 1$, $g = 3x^3 + 2x + 1$. *Then*

$$Res(f, g, x) = \det \begin{bmatrix} 1 & & & 3 & \\ 3 & 1 & & 0 & 3 \\ 1 & 3 & 1 & 2 & 0 \\ & 1 & 3 & 1 & 2 \\ & & 1 & & 1 \end{bmatrix} = -4,$$

so f *and* g *do not have a common factor in* $k[x]$. *That is,* $\gcd(f, g) = 1$.

REMARK 6.2. *The main problem with resultant computation is that when the size of the Sylvester matrix gets big, computation of determinant with numerical entries takes* $\mathrm{O}(n^3)$, *where* $n \times n$ *is the size of the matrix, while Euclid's algorithm for* gcd *of two numbers takes* $O(n)$ *time efficiency. Thus, computation of the resultant would take more time as* n *is increased although it has a polynomial time efficiency.*

On the other hand, there is an alternative way of computing the resultant which is similar to the Euclidean algorithm for $\gcd(f, g)$, *and most Computer Algebra Systems have a command for computing resultant that implements this algorithm.*

Then from the equation $Af + Bg = 1$ and Cramer's rule, we have:

PROPOSITION 6.2. *Let* $f, g \in k[x]$ *be two polynomials in* x *of positive degree in* x.

(1) *There exist polynomials* $A, B \in k[x]$ *such that*

$$Af + Bg = Res(f, g, x) \in (f, g) \cap k.$$

(2) *Moreover, the coefficients of* A *and* B *are integer polynomials of* $k[x]$ *in the coefficients of* f *and* g.

PROOF. If $Res(f,g,x)=0$, then there is nothing to prove. Thus, we may assume that $Res(f,g,x)\neq 0$. As before we write

$$\begin{aligned} f &= a_0x^p + a_1x^{p-1} + \cdots + a_p, \qquad a_0 \neq 0,\\ g &= b_0x^q + b_1x^{q-1} + \cdots + b_q, \qquad b_0 \neq 0,\\ A' &= c_0x^{q-1} + c_1x^{q-2} + \cdots + c_{q-1},\\ B' &= d_0x^{p-1} + d_1x^{p-2} + \cdots + d_{p-1}. \end{aligned}$$

After that, we want to solve the following equation for c_i and $d_i \in k$:

$$A'f + B'g = 1,$$

where the monomials x^j enables us to establish a system of $p+q$ linear equations in $p+q$ unknowns c_i and d_i:

$$\begin{array}{lllll} a_0c_0 & & + \quad b_0d_0 & & = 0 \text{ coefficient of } x^{p+q-1}\\ a_1c_0 & + \ a_0c_1 & \quad\ \ b_1d_0 & + \ b_0d_1 & = 0 \text{ coefficient of } x^{p+q-2}\\ \quad\ddots & & \quad\ddots & & \quad\vdots\\ & a_pc_{q-1} & + & b_qd_{p-1} & = 1 \text{ coefficient of } x^0. \end{array}$$

Thus, in matrix notation, we have a system of homogeneous linear equations:

$$Syl(f,g,x)\begin{bmatrix} c_0\\ \vdots\\ c_{q-1}\\ d_0\\ \vdots\\ d_{p-1} \end{bmatrix} = \begin{bmatrix} 0\\ \vdots\\ 0\\ 0\\ \vdots\\ 1 \end{bmatrix},$$

where the matrix of size $(p+q) \times (p+q)$

$$Syl(f,g,x) = \begin{bmatrix} a_0 & & & & b_0 & & & \\ a_1 & a_0 & & & b_1 & b_0 & & \\ a_2 & a_1 & \ddots & & b_2 & b_1 & \ddots & \\ \vdots & & \ddots & a_0 & \vdots & & \ddots & b_0 \\ & \vdots & & a_1 & & \vdots & & b_1 \\ a_p & & & & & & & \\ & a_p & & \vdots & & b_q & & \vdots \\ & & \ddots & & & & \ddots & \\ & & & a_p & & & & b_q \end{bmatrix}$$

is the same Sylvester matrix of f and g with respect to x as before.

Since we are assuming that $Res(f,g,x) \neq 0$, it has a unique answer, and then we just use Cramer's rule to get formulas for c_i's and d_i's for the unique solution $(c_0, c_1, ..., c_q, d_0, ..., d_p)^t$. For example,

$$\begin{aligned} c_0 &= \frac{1}{Res(f,g,x)} \cdot \det \begin{bmatrix} 0 & & & & b_0 & & & \\ 0 & a_0 & & & b_1 & b_0 & & \\ 0 & a_1 & \ddots & & b_2 & b_1 & \ddots & \\ \vdots & & \ddots & a_0 & \vdots & & \ddots & b_0 \\ & \vdots & & a_1 & & \vdots & & b_1 \\ 0 & & & & & & & \\ & a_p & & \vdots & & b_q & & \vdots \\ & & \ddots & & & & \ddots & \\ 1 & & & a_p & & & & b_q \end{bmatrix} \\ &= \frac{1}{Res(f,g,x)} \cdot (\text{an integer polynomial in } a_i\text{'s and } b_j\text{'s}). \end{aligned}$$

Similarly, we can have similar formulas for the other c_i's and the d_i's. We note that these formulas all have a common denominator,

$$Res(f,g,x) \in k[x],$$

so we just pull it out of both A' and B' to obtain:

$$\begin{aligned} A' &= \frac{1}{Res(f,g,x)} \cdot A, \\ B' &= \frac{1}{Res(f,g,x)} \cdot B, \end{aligned}$$

for some $A, B \in k[x]$. Therefore, with these A and B, we have finally established the claimed equality:

$$Af + Bg = Res(f,g,x).$$

□

Thus, the following corollary tells the exact relationship between the $\gcd(f,g)$ and $Res(f,g,x)$.

COROLLARY 6.2. *Let f and g be polynomials in $k[x]$ of positive degree. Suppose that $Res(f,g,x) \neq 0$. Then f and g have no common divisor if and only if there exist $A, B \in k[x]$ such that*

$$Af + Bg = 1 \cdot Res(f,g,x).$$

PROOF. Since we are in a *PID*, $k[x]$, $\gcd(f,g) = 1$ if and only if there exists $A', B' \in k[x]$ such that $A'f + B'g = 1$. Then by the proof of Proposition 6.2, this implies that there are $A, B \in k[x]$ such that

$$Af + Bg = 1 \cdot Res(f,g,x),$$

by clearing the common denominator $Res(f,g,x)$ of A' and B'. The reverse direction is obvious. □

1.3. Resultants in a *UFD*, $k[x,y]$. Before we move into the next section, we consider the following situation. Suppose that $f, g \in k[x,y]$ are polynomials in x, y of positive degree in x. Although $k[x,y]$ is not a *PID*, $k(y)[x]$ is a *PID* and certainly we have: $f, g \in k(y)[x]$. Thus, we have

COROLLARY 6.3. *Let $f, g \in k[x,y]$ be two polynomials of positive degree in x.*

(1) *There exist polynomials $A, B \in k[x,y]$ such that*

$$Af + Bg = Res(f,g,x) \in (f,g) \cap k[y].$$

(2) *Moreover, the coefficients of A and B are integer polynomials in $k[x,y]$ in the coefficients of f and g.*

PROOF. Just follow the proof of Proposition 6.2. □

REMARK 6.3. *Thus, $Res(f,g,x)$ is indeed in the elimination ideal $(f,g) \cap k[y]$ of the ideal (f,g) in $k[x,y]$. Hence, given a system of polynomial equations $f(x,y) = 0$ and $g(x,y) = 0$, we can eliminate the variable x using either the theory of Groebner basis (Theorem 5.1) or the theory of resultants.*

In the next section, we will see the resultant of $f, g \in k[x_1, ..., x_n]$, considering them as elements of $k(x_2, ..., x_n)[x_1]$ through exactly the same approach as we have seen in this subsection.

Exercises for Section 6.1

EXERCISE 6.1. *Let $f(x)$ be a polynomial of degree $p > 0$ and let $g(x)$ be a constant polynomial b_0. Prove that*

$$Res(f,g,x) = b_0^p.$$

EXERCISE 6.2. *Let $f(x)$ be a polynomial of degree $p > 0$ and $g(x)$ be a polynomial of $q > 0$. Prove that*

$$Res(f,g,x) = (-1)^{pq} Res(g, fx).$$

EXERCISE 6.3. *Let $f(x) = a_0x^p + \cdots + a_p$ be a polynomial in x over a field k with $a_0 \neq 0$, $p > 0$. Define the* ***discriminant*** *of f to be*

$$disc(f) = \frac{(-1)^{p(p-1)/2}}{a_0} Res(f, f', x),$$

where f' is the derivative of f with respect to x. Prove that f has a multiple root if and only if $disc(f) = 0$.

EXERCISE 6.4. *Let $f(x), g(x)$ be two polynomials over $\mathbb{C}$. Prove that they have a common root if and only if $Res(f, g, x) = 0$.*

2. Resultants and Elimination Theorem

The main purpose of this section is to prove the Elimination Theorem. Before proving it, we need to set up the current situation explicitly.

Let $f, g \in [x_1, ..., x_n]$ be polynomials of positive degree in x_1. We write

$$
\begin{aligned}
f &= a_0x_1^p + a_1x_1^{p-1} + \cdots + a_p, \qquad a_0 \neq 0, \\
g &= b_0x_1^q + b_1x_1^{q-1} + \cdots + b_q, \qquad b_0 \neq 0,
\end{aligned}
\tag{6.1}
$$

where $a_i, b_i \in k[x_2, ..., x_n]$. Under this setting, we have

DEFINITION 6.2. *The following determinant is called the **resultant**, $Res(f, g, x_1)$, of f and g with respect to x_1:*

$$
Res(f, g, x_1) = \det \begin{bmatrix}
a_0 & & & & b_0 & & & \\
a_1 & a_0 & & & b_1 & b_0 & & \\
a_2 & a_1 & \ddots & & b_2 & b_1 & \ddots & \\
\vdots & & \ddots & a_0 & \vdots & & \ddots & b_0 \\
 & \vdots & & a_1 & & \vdots & & b_1 \\
a_p & & & & & & & \\
 & a_p & & \vdots & & b_q & & \vdots \\
 & & \ddots & & & & \ddots & \\
 & & & a_p & & & & b_q
\end{bmatrix}.
$$

Using this definition, we can define a generalized resultant of more than two polynomials $k[x_1, ..., x_n]$ of positive degree in x_1:

DEFINITION 6.3. *Let $f_1, ..., f_s \in k[x_1, ..., x_n]$ be polynomials of positive degree in x_1, and let $u_2, ...u_s$ be new indeterminates. Then those h_α in the following resultant of f_1 and $f_2u_2 + \cdots + f_su_s$ with respect to x_1 are called the **generalized resultants** of $f_1, ..., f_s$ with respect to x_1:*

$$
Res(f_1, f_2u_2 + \cdots + f_su_s, x_1) = \sum_\alpha h_\alpha(x_2, ..., x_n)u^\alpha,
$$

where $u^\alpha = u_2^{\alpha 2} \cdots u_s^{\alpha 2}$ *are monomials and* $h_\alpha \in k[x_2, ..., x_n]$.

This definition may need an explanation. We consider the resultant

$$Res(f_1, f_2u_2 + \cdots + f_su_s, x_1) \in k[x_2, ..., x_n, u_2, ..., u_s],$$

of f_1 and $u_2f_2 + \cdots + u_sf_s$ in $k[x_1, ..., x_n, u_2, ...u_s]$ of positive degree in x_1 with respect to x_1; then we rearrange its terms with respect to those monomials $u^\alpha = u_2^{\alpha 2} \cdots u_s^{\alpha 2}$ in order to obtain the unique coefficients $h_\alpha(x_2, ..., x_n) \in k[x_2, ..., x_n]$ of these monomials, and these coefficients are called generalized resultants of $f_1, f_2, ..., f_s \in k[x_1, ..., x_n]$.

Thus, if $s = 2$, then the single $h_\alpha(x_2, ..., x_n)$ must be the same as $Res(f_1, f_2, x_1)$. We also note that generalized resultants are dependent on the list of polynomials $(f_2, ..., f_s)$. For instance,

$$Res(f_1, f_2u_2 + f_2u_3, x_1) \neq Res(f_1, f_3u_2 + f_2u_3, x_1).$$

We then recall that

PROPOSITION 6.3. *Let* $f, g \in k[x_1, ..., x_n]$ *have positive degree in* x_1. *Then*

(1) $Res(f, g, x_1)$ *is in the first elimination ideal* $(f, g) \cap k[x_2, ..., x_n]$.
(2) $Res(f, g, x_1) = 0$ *if and only if* f *and* g *have a common factor in* $k[x_1, ..., x_n]$ *which has positive degree in* x_1.

PROOF. The two statements are true because of Proposition 6.1, Proposition 6.2, and Corollary 6.1 □

COROLLARY 6.4. *For* $f, g \in \mathbb{C}[x]$, $Res(f, g, x) = 0$ *if and only if* f *and* g *have a common root.*

PROOF. Over $\mathbb{C}$ two polynomials have a common factor if and only if they have a common root. □

LEMMA 6.2. *Let* $(f_1, ..., f_s)$ *be an ideal of a ring* R. *Then for any positive integer* N *we have*

$$(f_1, f_2, ..., f_s) = (f_1 + f_2 + x_1^N f_1, f_3, ..., f_s).$$

PROOF. Since $f_2 = (f_2 + x_1^N f_1) + (-x_1^N) \cdot f_1$, we have the $\subseteq$ inclusion. The reverse inclusion is trivial. $\square$

Now, we are ready to prove a theorem, the Extension Theorem, which is geometrically very important. We note that its proof is based on the resultants of the Theory of Invariants.

THEOREM 6.1. ***(The Extension Theorem)*** *Let $I = (f_1, ..., f_s)$ an ideal of $\mathbb{C}[x_1, ..., x_n]$ be a polynomial ideal and let $I_1 = I \cap \mathbb{C}[x_2, ..., x_n]$, the first elimination ideal of I. For each $1 \le i \le s$, write*

$$f_i = g_i(x_2, ..., x_n)x_1^{N_1} + \text{ terms with } \deg(x_1) < N_i,$$

where each $N_i \ge 0$ and each $g_i \in \mathbb{C}[x_2, ..., x_n]$ is a non-zero polynomial. Then every partial solution $(c_2, ..., c_n) \in V(I_1) - V(g_1, ..., g_s)$ is extended to a total solution $(c_1, c_2, ..., c_n) \in V(I)$ for some $c_1 \in \mathbb{C}$.

REMARK 6.4. *This theorem is still true if we replace $\mathbb{C}$ with any other algebraically closed field k.*

PROOF. If $s = 1$, then $V(f_1) = V(f_1, f_1)$, and hence it is covered by the case when $s = 2$.

(Case 1) Suppose that $s = 2$. We write

$$\begin{aligned} f_1 &= a_0x_1^p + a_1x_1^{p-1} + \cdots + a_p, & a_0 \neq 0, \\ f_2 &= b_0x_1^q + b_1x_1^{q-1} + \cdots + b_q, & b_0 \neq 0, \end{aligned}$$

where $a_i, b_i \in k[x_2, ..., x_n]$ and

$$\begin{aligned} a_0 &= g_1(x_2, ..., x_n), \\ b_0 &= g_2(x_2, ..., x_n). \end{aligned}$$

We also write

$$\begin{aligned} f_1(x_1, c) &= f_1(x_1, c_2, ..., c_n), \\ f_2(x_1, c) &= f_2(x_2, c_2, ..., c_n). \end{aligned}$$

To have their resultant we write in detail as before,

$$\begin{aligned} f_1(x_1, c) &= a_0(c)x_1^p + a_1(c)x_1^{p-1} + \cdots + a_p(c), \qquad a_0(c) \neq 0, \\ f_2(x_1, c) &= b_0(c)x_1^q + b_1(c)x_1^{q-1} + \cdots + b_q(c), \qquad b_0(c) \neq 0, \end{aligned}$$

where

$$\begin{aligned} a_0(c) &= g_1(c), \\ b_0(c) &= g_2(c). \end{aligned}$$

If it occurs by our substitution of $c = (c_1, ..., c_n)$ to $(x_2, ..., x_n)$ that $b_0(c) = 0$, then by Lemma 6.2, we can replace the original f_2 with $f_2 + x_1^N f_1$ for a sufficiently large N such that the leading term in x_1 of $f_2 + x_1^N f_1$ is now $b_0(c)x_1^q$ but with $b_0(c) \neq 0$, and we can do such a replacement whenever it is necessary. Thus, we may assume that both $a_0(c) \neq 0$ and $b_0(c) \neq 0$. Thus,

$$Res(f\,_1(x_1, c), f_2(x_1, c), x_1)$$

is an element of $I_1 = I \cap \mathbb{C}[x_2, ..., x_n]$ by Proposition 6.3 and moreover it is equal to zero by hypothesis. Thus, by Corollary 6.4 $f_1(x_1, c)$ and $f_2(x_1, c)$ have a common root $(c_1, ..., c_n)$ for some $c_1 \in \mathbb{C}$.

(Case 2) Now, suppose $s \geq 3$. In order to have the generalized resultants, we consider the two polynomials f_1 and

$$f_2u_2 + \cdots + f_su_s,$$

in $\mathbb{C}[x_1, ..., x_n, u_2, ..., u_s]$ as before, where the u_i are new variables. Thus, with these two polynomials we can apply the proof of the case when $s = 2$, regarding $f_2u_2 + \cdots + f_su_s$ as f_2. In particular, we note that this is all possible because by Lemma 6.2, we can always assume that $a_0(c) \neq 0, b_0(c) \neq 0$ and moreover f_2 has degree in x_1 greater than $f_3, ..., f_n$.

Now, by Definition 6.3, we have

$$h = Res(f_1, u_2f_2 + \cdots + u_2f_2, x_1) = \sum_{\alpha} h_\alpha(x_2, ..., x_n)u^\alpha,$$

and in what follows we are going to show that all the generalized resultants $h_\alpha = h_\alpha(x_2, ..., x_n) \in \mathbb{C}[x_2, ..., x_n]$ are elements of the first elimination ideal I_1. For this, we are going to show that all $h_\alpha \in I$. Then each h_α does not have the variable x_1, we are done. Now, by Proposition 6.3,

$$Res(f_1, u_2f_2 + \cdots + u_2f_2, x_1) = Af_1 + B(u_2f_2 + \cdots + u_2f_2)$$

for some $A, B \in \mathbb{C}[x_1, ..., x_n, u_2, ..., u_n]$. To have a detailed comparison between terms with equal degree with resect to the monomials $u^\alpha = u_2^{\alpha_2} \cdots u_s^{\alpha_s}$, we write

$$\begin{aligned} A &= \sum_\alpha A_\alpha u^\alpha, \\ B &= \sum_\alpha B_\alpha u^\alpha, \\ u_2f_2 + \cdots + u_2f_2 &= \sum_{2\le i\le s} u^{e_{i-1}} f_i, \end{aligned}$$

where $e_i = (0, ..., 1, 0, ..., 0)$, 1 in the ith entry for $i = 1, ..., s$, and then we substitute them into the above expression to get

$$\begin{aligned} \sum_\alpha h_\alpha u^\alpha &= \left(\sum_\alpha A_\alpha u^\alpha\right) f_1 + \left(\sum_\alpha B_\alpha u^\alpha\right)\left(\sum_{2\le i\le s} u^{e_{i-1}} f_i\right) \\ &= \sum_\alpha (A_\alpha f_1) u^\alpha + \sum_\alpha \left(\sum_{\beta+e_{i-1}=\alpha} B_\alpha f_i\right) u^\alpha \\ &= \sum_\alpha \left(A_\alpha f_1 + \sum_{\beta+e_{i-1}=\alpha} B_\alpha f_i\right) u^\alpha \end{aligned}$$

Thus, by equating equal degree terms with respect to u^α, we see that, for each α,

$$h_\alpha = \left(A_\alpha f_1 + \sum_{\beta+e_{i-1}=\alpha} B_\alpha f_i\right) \in I,$$

which is what we wanted to show. Thus, we have shown that each $h_\alpha \in I_1$, so that $h_\alpha(c) = 0$ for each α by hypothesis. Thus,

$$h(c, u_2, ...u_s) = \sum_\alpha h_\alpha(x_2, ..., x_n) u^\alpha = 0,$$

where $h = Res(f_1, u_2f_2 + \cdots + u_2f_2, x_1)$.

On the other hand, since we can always assume that $g_1(c) = a_0(c) \neq 0$, $g_2(c) = b_0(c) \neq 0$ and moreover f_2 has degree in x_1 greater than $f_3, ..., f_n$, after substitution of $c = (c_2, ..., c_n)$ for $(x_2, ..., x_n)$ we still have the equality:

$$h(c, u_2, ...u_s) = Res(f_1(x_1, c), u_2f_2(x_1, c) + \cdots + u_2f_2(x_1, c), x_1),$$

that is, the expression on the right side is well-defined and hence, by definition of the resultant for two polynomials with respect to x_1, such an equality holds. Thus, putting these two together we see that

$$Res(f_1(x_1, c), u_2f_2(x_1, c) + \cdots + u_2f_2(x_1, c), x_1) = 0.$$

Thus, by Proposition 6.3 we know that there exists a common factor

$$F \in \mathbb{C}[x_1, ..., x_n, u_2, ..., u_s]$$

of positive degree in x_1 for $f_1(x_1, c)$ and $(u_2f_2(x_1, c) + \cdots + u_sf_s(x_1, c)$. But, since it divides $f_1(x_1, c)$, $F = F(x_1) \in \mathbb{C}[x_1]$. Then we moreover have an equality of polynomials,

$$F(x_1)A(x_1, u_2, ..., u_s) = u_2f_2(x_1, c) + \cdots + u_2f_2(x_1, c),$$

for some $A(x_1, u_2, ..., u_s) \in \mathbb{C}[x_1, u_2, ..., u_s]$. Thus, comparing the equal degree terms with respect to the monomials u^α, we conclude that $F(x_1)$ divides each $f_i(x_1, c)$, $i = 1, ..., s$. Now, let c_1 be a root of $F(x_1) = 0$. Then it is automatically a root of each $f_i(x_1, c)$, and hence $(c_1, ..., c_n)$ is indeed a total solution in $V(I)$ for any partial solution $(c_2, ..., c_n)$ in $V(I_1) - V(g_1, ..., g_s)$ for some $c_1 \in \mathbb{C}$. This finishes the proof of the Extension Theorem. □

REMARK 6.5. *In the proof of Emmy Noether's theorem, Theorem 4.4 of Chapter 4, we have also seen the introduction of new variables u_i, although their role is not the same as the one in the above proof.*

COROLLARY 6.5. *Let* $I = (f_1, ..., f_s) \subset \mathbb{C}[x_1, ..., x_n]$. *If there exists* i *such that*

$$f_i = cx_1^N + \text{ terms with } \deg(x_1) < N,$$

for some non-zero constant $c \in \mathbb{C}$ *and* $N > 0$. *Then every partial solution* $(c_2, ..., c_n)$ *in* $V(I_1)$ *is extended to a total solution* $(c_1, ..., c_n)$ *in* $V(I)$ *for some* $c_1 \in \mathbb{C}$.

PROOF. Since $g_i = c$ is a non-zero constant polynomial, no point in $\mathbb{C}^n$ can be a solution of $V(g_1, ..., g_n)$. □

The proof of Theorem 6.1 also establishes:

COROLLARY 6.6. *Let* $f(x), g(x)$ *be as in (6.1). If*

$$Res(f, g, x_1) \in \mathbb{C}[x_2, ..., x_n]$$

vanishes at $(c_2, ..., c_n) \in \mathbb{C}^{n-1}$, *then either*

(1) a_0 *or* b_0 *vanishes at* $(c_2, ..., c_n)$, *or*
(2) *there exists* $c_1 \in \mathbb{C}$ *such that* $f(x)$ *and* $g(x)$ *vanishes at* $(c_1, c_2, ..., c_n) \in \mathbb{C}^n$. □

Exercises for Section 6.2

EXERCISE 6.5. *Let* C *be a space curve defined parametrically by*

$$\begin{aligned} u &= \frac{t^2}{1+t^2}, \\ v &= \frac{t^3}{1+t^2}. \end{aligned}$$

(1) *Compute an implicit equation for* C *using the resultant expression in part (1) of Proposition 6.3.*
(2) *Apply Algorithm 25 to compute an implicit equation for* C.
(3) *Compare these two results. In particular, which method is more efficient?*

3. Resultants Plus Groebner Bases

3.1. The Geometry of Elimination and Extension. In this section, we are going to see that elimination in Algebra corresponds to projection of a variety onto a lower dimensional subspace. In what follows we will assume that our bottom field $k = \mathbb{C}$ as we have done it before, but the results are still true when k is any algebraically closed field.

Let $V = V(f_1, ..., f_s)$ be an affine variety in $\mathbb{C}^n$ and consider the projection map, for a fixed j,

$$\pi_j : \mathbb{C}^n \longrightarrow \mathbb{C}^{n-j},$$

defined by sending $(a_1, ..., a_n) \longmapsto (a_{j+1}, ..., a_n)$. Let

$$I_j = (f_1, ..., f_s) \cap \mathbb{C}[x_{j+1}, ..., x_n]$$

be the j-th elimination ideal of $I = (f_1, ..., f_s) \subset \mathbb{C}[x_1, ..., x_n]$.

LEMMA 6.3. *Under this setting, we have*

$$\pi_j(V) \subseteq V(I_1).$$

PROOF. We need to show that any polynomial f in the ideal I_j vanishes on the points of $\pi_j(V)$. Thus, let $(a_{j+1}, ..., a_n)$ be a point of $\pi_j(V)$. Then there exist $a_1, ..., a_j \in \mathbb{C}$ such that $(a_1, ..., a_n) \in V = V(I)$. But, we clearly have $f \in I_1 \subset I$, and hence

$$0 = f(a_1, ..., a_n) = f(a_{j+1}, ..., a_n),$$

simply because $f \in \mathbb{C}[a_{j+1}, ..., a_n]$. $\square$

We note that this lemma is nothing to do with the fact that $\mathbb{C}$ is algebraically closed.

DEFINITION 6.4. *Under this setting, the points of $V(I_1)$ are called* ***partial solutions*** *of $V(I)$ and the points of $V(I)$ are called* ***total solutions***.

Thus, what the above lemma really says is the following:

COROLLARY 6.7. *The elements of $\pi_j(V)$ are exactly those partial solutions that can be extended to total solutions in $V(I)$. That is,*

$$(a_{j+1}, ..., a_n) \in \pi_j(V)$$

if and only if there exists $a_1, ..., a_j \in \mathbb{C}$ such that $(a_1, ..., a_n) \in V(I)$.

The first question we may have is whether the equality holds in

$$\pi_j(V) \subseteq V(I_1),$$

and the following example says it is not the case in general.

EXAMPLE 6.2. *Let $V = V(xy - 1, xz - 1)$. Then a Groebner basis G for the ideal $I = 12y(xy - 1, xz - 1)$ under the lex order with $x > y > z$ is:*

$$G = \{xz - 1, y - z\},$$

and hence by the Elimination Theorem, Theorem 5.1 or its Corollary 5.2, we see that $y - z$ is the only generator for the 1st elimination ideal $I_1 = I \cap \mathbb{C}[x]$ of I, and thus $V(I_1)$ is the whole line $y = z$. But, we easily see that there is no preimage in $V(I)$ of the origin $(0,0)$ of the line under the projection

$$\pi_1 \colon \mathbb{C}^3 \longrightarrow \mathbb{C}^2,$$

and hence $(0,0) \notin \pi_1(V)$. Thus, we see that $\pi_1(V)$ is properly contained in $V(I_1)$.

COROLLARY 6.8. *Under the same setting as in Theorem 6.1, we have*

$$V(I_1) = \pi_1(V) \cup (V(g_1, ..., g_s) \cap V(I_1)) \in \mathbb{C}^{n-1},$$

where $\pi_1 \colon \mathbb{C}^n \longrightarrow \mathbb{C}^{n-1}$ is the projection of C^n onto the last $n-1$ components.

PROOF. Since $\pi_1(V)$ is a subset of $V(I_1)$, the $\supseteq$ inclusion is clear. For the reverse inclusion. let p be a point of $V(I_1)$. By Corollary 6.7, the points in $\pi_1(V)$ are exactly those which can be extended to total solutions in $V(I)$. Thus, if p does not belongs to $\pi_1(V)$, then it cannot be extended to a total solution in $V(I)$ and hence it should belong to $V(g_1, ..., g_s)$ in Theorem 6.1. Therefore, $p \in V(g_1, ..., g_s) \cap V(I_1)$. □

We have proved the following theorem while we are in Chapter 5: (Theorem 5.9):

THEOREM 6.2. *(**The Closure Theorem**) Let* $V = V(f_1, ..., f_s) \subset \mathbb{C}^n$, *and* $I_j = I \cap \mathbb{C}[x_{j+1}, ..., x_n]$ *be the jth elimination ideal of I. Then* $V(I_j)$ *is the smallest affine variety in* $\mathbb{C}^{n-j}$ *containing the image* $\pi_j(V)$ *of V under the projection map*

$$\pi_j \colon \mathbb{C}^n \longrightarrow \mathbb{C}^{n-j}.$$

3.2. An Example of Elimination and Extension. In Example 3.1 of Chapter 3, we have seen the role of the dictionary ordering, i.e., the lex order when we solve a system of polynomial equations or when we want to find an implicit equation for a parametrized surface: By computing a Groebner basis of the associated ideal under the lex order, we are able to handle one equation at one time. That is, among the elements of the computed Groebner basis G the variables are now separated so that we can solve each one of them one by one, starting from the last one and then moving to the next last one, etc. In the case of Example 3.1, the computed Groebner basis was

$$2x - 1,$$
$$12y - 4z + 1,$$
$$160z^2 - 8z - 107,$$

so we first solve the last equation, $160z^2 - 8z - 107 = 0$ for the unknown z, and then with the two solutions for $z \in \mathbb{C}^1$, we then solve the next last one, $12y - 4z + 1 = 0$ for the unknown y, and then with the two solutions $(y, z) \in \mathbb{C}^2$, we solve the next last one, i.e., the first equation in this case, $2x - 1 = 0$ for the unknown x, and hence we obtain two solutions $(x, y, z) \in \mathbb{C}^3$.

Moreover, if z is a real solution of the third equation, then the extended partial solution $(y, z) \in \mathbb{C}^2$ of the real partial solution $(z) \in \mathbb{R}^1$ is again a real solution in $\mathbb{R}^2$, and the extended total solution $(x, y, z) \in \mathbb{C}^3$ of the real partial solution $(y, z) \in \mathbb{R}^2$ is again a real solution in $\mathbb{R}^3$. In a later subsection, we are going to see what this means geometrically.

In this process of solving the given system of polynomial equations by backward substitution, the intermediate solutions are called *partial solutions* and the final solutions (x, y, z) are called the *total solutions.* This process is possible only when we can "separate variables" x, y, and z, *and* at each step we should be able to solve the equation for a single unknown, for example, it is the case if our bottom field k for the coefficients is an algebraically closed field.

If we compare the above computed Groebner basis for the ideal I with the given generating polynomials for the same ideal I on the left sides of these three equalities, we clearly see that we can not solve these original equations one by one separately, as we have just seen above with the computed Groebner basis for I under the lex order with $x > y > z$..

$$\begin{cases} x^2 + y^2 + z^2 - 1 &= 0, \\ x^2 + y^2 + z^2 - 2x &= 0, \\ 2x + 3y - z &= 0. \end{cases}$$

In other words, computing a Groebner basis for a polynomial ideal under the lex order guarantees to eliminate variables one by one, so that the variables are separated, and hence this process is usually called the "*Elimination Step*", and then the backward substitution process, like the one we have just seen above, is called the "*Extension Step*". In the case of Example 3.1, we are very lucky, because all the partial solutions, (z) and (y, z) are successfully extended to the total solutions, i.e., to those solutions for the given system of polynomial equations.

We note that the subjects we have studied in Chapter 5 covers the whole Elimination Step, and the subjects we have studied in this chapter covers the Extension Step.

Mor precisely, the last equation

$$160z^2 - 8z - 107$$

in the computed Groebner basis for I under the lex order with $x > y > z$,

$$\begin{aligned} &2x - 1, \\ &12y - 4z + 1, \\ &160z^2 - 8z - 107, \end{aligned}$$

does not have the variable y so that this y variable is eliminated while moving from the previous equation $12y - 4z + 1$ to the third equation. Then since the coefficient of the eliminated variable y in this polynomial is a non-zero constant 12, by Corollary 6.5 the two solutions

$$\begin{aligned} z &= \frac{3\sqrt{119}+1}{40}, \\ z &= \frac{-3\sqrt{119}+1}{40}, \end{aligned}$$

of

$$160z^2 - 8z - 107 = 0$$

are extended to the two solutions (y, z) of $12y - 4z + 1 = 0$,

$$(\frac{-\sqrt{119}-3}{40}, \frac{-3\sqrt{119}+1}{40}), \ (\frac{\sqrt{119}-3}{40}, \frac{3\sqrt{119}+1}{40}).$$

And then we clearly see that the x variable is eliminated while moving from the first equation to the second equation. Then since the coefficient of this eliminated variable x of the first equation $2x - 1$ is a non-zero constant 2, these two partial solutions are again by Corollary 6.5 extended to the two total solutions,

$$(\frac{1}{2}, \frac{-\sqrt{119}-3}{40}, \frac{-3\sqrt{119}+1}{40}), \ (\frac{1}{2}, \frac{\sqrt{119}-3}{40}, \frac{3\sqrt{119}+1}{40}).$$

This concrete example explains the role of the Extension Theorem, Theorem 6.1, very explicitly.

REMARK 6.6. *We do not know whether we can prove the Extension Theorem without using the Theory of Resultants in the Invariant Theory of the 19th Century.*

Exercises for Section 6.3

EXERCISE 6.6. *Apply the Extension Theorem, Theorem 6.1, to the following system:*

$$\begin{aligned} x^2 + y^2 + z^2 - 1 &= 0, \\ xyz - 1 &= 0, \end{aligned}$$

as we have done above.

4. Implicitization and Parametrization

As we have explained in Section 5 of Chapter 3, implicitization has a long history in mathematics. The basic problem in implicitization is the following: Given a parametric expression for an affine variety, we want to eliminate those parameters and want to obtain a defining equation for the variety, and the name "implicitization" comes from the name for defining equations, "implicit representation" of the variety. Thus, moving into one direction is called *parametrization* and moving into another direction is called *implicitization.*

Our basic example for implicitization in Section 5 of Chapter 3 was the implicitization of the surface tangent to the twisted cubic. (See Example 3.11).

In what follows we will first deal with two cases, *polynomial parametrization* and *rational parametrization.* We will soon see that we have already studied all the necessary subjects in Algebra while we are in Chapter 5 and almost all of the subjects in Geometry while we are in both Chapter 5 and this chapter.

4.1. Implicitization of a Polynomial Parametrization. The only thing we need to examine is about the relationship between the graph of a polynomial mapping and its projection onto a subspace. In fact, we have also discussed this while we are in Section 3 of Chapter 5. Suppose we are given a polynomial parametrization:

$$
\begin{aligned}
y_1 &= f_1(x_1, ..., x_n), \\
&\vdots \\
y_m &= f_m(x_1, ..., x_n),
\end{aligned}
\tag{6.2}
$$

where $f_i(x_1, ..., x_n)$ are polynomials in $k[x_1, ..., x_n]$, and hence these variables or parameters $x_1, ..., x_n$ are the objects of elimination. Then as we have discussed in Section 3 of Chapter 5, we can think of this geometrically as a function

$$f : k^n \longrightarrow k^m,$$

defined by sending $(x_1, ..., x_n) \longmapsto (f_1(x_1, ..., x_n), ..., f_m(x_1, ..., x_n)$, and this function is called a *polynomial mapping.* It is called a polynomial mapping because it is defined by a set of polynomial functions.

Then the image $f(k^n)$ of k^n under f is the subset of k^m parametrized by the equations in (6.2), and as we have seen before, $f(k^n)$ is not in general an affine variety, so the next thing we would like to know is the smallest affine variety V containing this image $f(k^n)$. Once we are able to figure out this, then the next question is whether V and $f(k^n)$ are in fact the same or not.

The traditional way of relating implicitization to elimination is to move into the Cartesian product of the domain and the target, k^{n+m}. In there, there is the graph $\Gamma(f)$ of the polynomial mapping f,

$$\Gamma(f) = V(I) \subset k^{n+m},$$

where $I = (y_1 - f_1, ..., y_m - f_m)$ is the ideal in $k[x_1, ..., x_n, y_1, ..., y_m]$. By definition, it is an affine variety in k^{n+m} defined by the m polynomials $y_i - f_i \in k[x_1, ..., x_n, y_1, ..., y_m]$. Then we have two other associated mappings: The fist map is the injection,

$$i : k^n \hookrightarrow k^{n+m},$$

which sends

$$(x_1, ..., x_n) \longmapsto (x_1, ..., x_n, f_1(x), ..., f_m(x)),$$

where $x = (x_1, ..., x_n)$ and the second map is the projection,

$$\pi_n : k^{n+m} \longrightarrow k^m,$$

which sends

$$(x_1, ..., x_n, f_1(x), ..., f_m(x)) \longmapsto (f_1(x), ..., f_m(x)),$$

and hence their composition is exactly the same as f, that is,

$$\pi_n \circ i = f.$$

Then since

$$i(k^n) = \Gamma(f),$$

Thus, under this setting, we have established:

COROLLARY 6.9.

$$f(k^n) = (\pi_n(i((k^n)) = \pi_n(V) = \pi_n(\Gamma(f)). \tag{6.3}$$

That is, the image $f(k^n)$ of the parametrization, i.e., the polynomial mapping f, is the same as the projection $\pi_n(\Gamma(f))$ of its graph $\Gamma(f)$.

Again under this setting, we have:

THEOREM 6.3. ***(Polynomial Implicitization)*** *Suppose that k is an infinite field. Let $f : k^n \longrightarrow k^m$ be the polynomial mapping defined parametrically by polynomial functions in (6.2). Let*

$$I = (y_1 - f_1, ..., y_m - f_m) \subset k[x_1, ..., x_n, y_1, ..., y_m]$$

be the defining ideal for the graph $\Gamma(f)$ of f and let $I_n = I \cap k[y_1, ..., y_m]$ be the nth elimination ideal of I. Then $V(I_n)$ is the smallest affine variety in k^m containing $f(k^n)$.

PROOF. If $k = \mathbb{C}$, the field of complex numbers, we have proved this in Theorem 5.9 in Chapter 5. Everybody knows nowadays that $\mathbb{C}$

is an algebraically closed field. (The fundamental Theorem of Algebra, Theorem 1.3 of Chapter 1).

On the other hand, if k is not a subfield of $\mathbb{C}$ then one can prove that it is a subfield of some algebraically closed field K. Then since the Closure Theorem, Theorem 5.9, still holds over any algebraically closed field, the theorem in that case will follow from what we are going to prove now:

Suppose that k is a subfield of $\mathbb{C}$. Since the characteristic of $\mathbb{C}$ is zero, $\mathbb{Q}$ is also a subfield of k, and hence k is automatically an infinite field. We will use the subscript k to keep track of which field we are working with, so that $V_k(I_n)$ is the affine variety in k^m and $V(I_n)$ is the affine variety in $\mathbb{C}^m$.

REMARK 6.7. *Of course, $V_k(I_n)$ could be a proper subset of $V(I_n)$, but the ideal I_n is never changed by moving into the larger field, because the ideal I is by its definition, only affected by the bottom field k, and the algorithm for computing the elimination ideal is not affected by changing from k to $\mathbb{C}$.*

We have to prove that $V_k(I_n)$ is the smallest affine variety in k^m containing $f(k^n)$. First, we note that

$$f(k^m) = \pi_n(V_k) \subseteq V_k(I_n),$$

by Equation (6.3) and Lemma 6.3. So let

$$W_k = V_k(g_1, ..., g_s)$$

be an affine variety in k^m for some polynomials $g_i \in k[y_1, ..., y_m]$ such that

$$f(k^n) \subseteq W_k$$

Then, each polynomial = polynomial function g_i (because we assume that k is an infinite field) finishes on the points of $f(k^n)$; hence its composition with f, $g_i \circ f = g_i(f_1, ..., f_m) \in k[x_1, ..., x_n]$ vanishes on the whole k^n. Thus by Proposition 1.5 of Chapter 1, the polynomial $g_i(f_1, ..., f_m)$ is the zero polynomial or equivalently the zero function, since k is an infinite field. In particular, this zero polynomial $g_i(f_1, ..., f_m) \in \mathbb{C}[x_1, ..., x_n]$ obviously vanishes on $\mathbb{C}^n$; hence we have

$$0 = g_i(f_1, ..., f_m)(\mathbb{C}^n) = g_i(f(\mathbb{C}^n)).$$

Therefore, each g_i vanishes on the image $f(\mathbb{C}^n)$; consequently $V(g_1, ..., g_s)$ contains $\pi_n(V)$. But, this just means that

$$V(I_n) \subseteq V(g_1, ..., g_s)$$

in $\mathbb{C}^m$ by the Closure Theorem, Theorem 5.9, and hence

$$V_k(I_n) \subseteq V_k(g_1, ..., g_s),$$

and this finishes the proof. □

This result gives us an algorithm, which is similar to Algorithm 21.

ALGORITHM 24. *(Implicitization Algorithm for Polynomial Parametrization) Let k be an* infinite *field. For a given polynomial parametrization in (6.2), we do the following to have its polynomial implicitization.*

(1) *Compute a Groebner basis G for the ideal*

$$I = (y_1 - f_1, ..., y_m - f_m) \subset k[x_1, ..., x_n, y_1, ..., y_m]$$

under an elimination order with the x variables bigger than the y variables.

(2) *By Theorem 5.1, $G \cap k[y_1, ..., y_m]$ is a Groebner basis for the n-th elimination ideal $I_n = I \cap k[y_1, ..., y_m]$.*

(3) *By Theorem 6.3, $V(I_n)$ is the smallest affine variety in k^m containing the image $f(k^n)$ of k^n under f.*

4.2. An Example of Implicitization of a Polynomial Parametrization. In this subsection, we are going to see the usefulness of both Groebner bases and the Extension Theorem in a delicate geometric question of whether the Zariski closure of the image of a polynomial mapping is a subset of the image. We are going to see that the first half of the whole arguments is basis on the Elimination Theorem, Theorem 5.1 of Chapter 5 due to the theory of Groebner bases, and the second half is based on the Extension Theorem, Theorem 6.1, due to the theory of resultants.

For this, let's examine the computed Groebner basis for the implicitization problem in Example 3.11. The problems was to figure out an implicit defining equation for the surface tangent to the twisted cubic of which parametric equation is given by

$$\begin{aligned} x &= t+u, \\ y &= t^2+2tu, \\ z &= t^3+3t^2u, \end{aligned} \tag{6.4}$$

where $t, u \in \mathbb{R}$. We have seen that a Groebner basis for the associated ideal $I = (x-(t+u), y-(t^2+2tu), z-(t^3+3t^2u))$ under the lex order with $t > u > x > y > z$ is the following:

$$t+u-x,$$

$$\begin{aligned} &u^2-x^2+y, \\ &2ux^2-2uy-2x^2+3xy-z, \\ &uxy-uz-x^2y-xz+2y^2, \\ &2uxz-2uy^2+2x^2z-xy^2-yz, \\ &2uy^3-2uz^2-4x^2yz+xy^3-2xz^2+5y^2z, \end{aligned}$$

$$4x^3z-3x^2y^2-6xyz+4y^3+z^2.$$

Then the combination of the Elimination Theorem, Theorem 5.1 of Chapter 5 and the Closure Theorem of Zariski, Theorem 5.9 of Chapter 5, tells us that the variety defined by the last equation is the smallest affine variety containing the image of the polynomial mapping defined by the given polynomial parametrization in (6.4).

Here, we have the three different groups of equations: The first group consists of only one equation with variables t, u, and x. The second group consisting of 5 equations does not have the first variable t, that is, the variable t is eliminated, and the third group consisting of just one equation does not have the next variable u, that is the variable u is eliminated. Let S' be the solution set of the last equation. Then it is an affine variety in k^3

In fact the current situation we have is the following: We are given a polynomial mapping f: $\mathbb{R}^2 \longrightarrow \mathbb{R}^3$ defined by (6.4):

$$(t, u) \longmapsto (t + u, t^2 + 2tu, t^3 + 3t^2u),$$

so that the affine variety

$$V = V(x - (t + u), y - (t^2 + 2tu), z - (t^3 + 3t^2u)) \subset k^{2+3}$$

is the graph $\Gamma(f)$ of f, where

$$I = (x - (t + u), y - (t^2 + 2tu), z - (t^3 + 3t^2u))$$

is the defining ideal in $k[t, u, x, y, z]$ for the graph $\Gamma(f)$ of f. Thus, the projection of $V(I)$ onto k^3 is the image $f(\mathbb{R}^2)$ in $\mathbb{R}^{2+3}$ of the domain $\mathbb{R}^2$ under f. We note that $f(\mathbb{R}^2)$ is usually not an affine variety, but by Theorem 6.3, which is a generalization of the Closure Theorem of Zariski, Theorem 5.9, we know that

$$S_{\mathbb{R}} = V_{\mathbb{R}}(4x^3z - 3x^2y^2 - 6xyz + 4y^3 + z^2)$$

is the smallest affine variety in $\mathbb{R}^3$ containing $f(\mathbb{R}^2)$. That is,

$$f(\mathbb{R}^2) \subseteq S_{\mathbb{R}}. \tag{6.5}$$

Then, our main concern is shifted to the question of whether a real solution $(x, y, z) \in \mathbb{R}^3$ of the last equation for $S_{\mathbb{R}}$ can be extended to a real solution $(u, x, y, z) \in \mathbb{R}^4$ of the second group of the equations, and finally this can also be extended to a real solution $(t, u, x, y, z) \in \mathbb{R}^5$ of the first equation, because this will answer whether we have an equality or not in the inclusion in (6.5).

For the following arguments, we definitely need the Extension Theorem, Theorem 6.1, and its corollary, Corollary 6.5, of this chapter, and the answer goes as follows. First, we note that the same assignment

$$(t, u) \longmapsto (t + u, t^2 + 2tu, t^3 + 3t^2u)$$

defines a polynomial mapping

$$f : \mathbb{C}^2 \longrightarrow \mathbb{C}^3.$$

Then since the leading coefficient of the first equation in the second group is a non-zero constant 1, we can extend a complex solution $(x, y, z) \in \mathbb{C}^3$ of the last equation

$$4x^3z - 3x^2y^2 - 6xyz + 4y^3 + z^2 = 0$$

to a complex solution $(u, x, y, z) \in \mathbb{C}^4$ of the second group of 5 equations. Then again because of the similar reason, the leading coefficient of the first equation is a non-zero constant 1, a complex solution $(u, x, y, z) \in \mathbb{C}^4$ of the second group of equations

$$\begin{aligned}
&u^2 - x^2 + y, \\
&2ux^2 - 2uy - 2x^2 + 3xy - z, \\
&uxy - uz - x^2y - xz + 2y^2, \\
&2uxz - 2uy^2 + 2x^2z - xy^2 - yz, \\
&2uy^3 - 2uz^2 - 4x^2yz + xy^3 - 2xz^2 + 5y^2z,
\end{aligned} \tag{6.6}$$

is extended to a total complex solution $(t, u, x, y, z) \in \mathbb{C}^5$ of the first equation. But, this total solution is just a point of $V(I)$ in $\mathbb{C}^{2+3}$, where this time,

$$I = (x - (t + u), y - (t^2 + 2tu), z - (t^3 + 3t^2u)) \subset \mathbb{C}[t, u, x, y, z].$$

What does this all mean geometrically? We in fact have verified, thanks to the Extension Theorem, Theorem 6.1, that

$$\begin{aligned}
V(I_2) &\subseteq \pi_1(V(I_1)), \\
V(I_1) &\subseteq \pi_1(V(I)),
\end{aligned}$$

and consequently the following equalities hold:

$$S = V(I_2) \subseteq \pi_1(V(I_1)) \subseteq \pi_1(\pi_1(V(I))) = \pi_2(V(I)) = f(\mathbb{C}^2),$$

where I_1 is the ideal generated by those polynomials in (6.6) *and* $4x^3z - 3x^2y^2 - 6xyz + 4y^3 + z^2$ over $\mathbb{C}$, and I_2 is generated by the single polynomial $4x^3z - 3x^2y^2 - 6xyz + 4y^3 + z^2$ over $\mathbb{C}$. Then since we already know the reverse inclusions hold by Theorem 6.3, when the bottom field is $\mathbb{C}$, we have verified that $f(\mathbb{C}^2)$ itself is an affine variety in $\mathbb{C}^3$ defined by a single equation:

$$4x^3z - 3x^2y^2 - 6xyz + 4y^3 + z^2 = 0.$$

Now, let $(x, y, z) \in V_{\mathbb{R}}(I_2) \subset \mathbb{R}^3$ be a real point with real coordinates $x, y, z \in \mathbb{R}$, then the extended partial solution (u, x, y, z) in $\mathbb{C}^4$ of this real point (x, y, z) is a real point in $\mathbb{R}^4$, for instance, because u is a real number by the second equation,

$$2ux^2 - 2uy - 2x^2 + 3xy - z = 0,$$

if either x or y is a non-zero real number and u is zero by the first equation

$$u^2 - x^2 + y = 0,$$

if both x and y are zero. Then the extended total solution (t, u, x, y, z) in $\mathbb{C}^5$ of this real point (u, x, y, z) in $\mathbb{C}^4$ through the equation,

$$t + u - x = 0,$$

is again a real point in $\mathbb{R}^5$. Thus, we have

$$S_{\mathbb{R}} = V_{\mathbb{R}}(I_2) \subseteq \pi_1(V_{\mathbb{R}}(I_1)) \subseteq \pi_1(\pi_1(V_{\mathbb{R}}(I))) = \pi_2(V_{\mathbb{R}}(I)) = f(\mathbb{R}^2).$$

Then since we have the reverse inclusions by Theorem 6.3, the real surface $S_{\mathbb{R}}$ in $\mathbb{R}^3$ defined parametrically by (6.4), namely a real surface tangent to the twist cubic defined parametrically

$$\begin{aligned} x &= t, \\ y &= t^2, \\ z &= t^3. \end{aligned}$$

in $\mathbb{R}^3$, is exactly the same as the affine variety in $\mathbb{R}^3$ defined by a single polynomial equation:

$$4x^3z - 3x^2y^2 - 6xyz + 4y^3 + z^2 = 0.$$

4.3. Implicitization of a Rational Parametrization. In this section, we are going to study a rational mapping, that is, a map

$$f : k^n \longrightarrow k^m$$

defined by sending

$$(x_1, ..., x_n) \longmapsto \left(\frac{f_1(x_1, ..., x_n)}{g_1(x_1, ..., x_n)}, ..., \frac{f_m(x_1, ..., x_n)}{g_m(x_1, ..., x_n)}\right), \tag{6.7}$$

where each $f_i = f_i(x_1, ..., x_n)$ and each $g_i = g_i(x_1, ..., x_n)$ are polynomials in n variables $x_1, ..., x_n$ over the bottom field k and each g_i *never* vanishes on the domain of the map. We write

$$W = V(g_1, ..., g_m) \subset k^n.$$

Then outside of W, is the map f well-defined. As we have done for a polynomial mapping, we move into the Cartesian product k^{n+m+1} and consider the graph of the rational mapping f. For this, we introduce new coordinates $y_1, ..., y_m$ and express the map as

$$\begin{aligned} y_1 &= \frac{f_1(x_1, ..., x_n)}{g_1(x_1, ..., x_n)}, \\ &\vdots \\ y_m &= \frac{f_m(x_1, ..., x_n)}{g_m(x_1, ..., x_n)}. \end{aligned}$$

Then clearing the denominators we obtain $m + 1$ equations:

$$\begin{cases} g_1y_1 - f_1 &= 0, \\ \vdots & \\ g_my_m - f_m &= 0, \\ 1 - g_1 \cdots g_m \cdot y &= 0, \end{cases}$$

where we need to add the last equation,

$$1 - g_1 \cdots g_m \cdot y = 0,$$

once and for all in order to prevent the denominators of the rational mapping f from becoming zero. These equations are all polynomial equations, so let I be an ideal in $k[y, x_1, ..., x_n, y_1, ..., y_m]$ defined by

$$I = (1 - g_1 \cdots g_m \cdot y, g_1 y_1 - f_1, ..., g_m y_m - f_m).$$

Then $V(I)$ is certainly an affine variety in the affine $(n+m+1)$-space k^{n+m+1}, and moreover, those polynomials in I never vanish on

$$W = V(g_1 \cdots g_m).$$

Thus, we also have, as before, have the injection map

$$i : k^n - W \longrightarrow k^{n+m+1}$$

defined by sending

$$(x_1, ..., x_n) \longmapsto \left(\frac{1}{g_1 \cdots g_m}, x_1, ..., x_n, \frac{f_1(x_1, ..., x_n)}{g_1(x_1, ..., x_n)}, ..., \frac{f_m(x_1, ..., x_n)}{g_m(x_1, ..., x_n)} \right),$$

and we have the projection map

$$\pi_{n+1} : k^{n+m+1} \longrightarrow k^m,$$

defined by sending

$$(y, x_1, ..., x_n, y_1, ..., y_m) \longmapsto (y_1, ..., y_m).$$

Then, we obviously have

$$\pi_{n+1} \circ i = f$$

because

$$\begin{aligned} (\pi_{n+1} \circ i)(k^n - W) &= \pi_{n+1}(i(k^n - W)) \\ &= \left(\frac{f_1(x_1, ..., x_n)}{g_1(x_1, ..., x_n)}, ..., \frac{f_m(x_1, ..., x_n)}{g_m(x_1, ..., x_n)} \right) = f(k^n - W). \end{aligned}$$

LEMMA 6.4. *Under this setting, we have*

$$i(k^n - W) = V(I).$$

PROOF. The inclusion $i(k^n - W) \subseteq V(J)$ is clear from the definitions. For the reverse inclusion, let $p = (y, x_1, ..., x_n, y_1, ..., y_m)$ be a point of $V(I)$. Then, $(g_1 \cdots g_m)y = 1$ implies we can solve it for y:

$$y = \frac{1}{g_1 \cdots g_m},$$

and hence none of the g_i can vanish at p, and hence each

$$g_i y_i - f_i = 0$$

can be solved for y_i:

$$y_i = \frac{f_i(x_1, ..., x_n)}{g_i(x_1, ..., x_n)},$$

for each i. We note that all these arguments are possible only for those points $(x_1, ..., x_n) \in k^n - W$. Thus, $p \in i(k^n - W)$. □

Thus, we have

$$f(k^n - W) = \pi_{n+1}(i(k^n - W)) = \pi_{n+1}(V(I)). \tag{6.8}$$

Thus under this setting, we have established:

COROLLARY 6.10. *The image $im(f)$ of the rational parametrization f given in (6.7) is the same as the projection $\pi_{n+1}(V(I))$ in k^m of the affine variety $V(I)$ in $k^{n+m=1}$.*

Again under this setting, we have:

THEOREM 6.4. ***(Rational Implicitization)*** *Suppose that k is an* infinite *field. Let $f : k^n - W \longrightarrow k^m$ be a rational mapping defined parametrically by rational functions in (6.7), where $W = V(g_1 \cdots g_m)$. Let I be the ideal*

$$I = (1 - g_1 \cdots g_m \cdot y, g_1 y_1 - f_1, ..., g_m y_m - f_m) \subset k[y, x_1, ..., x_n, y_1, ..., y_m],$$

and let $I_{n+1} = I \cap k[y_1, ..., y_m]$ be the $(n+1)$st elimination ideal of I. Then $V(I_{n+1})$ is the smallest affine variety in k^m containing $f(k^n - W)$.

PROOF. The proof of this theorem is very similar to that of Theorem 6.3. Here, we need to use Equation (6.8) instead of Equation (6.3). In the proof of this theorem, we also need to use the fact that a polynomial h vanishing on $k^n - W$ must be the zero polynomial, but it is easy to verify: If h vanishes on $k^n - W$ then $h \cdot g_1 \cdots g_m$ vanishes on k^n and hence $h \cdot g_1 \cdots g_m \equiv 0$, the zero polynomial = the zero function, since k is an infinite field. Then since each g_i never vanishes on $k^n - W$, h must be the zero polynomial. $\square$

ALGORITHM 25. *(Implicitization Algorithm for Rational Parametrization) Let k be an infinite field. For a given polynomial parametrization in (6.2), we do the following to have its rational implicitization.*

(1) *Compute a Groebner basis G for the ideal*

$$I = (1 - g_1 \cdots g_m \cdot y, y_1 - f_1, ..., y_m - f_m) \subset k[y, x_1, ..., x_n, y_1, ..., y_m]$$

under an elimination order with both the y variable and the x variables bigger than the variables $y_1, ..., y_m$.

(2) *By Theorem 5.1, $G \cap k[y_1, ..., y_m]$ is a Groebner basis for the $(n+1)$st elimination ideal $I_{n+1} = I \cap k[y_1, ..., y_m]$.*

(3) *By Theorem 6.4, $V(I_{n+1})$ is the smallest affine variety in k^m containing the image $f(k^n - W)$ of k^n under f.*

4.4. An Example of Rational Implicitization. Let

$$f : \mathbb{R} \longrightarrow \mathbb{R}^2$$

be a rational mapping defined by a rational parameterization:

$$\begin{aligned} x_1 &= \frac{1 - t^2}{1 + t^2}, \\ x_2 &= \frac{2t}{1 + t^2}. \end{aligned}$$

To obtain the smallest affine variety in $\mathbb{R}^2$ containing the image of this map, let

$$I = (1 - (1 + t^2) \cdot y, 1 - t^2 - (1 + t^2)x_1, 2t - (1 + t^2)x_2) \subset \mathbb{R}[y, t, x_1, x_2],$$

and compute a Groebner basis for I under the lex order with $y > t > y_1 > y_2$. For this, we use the REDUCE:

```
off echo$
off nat$
load_package groebner$
torder({y,t,x1,x2},lex)$
f1:=1-(1+t^2)*y$
f2:=1-t^2-(1+t^2)*x1$
f3:=2*t-(1+t^2)*x2$
gb:=groebner({f1,f2,f3});
gb := {2*y - x1 - 1,
t*x1 + t - x2,
t*x2 + x1 - 1,
x1**2 + x2**2 - 1}$
```

Thus, the smallest affine variety containing the image of f is the circle,

$$x_1^2 + x_2^2 = 1,$$

with radius one centered at the origin. Then since the second and the third equations are

$$(x_1 + 1)t - x_2,$$
$$x_2 t + x_1 - 1,$$

the point on the circle,

$$(-1, 0),$$

which is the only common zero of $x_1 + 1 = 0$ and $x_2 = 0$, is the only point which can not be extended to a partial solution (x_1, x_2). Except for this point, all other points (x_1, x_2) of the circle are extended to partial solutions (t, x_1, x_2) for some non-zero t and then they are all extended to (y, t, x_1, x_2) for some y. Thus, the image of f is *properly* contained in the circle, because the point $(-1, 0)$ of the circle is not in the image of f.

4.5. An Example of Rational Parametrization. We note that it is usually not easy to derive a parametric representation of a given affine variety defined implicitly, i.e., defined by a set of polynomial equations.

But, the above "rational parametrization of the circle" is in fact a special case and it provides us with a very powerful method for integration of a rational function of trigonometric functions. The following example is a prototypical example of *Projective Algebraic Geometry.*

Let C be a circle of radius one centered at the origin $(0,0)$ and let L be a line passing through the given point $(-1,0)$. Then for each slope t of the line L, except for the vertical line, we have only another intersection point at the circumference of the circle, and conversely, for a given point on the circumference of C we have only one line passing through the fixed point $(-1,0)$, and hence only one slope t of the line L.

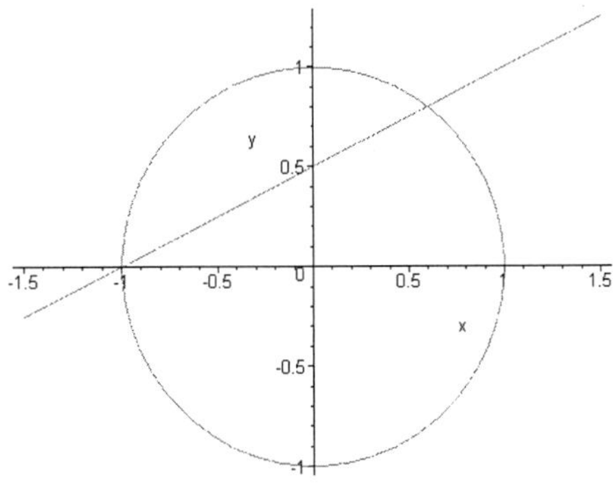

Say, (x_1, x_2) is a point of C. Then the slope t is given by

$$t = \frac{x_2 - 0}{x_1 - (-1)} = \frac{x_2}{x_1 + 1}.$$

Thus, $x_2 = t(x_1 + 1)$. If we substitute this value for x_2 in the equation for the circle

$$x_1^2 + x_2^2 = 1,$$

we get a quadratic equation.

$$(1+t^2)x_1^2 + 2t^2x_1 + t^2 - 1 = 0.$$

Since $(-1,0)$ is the fixed intersection point, this equation is rewritten as

$$(x_1+1)((1+t)x_1 - (1-t^2)) = 0,$$

and from this, we obtain a rational parametric representation of the circle, except for one point, $(-1,0)$:

$$\begin{aligned} x_1 &= \frac{1-t^2}{1+t^2}, \\ x_2 &= \frac{2t}{1+t^2}. \end{aligned}$$

Now, if we introduce an angle θ between x_1-axis and the point (x_1,x_2), everybody knows that the angle between the line L and the x_1-axis is $\frac{1}{2}\theta$, and moreover we have,

$$\begin{aligned} \tan\frac{\theta}{2} &= t, \\ \cos\theta &= \frac{1-t^2}{1+t^2}, \\ \sin\theta &= \frac{2t}{1+t^2}, \\ \tan\theta &= \frac{2t}{1-t^2}. \end{aligned} \tag{6.9}$$

This substitution allows us to integrate,

$$\int \frac{p(\theta)}{q(\theta)} d\theta.$$

Thus, by changing the variable from θ to t, the integrand $\frac{p(\theta)}{q(\theta)}$, where θ is an angle, is expressed as a rational function of t, and hence we can integrate it with respect to t.

REMARK 6.8. *It was known that this powerful method of integration,*

$$\int \frac{p(\theta)}{q(\theta)} d\theta,$$

where θ is an angle, was first introduced by Karl Weierstrass (1815 - 1897).

Exercises for Section 6.4

EXERCISE 6.7. *Let C be a space curve in $\mathbb{R}^3$ defined parametrically by*

$$\begin{aligned} u &= s^2, \\ v &= st, \\ w &= t^2. \end{aligned}$$

Apply Algorithm 24 to compute an implicit equation for C.

EXERCISE 6.8. *Let C be a curve in $\mathbb{R}^2$ defined parametrically by*

$$\begin{aligned} u &= \frac{t^2}{1+t^2}, \\ v &= \frac{t^3}{1+t^2}. \end{aligned}$$

(1) *Compute an implicit equation for C using the resultant expression in part (1) of Proposition 6.3.*
(2) *Apply Algorithm 25 to compute an implicit equation for C.*
(3) *Compare these two results. In particular, which method is more efficient?*

EXERCISE 6.9. *Without using a Computer Algebra System (CAS), compute*

$$\int \frac{\sin\theta}{1+\sin\theta} d\theta.$$

[Hint: Use Equations in (6.9)].

APPENDIX A

An Introduction to Algebraic Analysis

1. Tools in Mathematics

The whole body of Mathematics consists of many different kinds of knowledge, or results and procedures which produce such results, most of which are verified through the process of the so-called "proof". Those proved and somewhat useful or important are usually called **theorems**. Unproven ones are formally called **conjectures**. The place for most mathematicians should be between these two words, because we are born to establish reasonable conjectures and then prove them. When one tries to prove something, something itself must be first recognized as a reasonable mathematical object; discovering such object could be quite personal.

In proving reasonable conjectures, one generally uses two tools, a special type of logics, called **mathematical logic**, and **computation,** definitely guided by his own mathematical intuitions or experiences. Since they are applied as tools in the process of the so-called **proof**, they themselves cannot be called mathematics. However, the kinds of the methods employed certainly put some unavoidable limitations on the character of the consequences. These two tools, nevertheless, do not hurt each other. In contrast, they help each other, and these two are the most powerful tools we, mathematicians, have. Depending on the degree of putting more emphasis on any of these two, one usually has two basic tendencies in doing mathematics, axiomatic or computational.

1.1. Axiomatic Nature. Although it is very difficult to figure out what constitutes mathematics, one may describe its axiomatic nature as follows: Once a reasonable mathematical conjecture is recognized, one accepts, without questioning their existence or validity, a finite number of undefined terminologies and that of postulates (i.e., axioms or principles) which describe some relations between the undefined terminologies, and one applies a finite number of steps of mathematical logic directly to them or to some of their consequences until one finally reaches the desired positive or negative conclusions. As a demonstrative science started even before Greek period, mathematics has shown this axiomatic nature for a long time. One can find such an evidence in, for instance, Euclid's Elements [**E**] and other works of Greek mathematicians and philosophers. In fact, one can even trace back to their works to see that most of the formal terminologies used in mathematics, something like definitions, propositions, corollaries, etc., were already used in his Elements and other Greek literatures. The meaning and usage of these terminologies and other metamathematical concerns are also well explained in the introduction of A. Whitehead and B. Russell's famous work, *Principia Mathematica* [**WR**], which was originally published in 1910.

1.2. Computational Nature. When computations in general are concerned, we know that there are basically two different kinds, namely, exact computation and approximation. **Approximation** is done through the limiting process mostly in terms of inequalities, where as **exact computation or symbolic computation** is done through the so-called four binary operations: addition, subtraction, multiplication and acceptable division. These operations are applied not only to numbers but also to a mixture of numbers and **symbols**.

Computations are very important to us at least for the following three reasons. First, they usually show up in the middle of the process called proof. This is almost unavoidable to every mathematician. Secondly, a decent amount of computational experiments quite often reveals to us what's going on in a very clean manner, while simple-minded 'what if ...' ways of logical questioning cannot. Thirdly, we, mathematicians, can communicate with other branches of sciences very efficiently and successfully, through computational results and their usefulness. Considering that mathematics has so far grown up to this much within the accumulation of our civilization, this point of view is also very important, because isolation only hurts oneself till one's starvation.

Up to the end of the 19th century or the beginning of the 20th century, some important mathematical works discovered and proved were based on individual mathematician's heavy hand-computational experiments. However, as was the case in Invariant Theory of the 19th century, mathematicians were practically forced out to give up heavy computations when the number of unknowns and the size of numbers are substantially increased so that hand-calculation obviously becomes terrible. Any attempt based on computational experiments may be called a **computational approach**.

2. Overemphasis on Axiomatic Approach in the 20th Century

The revival of the so-called **axiomatic approach** after the Greek period (namely, Euclid's Elements) is clearly evident in the Lecture Notes of E. Artin and E. Noether, written up by B.L. van der Waerden in 1930s [**BW**]. This approach, especially well-supported by precise notations adapted from the subject, Set Theory, was so good in pursuing a particular-minded concept or property, that it rapidly pushed out the old tradition of computational approaches, which was quite common up to the first quarter of 20th century, from the mathematical world. In addition, some eminent mathematicians, like Banach in Analysis or later E. Spanier in Algebraic Topology [**ES**], tried to figure out and establish sufficiently small number of axioms but still good enough to explain successfully the rest of the theory they know. In order to fulfill such tasks, they examined many computed examples of different kinds and pulled out some common underlying axioms which replace too much involved terrible computations. The success of this attempt is quite evident in almost every part of current mathematics in both research articles and textbooks. It is also around this period that most mathematicians officially but still reluctantly began to acknowledge the usefulness of the non-traditional mathematical axiom, i.e., the axiom of choice, or its equivalence, the Zorn's Lemma.

The point of views expressed in this article could be quite personal until they get more acceptance from general audience. However, the readers will not be hurt by the author's viewpoints on mathematics expressed in this article but it will supply a fresh taste or sense to our main meal, mathematics.

2.1. Danger of Too Much Emphasis on Axiomatic Approach. One should note that those mathematicians, who introduced an appropriate set of axioms for the knowledge they already know, were also very good at computation, since they were trained as, called from now on, the "first" generation mathematicians, by their teaches most of whom were under the strong influence of computational approach of the 19th century. On the other hand, the "second" generation mathematicians, i.e., those who were mathematically trained by the first generation mathematicians, tended to start their mathematics not from heavy computational experiments but from a small amount of axioms, and they were more and more away from computational experiences. The lack of computational experiences is more significant in the "third" generation mathematicians, i.e., those trained by the second generation mathematicians.

Nevertheless, the strength of axiomatic approach had been fully demonstrated by the monumental works of a single mathematician, Alexander Grothendieck (1928 - 2014), on the Foundation of Algebraic Geometry, [**AG1**], FGA [**AG2**], EGA's [**AG3**], [**AG4**], and SGA's [**AG5**] during 1950s up until the beginning of 1970s. His works in turn simply recall the works of a group of mathematicians, called Bourbaki [**B2**] and [**B1**], as elementary results. Anyone who is very much interested in the scope of axiomatic approach in mathematics explored during the 20th century is strongly encouraged to study and digest the above mentioned Grothendieck's works, of which the basic concepts and results are described in terms of categorical languages, called the **language of schemes**.

In particular, the tendency of the results obtained under a strong influence of axiomatic approach is too much conceptual, and hence becomes more and more **soft** and hence fragile, as time goes. In this case, a good-minded mathematician should worry about whether his and/or her results will still survive after, say, 1,000 years from now. A prototypical example is Euclid's algorithm for computing the greatest common divisor of two numbers. His method is still known to us because it is about computation of certain mathematical objects in an efficient manner. Moreover, lack of computability only speeds up the isolation of Mathematics from other branches of Sciences. This is in no way acceptable to us.

For an elegant axiomatic treatment but still with a strong emphasis on computational point of views, of Algebraic Geometry, one may refer to Prof. George R. Kempf's two books [**GK1**] and [**GK2**].

3. A New Computational Tool: Computers

3.1. Personal Computers and High Performance Servers. Starting at around the beginning of the fourth quarter of the 20th century, computer industry has grown up with an unexpected rate of success; this nowadays allows us to compute very big numbers using a Personal Computer or even a Notebook Computer. Recently, inexpensive High Performance Cluster Servers are also available to us. As a tool for actual computation, computers are already playing a significant role and such role will surely change the mathematical world of the 21-st century. Here are some examples.

EXAMPLE A.1. *Here, $p(n)$ denotes, by definition, the number of ways of writing a given natural number n as a sum of others (called components) ignoring the order among the components. One starts with a very famous theorem:*

THEOREM A.1. *(Rademacher [**GA**]) Given a natural number n, one has*

$$p(n) = \frac{1}{\pi\sqrt{2}} \sum_{k=1}^{\infty} A_k(n)\sqrt{k}\frac{d}{dn}\frac{\sinh((\pi/k)\sqrt{(2/3)(n-\frac{1}{24})})}{\sqrt{n-\frac{1}{24}}}, \tag{A.1}$$

where

$$A_k(n) = \sum_{\substack{0 \le h < k \\ (h,k)=1}} \varpi_{h,k} \exp(-2\pi inh/k),$$

where

$$\varpi_{h,k} = \exp(\pi is(h,k)),$$

is a 24kth root of unity and where

$$s(h,k) = \sum_{\mu=1}^{k-1} \left(\frac{\mu}{k} - \left[\frac{\mu}{k}\right] - \frac{1}{2}\right)\left(\frac{h\mu}{k} - \left[\frac{h\mu}{k}\right] - \frac{1}{2}\right)$$

is the so-called Dedekind sum.

*It turns out that the value of $p(20,000,000)$, when computed it using this theorem, requires more than 4,900 number of decimal digits (Cf. [**STS1**]). Thus the final result is really a super-ultra big number, which is one point. Moreover, another point is that it only took about 17 Minutes with already-done pre-calculations of the so-called Dedekind sums, under the same hardware and software limitations as in the next example. These two points clearly show us the computing capability of current PC's.*

REMARK A.1. *One should note that computation of such a super-ultra number eventually raises a serious concern whether the computed result is absolutely correct. This is certainly a new kind of headaches to mathematicians.*

On the other hand, the author recently discovered an exact algebraic formula for a graded partition function $p_{\mathbf{M}}(n)$ over a multiset $\mathbf{M}$ and an associated quasi-recursive formula for $p_{\mathbf{M}}(n)$, both of which never *require approximation at all (Cf. [**STS2**]). One should note that it is neither quite* *possible to obtain an analytically exact formula for* $p_{\mathbf{M}}(n)$ *nor to obtain a recursive formula for it, something like either Rademacher's formula or the well-known Euler's recursive formula for* $p(n)$*, unless there is a significant symmetry among the elements of* $\mathbf{M}$*. However, one version of the author's formulas computes all the values of* $p_{\mathbf{M}}(i)$ *for* $1 \leq i \leq n$ *with time efficiency of* $\mathrm{O}(n^2)$ *for a general multiset* $\mathbf{M}$ *(Cf.* [**STS3**]*).*

3.2. Computer Algebra Systems. Of course, having just a very fast computing machine of which hardware is very reliable is not enough. One also needs a reliable software which guarantees several facilities. Among them, the most urgent requirements are, for instance, ability to handle numbers of any sizes and arbitrary precision in representing fractions, each of which absolute reliability is definitely required in any computation with computers.

EXAMPLE A.2. *The following Table shows the actual computation of the value of the classical partition function* $p(n)$*, for* $n = 200$*, using the Rademacher's formula in Eq (A.1). Here, by adding the first* 8 ***non-zero*** *terms of fractional real numbers, one gets a fractional real number,*

$$3972999029388.001346332,$$

which is very close to the true integer value $p(200) = 3972999029388$.

$$\begin{array}{rcr}
t(1) & = & +3.97299899318589575787 2E+12 \\
t(2) & = & +36282.9784854411847056 \\
t(3) & = & -87.58391287399 \\
t(4) & = & +5.146534584 \\
t(5) & = & +1.424217569 \\
t(6) & = & +0.07102143214 \\
t(7) & = & 0 \\
t(8) & = & +0.04357655366 \\
t(9) & = & +0.0256657539 \\
[t(10) & = & +0.008603641042] \\
\hline
\text{Up to 9th Partial Sum} & = & +3972999029388.001346332
\end{array}$$

When it is about computation of an exact integer value in this way of approximation, there should not be any approximation-related errors in all of the intermediate steps. Do not expect that an error-free computation of such a kind is quite easy (Cf. [**STS1**]*), unless one uses very reliable Computer Algebra Systems.*

REMARK A.2. *This example, which is prepared by the author of this book, is especially interesting because it is a direct computational verification of the famous example computed by Hardy and Ramanujan (p. 284 of* [**HR**]*) and quoted by Andrews (p. 70 of* [**GA**]*). Indeed, the original calculation in Hardy's own handwriting occurs on the dust jacket of the 1969 edition of "A Mathematician's Apology",* [**H**]*. To our surprise, the calculation by Hardy and Ramanujan contains an error. The corrected calculation as in Example A.2 turns out to be much closer to* $p(200)$ *than the original incorrect version. Therefore, the third term,* -87.555*, in the original computation table for* $p(200)$ *in p. 284 of* [**HR**] *and later quoted in p. 70 of* [**GA**]*, needs to be replaced with* -87.584*.*

Any software package lacking these two capabilities should not be our choice for a tool for computational experiments. In addition to them, it is also required to have an ability to solve a given system of polynomial equations, i.e., non-linear equations over, say, the field of complex numbers. This means that it must have an ability to handle symbols as well as numbers. The theory for such computation had not fully understood until Bruno Buchberger (1942 -) discovered his famous algorithm for Groebner bases in 1965. Any reliable software package with at least these three facilities could be called a **Computer Algebra System**.

There are two types of Computer Algebra Systems, commercially available or not. Some of the commercially available ones are, MAPLE, MACSYMA, REDUCE, etc. Some of non-commercially available ones are, MACAULAY 2, CoCoA, etc. There should be more than these.

3.3. Behind Computational Experiments. One first notes that there are two kinds of computations one can do with computers, **exact computation** and **approximation**. Exact computation only needs addition, subtraction, multiplication, and allowed division depending on given conditions, on a given mixture of known symbols, unknown symbols, and numbers. In contrast to this, approximation allows us to apply non-exact division and limiting process, as well as addition, subtraction, multiplication and division on given numbers only. Thus, exact computation allows us to manipulate not only numbers but also **symbols**, whereas approximation only allows us to manipulate numbers. This tells us why exact computation is also called **Algebraic Analysis** or **Symbolic Analysis**, while approximation is called **Numerical Analysis**. Examples of Numeric Analysis are numerous. For instance, FEM (Finite Element Method) is just one of them. Also, see Example A.2 above.

These two approaches in computational methods do not hurt each other. In fact, they help each other. The only general trouble is that Algebraic Analysis does require some amount of elementary Algebra, mostly about polynomial rings in more than one unknowns, as well as some amount of Modern Algebraic Geometry, developed by Oscar Zariski [**OZ**] and J. P. Serre [**JS**], and a few others (For beginners, this amount of accounts on the subject may be good enough).

A proper sequence of computational approach is as follows:

(1) to apply appropriate exact methods in Algebraic Analysis until all available ones are exhausted, and then, as a final step,
(2) to apply appropriate approximation methods in Numeric Analysis.

The subject of Numeric Analysis has been well known to us after Sir Isaac Newton. However, Algebraic Analysis is very new to us. The most important theoretical background for recent revival of computational approach is the Buchberger's astonishing discovery of the algorithm for Groebner bases for polynomial ideals in more than one indeterminates. He figured out the method while he was working on his Ph.D. thesis, and it was first recognized by mathematicians oriented to computer science somewhat before or around 1975. I myself learned this remarkable algorithm in mid 1980s.

3.4. Danger of Too Much Emphasis on Computational Approach. If one puts too much emphasis on computational approach, he will soon find that he does not know why the method works and eventually he will get lost, because the arithmetic operations themselves never tell him what's going on, unless he knows *in advance* what's really going on. Unfortunately, it is obviously impossible most of the time, especially when one works on some mathematical phenomena not yet well understood to him. This means that one really needs a well-developed theory for exact computation or Algebraic Analysis.

There was no general theory for exact computation with computers until B. Buchberger discovered a very important algorithm. The mixture of his algorithm with the division algorithm produces very important algorithms for polynomial ideals in more than one variables. It is very urgent to understand this theory for hard but honest and diligent computation. The readers of this article should not underestimate the strength and usefulness of this theory.

4. Expected Trends of the 21st Century Mathematics

4.1. Computer Age. It is well known that there were two types of computers, analogue and digital, in the past. Nowadays, however, computers usually mean **digital computers** or their variants for better performance, since performance of digital computers (for instance, signal to noise ratio) has already proved to be a lot better than analogue types. It is also well known that the theoretical creation of digital computers was done by a famous mathematical logician, von Neumann, who invented Class Theory to answer, for instance, the question like whether the set of all sets is a set or not. After theoretically creating digital computers, he immediately developed a good deal of approximation theory for computing the values of non-polynomial functions with his theoretical machines.

Although Electric Engineering, Computer Engineering, and Computer Science are respectively needed to establish hardware, architecture, and operating system and software, the final combination of them is a system, called, computer, and the main reason for developing such a system is to compute or calculate something in any possible senses. The point we, mathematicians, should not miss is that **computation belongs to mathematics**. Thanks to computers, paperless computation of huge numbers is already available to those who know how to use some of the commercially or non-commercially available Computer Algebra Systems, so that the scope of Algebra and Algebraic Geometry in both pure and applied directions has already been expanded a lot.

4.2. Computers Are for Computation Related Human Activities. Since computation solely belongs to mathematics, it is mathematician's due obligation to examine and develop the needed mathematical algorithms in order to obtain from the machine the desired results with absolute correctness. In addition to this, our second obligation is to make it as efficient as possible. If it takes too much computation time, then such a computational approach will never buy other people's approval at all. To my regret, these two can't be easily accomplished by those who were not trained as mathematicians.

Unlike any other centuries, existence of so many mathematical objects have been proved during the 20th century, mostly thanks to active application of axiomatic methods influenced by Set Theory or Category Theory. These results are all inputs for further investigation that which of them are computationally constructible, applying the methods of Algebraic Analysis.

I should mention one dangerous aspect of the Zorn's Lemma or its equivalent axioms. If one used the Zorn's Lemma in some place of his proof of the existence of certain mathematical object, then his proof is computationally useless in the sense that it never really tells us how to construct such an object only using the four arithmetic operations and the limiting process. This is really a disappointing aspect of the Zorn's Lemma, but is, as long as **algorithmic construction** is concerned, still important enough to keep in mind in order for us not to use it in our proof as possible as we could.

Computers have already become a convenient and very capable tool for computations in any part of mathematics. We do not necessarily require to have an access to Supercomputers, Servers, or even Workstations for computational experiments. We only need to have Personal Computers for most computational experiments. Since Personal Computers are nowadays available to everyone and since they are already too fast and too capable of, most computational experiments could be done with them without spending too much time and money. Moreover, the hardware-related performance of these machines are nowadays very much reliable. The author strongly believes that such a very reliable computational environment, already available to everybody, sooner or later changes the whole atmosphere of mathematics of the 21-st century. For instance, the so-called **internet computing** will soon become very common.

One of the main ingredients of mathematics is perhaps its **exactness** in the process of both **proof** and **computation.** The point is that both of them require quite different types of exactness. This may be also the main reason why no other branches of Sciences even try to compete with mathematics, but only to rely on the results produced by mathematicians. Consequently mathematics has its own everlasting life. We should not give up any of these two powerful tools, proof and computation, to other branches of Sciences. On the other hand, we have been silently forced to ignore the significance of computational aspects for at least a century, mostly because of our inability in exactly hand-computing mixtures of big numbers and many parameters and also partly because of strong influences of axiomatic approach of the 20th century. Isn't it a time for us to revive computation related mathematical activities with a new tool, computers, to balance ourselves between them?

5. About Our InetCompu Service

We have recently started an E-mail based Internet Computing Service, called InetCompu.

If one visits

http://trinitas.mju.ac.kr/intro2InetCompu.html[1]

the detailed information and several demonstrative services are provided:

5.1. Our Concept and Terminology. If a person, for instance, using PC, PDA, or Mobile Phone, can perform the kinds of computations he wants on a remote Server, where he is not even required to have his own account, and is also allowed to get back the computed results immediately, then we call such a service, **InetCompu** (Internet Computing, Copyrighted) or **GridCompu** (Grid Computing, Copyrighted).

5.2. What We Have Done So Far. Several years ago, the author of this book figured out a practical method to actually realize the needed Software Interface (called General Purpose Grid Middleware Interface) for such an InetCompu Service, and he finished the development of the first primitive but working version of the Interface at around August 16, 2001. Later on, the Interface for InetCompu was significantly upgraded by him to such a degree that it could handle the incoming data more neatly and smartly. The final version currently in usage also passed a rigorous local fields test successfully.

As of August 28, 2001, we began to provide an InetCompu Service, i.e., whoever sends an E-mail to a particular E-mail address of our Server, for example, containing

(1) either an input data for an application program in our Server for a certain mathematical formula (for instance, input: n=1234567 to numbpart@trinitas.mju.ac.kr in plain text format

(2) or his own program, for instance, written in Reduce commands,

then he will soon receive the computed results from our Server.

[1] an archive site: http://ch.catholic.or.kr/pundang/archive/4/intro2InetCompu.html

5.3. An Example of InetCompu. In particular,

http://trinitas.mju.ac.kr/intro2numbpart.html[2]

explains how one can use our InetCompu Service to compute the value of $p(n)$ for a given input n, which utilizes Rademacher's formula for $p(n)$ in Theorem A.1.

[2]an archive site: http://ch.catholic.or.kr/pundang/archive/4/intro2numbpart.html

Bibliography

[AL] W.W. Adams and P. Loustaunau, *An Introduction to Gröbner Bases,* GSM Vol. 3, AMS, 1994.

[GA] G.E. Andrews, *The Theory of Partitions,* Reading, 2nd printing,1981, Addison-Wesley

[BW] T. Becker and V. Weispfenning, *Groebner Bases: A Computational Approach to Commutative Algebra*, Springer Verlag, 1993.

[B1] N. Bourbaki, *Algebra,* Part I and II (in English), Springer-Verlag, 1970, 1981.

[B2] N. Bourbaki, *Commutative Algebra* (in English), Springer-Verlag, 1985.

[CLO] D. Cox, J. Little, and D. O'Shea, *Ideals, Varieties, and Algorithms, An Introduction to Computational Algebraic Geometry and Commutative Algebra,* 2nd Ed., Springer-Verlag, 1997.

[ER] D. Eisenbud and L. Robbiano, *Computational Algebraic Geometry and Commutative Algebra,* Symposia Mathematica Vol. XXXIV, Istituto Nazionale di Alta Mathematica Francesco Severi, Cambridge Univ. Press, 1993.

[DE] D. Eisenbud, *Commutative Algebra with a View Toward Algebraic Geometry,* Springer-Verlag, 1995.

[E] Euclid, *The Elements,* Vol. 1 - 3 (English translation), 2nd Ed., Dover, 1956.

[AG1] A. Grothendieck, *Algébre Homologique,* Secrétariat mathématique, 11 rue Pierre Curie, PARIS 5e, 1958.

[AG2] A. Grothendieck, *Fondements de la Géométrie Algébrique* (Extracted from Bourbaki Seminar 1957-1962), Secrétariat mathématique, 11 rue Pierre Curie, PARIS 5e, 1962.

[AG3] A. Grothendieck and J.A. Dieudonné, *Eléments de Géométrie Algébrique,* Vol. I-IV, IHES, 1960-1967.

[AG4] A. Grothendieck and J.A. Dieudonné, *Eléments de Géométrie Algébrique I,* Springer-Verlag, 1970.

[AG5] A. Grothendieck, *Séminaire de Géométrie Algébrique,* Vol. 1-7, IHES, 1960-1962.

[HR] G. H. Hardy and S. Ramanujan, *Asymptotic Formulae in Combinatory Analysis*, Proc. London Math. Soc. (2), Vol. 17 (1917), pp. 75-115. (Reprinted in

Collected Papers of S. Ramanujan, pp. 276-309, Cambridge University Press, London, 1927.)

[H] G.H. Hardy, *A Mathematician's Apology*, Cambridge University. Press, 1969.

[GK1] G. Kempf, *Algebraic Varieties*, London Math. Soc. LNS 172, Cambridge Univ. Press, 1993.

[GK2] G. Kempf, *Algebraic Structures,* Friedrich Vieweg & Sohn, 1995.

[DM] D. Mumford, *Algebraic Geometry I, Complex Projective Varieties,* Springer-Verlag, 1976.

[JS] J. P. Serre, *Faisceaux Algébriques Cohérents*, Annals of Mathematics, Vol 61, No. 2, March, 1955.

[STS1] S.T. Soh, *A Reliable Implementation of the Formula of Rademacher for the Partition Function* $p(n)$, preprint

[STS2] S.T. Soh, *A Diophantine Construction of an Exact Algebraic Formula for Graded Partition Functions*, J. Korean Math. Soc., Vol 36 (1999), No. 2, pp. 267-298

[STS3] S.T. Soh, *Efficient (Quasi-) Recursive Formulas for Graded Partition Functions,* preprint

[ES] E.H. Spanier, *Algebraic Topology,* Springer-Verlag, 1966.

[BS] B. Sturmfels, *Algorithms in Invariant Theory,* Texts and Monographs in Symbolic Computation, Springer-Verlag, 1993.

[BW] B.L. van der Waerden, *Algebra,* Vol. 1 and 2, Frederick Ungar Pub., 1970.

[WR] A.N. Whitehead and B. Russell, *Principia Mathematica*, Cambridge Univ. Press, 1910.

[OZ] O. Zariski, *Collected Papers,* Vol. 1 - 4., MIT Press, 1972.

[ZP] O. Zariski and P. Samuel, *Commutative Algebra,* Vol.1 and 2, Springer-Verlag, 1958.

APPENDIX B

Computer Algebra Systems

In recent years, several Computer Algebra Systems were developed and they are currently used by many scientists and engineers. Among them, MAPLE, MATHEMATICA, etc. are commercially available. A CAS, REDUCE, is now a well-known free CAS available from

http://ch.catholic.or.kr/pundang/archive/4/reduce-algebra.htm

In this section, we discuss very briefly several basic commands of REDUCE.

1. A Groebner Basis Package in REDUCE

To load a package for Groebner basis computation within REDUCE, one needs to type in the following line at Reduce prompt:

load_package groebner; (and then press Enter key)

REMARK B.1. *Here are quick explanations on those commands that are used in our codes for REDUCE in many Examples of this book:*

; displays the result of the command.
$ does not display the result of the command.
bye$ finishes the current reduce session.

% begins a one line comment.
comment begins a multiple line comment until ; or $
off echo$ disables the echo from the system.
off nat$ allows us to use the output as an input.
out t$ redirects the output to the terminal. (Default).

---

sqrt(2) displays square root of 2 knowing that its square is 2.

i displays $\sqrt{-1}$ knowing that its squre is -1.

mat((a,b),(c,d))$ assigns a 2 by 2 matrix.

det(A)$ computes the determinant of A.

---

in "filename"$ reads the contents of the file, filename, into REDUCE.

out "filename.out"$ redirects the output to the file, filename.out

write "comment"$ writes your comment to the output file, filename.out .

shut "filename.out"$ closes the file, filename.out .

---

sub({x=3},y=x)$ substitutes 3 for x in the expression y=x .

mkid(result,3)$ creates a new variable "result3".

solve({f1,f2,f3})$ solves a given system of equations f1=f2=f3=0 .

on fullroots$ attempts to express very explicitly.

length(result)$ reports the number of elements of he list, result .

part(result,j)$ takes the j-th element out of the list, result .

coeffn(f,j)$ takes the j-th coefficient of the polynomialf in one variable.

---

setmod 3$ on modular$ computes in modulo 3.

off modular$ turns off modular computation.

---

on rounded$ starts numerical computation.

precision 50$ adjusts the number of digits for numerical computation to 50.

off rounded$ ends numberical computation. (Default).

REMARK B.2. *The following commands are needed for any computation related to Groebner basis of a polynomial ideal.*

load_package groebenr$ starts the Groebner package.

torder({x,y,z}, lex)$ assigns the lex order. (Default).

torder({x,y,z}, gradlex)$ assigns the graded lex order.

torder({x,y,z}, revgradlex)$ assigns the reverse graded lex order.

groebner({f1,f2,f3}); compute the reduce Groebner basis for the ideal generated by f1, f2, and f3.

---

groebnert({given_f1=f1,given_f2=f2,given_f3=f3}); not only computes the reduced Groebner basis but also keep track of all the intermediate division process so that the output of this command allows us to figure out the quotients of the division by f1, f2, and f3. See Program 2 and Example 3.4.

groebprotfile; contains all the intermediate division results.

---

greduce(f,{f1,f2,f3}); computes the unique remainder of the division of f by the reduced Groebner basis for the ideal generated by f1, f2, and f3.

preduce(f,{f1,f2,f3}); computes a non-unique remainder of the division of f by f1, f2, and f3.

REMARK B.3. *The following table is sometimes very useful.*

Built In Infix	**Equivalent Expression**
:=	setq
=	equal
>=	geq
>	greaterp
<=	leq
<	lessp
+	plus
-	difference (if unary, minus)
*	times
/	quotient (if unary, recip)
^ or **	expt (raising to a power)
.	cons

REMARK B.4. *For detailed usage informations on these commands, one may visit the following homepage:*

http://ch.catholic.or.kr/pundang/archive/4/reduce-algebra_sourceforge_io.htm

where one may easily find several manuals for the Computer Algebra System, REDUCE.

APPENDIX C

Suggestions for Further Reading and References

W. Adams and P Loustaunau, *An Introduction to Groebner Bases*, Graduate Studies in Mathematics, Vol. 3, AMS, 1994.

M. Artin, *Algebra*, Prentice Hall, 1991.

T. Becker and V. Weispfenning, *Groebner Bases: A Computational Approach to Commutative Algebra*, Springer Verlag, 1993.

G. Birkhoff and S. MacLane, *A Survey of Modern Algebra*, 4th Edition, Macmillan, New Nork, 1997.

D. Cox, J. Little, and D. O'Shea, *Ideals, Varieties, and Algorithms*, 2nd Edition, UTM, Springer-Verlag, 1997.

D. Cox, J. Little, and D. O'Shea, *Using Algebraic Geometry*, GTM, Springer-Verlag, 1998.

D. Cox, B. Sturmfels (Editors), *Applications of Computational Algebraic Geometry*, PSAPM 53, AMS, 1998.

D. Eisenbud and L. Robbiano, *Computational Algebraic Geometry and Commutative Algebra,* Symposia Mathematica Vol. XXXIV, Istituto Nazionale di Alta Mathematica Francesco Severi, Cambridge Univ. Press, 1993.

D. Eisenbud et al, *Computations in Algebraic Geometry with Macaulay 2*, Springer-Verlag, 2002.

G.M. Greuel and G. Pfister, *A Singular Introduction to Commutative Algebra*, Springer-Verlag, 2002.

A.C. Hearn et al, *REDUCE User's Contributed Package Manual*, Version 3.8, Konrad-Zuse-Zentrum, Berlin, 2004.

M. Kreuzer and L. Robbiano, *Computational Commutative Algebra 1*, Springer-Verlag, 2000.

M. Kreuzer and L. Robbiano, *Computational Commutative Algebra 2*, Springer-Verlag, 2005.

G. Kempf, *Algebraic Varieties*, London Math. Soc. LNS 172, Cambridge Univ. Press, 1993.

G. Kempf, *Algebraic Structures,* Friedrich Vieweg & Sohn, 1995.

M. Kofler, *Maple: An Introduction and Reference*, Addison-Wesley, 1997.

S. MacLane and G. Birkhoff, *Algebra*, 2nd Edition, Macmillan, New York, 1979.

S.T. Soh, *Computational Algebra*, Kyowoo, Seoul, 2006.

B. Sturmfels, *Algorithms in Invariant Theory*, Texts and Monographs in Symbolic Computations, Springer-Verlag, 1993.

B. Sturmfels, *Solving Systems of Polynomial Equations*, CBMS 97, AMS, 2002.

B.L. van der Waerden, *Modern Algebra*, Ungar, New York, 1970.

O. Zariski and P. Samuel, *Commutative Algebra,* Vol.1 and 2, Springer-Verlag, 1958.

http://trinitas.mju.ac.kr/intro2maple.html

http://trinitas.mju.ac.kr/intro2reduce.html

http://trinitas.mju.ac.kr/intro2InetCompu.html

http://ch.catholic.or.kr/pundang/archive/4/soh/2002_InetCompu

http://www-groups.dcs.st-and.ac.uk/~history/

http://www.reduce-algebra.com/docs/Groebner.pdf

http://www.reduce-algebra.com/docs/reduce.pdf

Index